Lecture Notes in Management and Industrial Engineering

Series editor

Adolfo López-Paredes, Valladolid, Spain

This bookseries provides a means for the dissemination of current theoretical and applied research in the areas of Industrial Engineering & Engineering Management. The latest methodological and computational advances that both researchers and practitioners can widely apply to solve new and classical problems in industries and organizations constitute a growing source of publications written for and by our readership.

The aim of this bookseries is to facilitate the dissemination of current research in the following topics:

- Strategy and Enterpreneurship
- Operations Research, Modelling and Simulation
- Logistics, Production and Information Systems
- Quality Management
- Product Management
- Sustainability and Ecoefficiency
- Industrial Marketing and Consumer Behavior
- Knowledge and Project Management
- Risk Management
- Service Systems
- Healthcare Management
- Human Factors and Ergonomics
- Emergencies and Disaster Management
- Education

More information about this series at http://www.springer.com/series/11786

Marlene Amorim · Carlos Ferreira
Milton Vieira Junior · Carlos Prado
Editors

Engineering Systems and Networks

The Way Ahead for Industrial Engineering and Operations Management

Springer

Editors
Marlene Amorim
Departamento de Economia Gestão
Engenharia Industrial e Turismo
Universidade de Aveiro
Aveiro
Portugal

Carlos Ferreira
Departamento de Economia Gestão
Engenharia Industrial e Turismo
Universidade de Aveiro
Aveiro
Portugal

Milton Vieira Junior
Universidade Nove de Julho - UNINOVE
São Paulo
Brazil

Carlos Prado
Escuela de Ingeniería Industrial
Universidad de Vigo
Vigo
Spain

ISSN 2198-0772 ISSN 2198-0780 (electronic)
Lecture Notes in Management and Industrial Engineering
ISBN 978-3-319-45746-8 ISBN 978-3-319-45748-2 (eBook)
DOI 10.1007/978-3-319-45748-2

Library of Congress Control Number: 2016949617

Printed on acid-free paper

This Springer imprint is published by Springer Nature
The registered company is Springer International Publishing AG
The registered company address is: Gewerbestrasse 11, 6330 Cham, Switzerland

Preface

We live in an interconnected world where increasing flows of goods, information and individuals strengthen the links among companies and nations at an unprecedent pace. This scenario is reinforced by the rising prosperity and participation of emerging economies, and by the dissemination of digital technologies. Aiming at an effective participation in today's interconnected production contexts requires the development of specific knowledge that can inform managerial practice for taking full advantage of the existing opportunities.

This volume presents the best papers from the ICIEOM-CIO-IIIE 2015, "XXI International Conference on Industrial Engineering and Operations Management", "9th International Conference on Industrial Engineering and Industrial Management" (XIX Congreso de Ingeniería de Organización) and "International IIE Conference 2015". The moto of ICIEOM-CIO-IIIE 2015 International Conference was "Engineering Systems and Networks: the way ahead for industrial engineering and operations management". The conference provided a unique forum to disseminate, to all branches of industry, information on the most recent and relevant research, theories and practices in industrial engineering, management and operations. The chapters gathered in this volume offer a multidisciplinary and timely overview of the key themes that drive the current research in the field of industrial engineering and management, and that are driving the debate between researchers and practitioners from different industry sectors.

The ICIEOM-CIO-IIIE 2015 received a total number of 353 submissions, addressing a multidisciplinary range of industrial engineering and operations management topics, and creating a rich setting for three conference days of vibrant debate. A total of 200 of such submissions were selected for presentation at the conference. This volume presents a selection of 39 of the presented papers.

The Organizing and Scientific Committees of ICIEOM-CIO-IIE 2015 express their gratefulness to all the authors, invited speakers and the members of the Program Committee who have generously committed their time and expertise in the rigorous process of revision of the submitted manuscripts and provided key ingredients to set up a conference of very high standards, built on the experience of previous editions of ICIEOM, CIO and IIIE.

Aveiro, Portugal Marlene Amorim

Contents

Part I OR, Modelling and Simulation

A Simulation-Based Analysis of a Cork Transformation System 3
J. Teles, R.B. Lopes and A.L. Ramos

Using Overall Equipment Effectiveness (OEE) to Predict Shutdown Maintenance 13
Rolando Jacyr Kurscheidt Netto, Eduardo Alves Portela Santos, Eduardo de Freitas Rocha Loures and Rodrigo Pierezan

Forecasting Cloud Computing: Producing a Technological Profile 23
Iñaki Bildosola, Rosa Rio-Bélver, Ernesto Cilleruelo and Javier Gavilanes

Modelling the Strategies Alignment Process in the Collaborative Network Context 33
Beatriz Andres and Raúl Poler

The Role of Complexity and Flexibility of the Instance in the Joint Solution Approach 43
Raul Pulido, Álvaro García-Sánchez, Alessandro Brun and Miguel Ortega-Mier

Towards Increasing Sustainability in Large Urban Mobility Attractors 53
Jesús Muñuzuri, Luis Onieva, José Guadix and Elena Barbadilla

Stock Market Firm Value Effects of Research and Development Expenditures in the Oil and Gas Industry 61
Eduardo Pontual Ribeiro, Willian Freitas de Almeida and Rosemarie Bröker Bone

Part II Logistics, Production and Information Systems

A Decision Support Framework for Production Flow Coordination Using Supply Chain Management Practices, Ordering Systems and Modeling Techniques 71
Wagner de Barros Neto, Laisa Caroline de Paiva Gomes, Maico Roris Severino and Moacir Godinho Filho

A Greedy Primal-Dual Type Heuristic to Select an Inventory Control Policy 79
Nazanin Esmaili, Bryan A. Norman and Jayant Rajgopal

A Model that Integrates Direct and Reverse Flows in Omnichannel Logistics Networks 89
Javier Guerrero-Lorente, Eva Ponce-Cueto and Edgar E. Blanco

A Nonlinear Integer Programming Model for Warehousing Sustainable Logistics 99
Francesco Boenzi, Salvatore Digiesi, Francesco Facchini, Giorgio Mossa and Giovanni Mummolo

A Reference Framework to Design Inventory Policies Using a Fill Rate Criterion in Lost Sales Contexts 109
Eugenia Babiloni, Ester Guijarro and Manuel Cardós

Contribution of Lean Principles in the Information Systems Development: An Experience Based on a Practical Case 117
Joana Pereira and Leonor Teixeira

Data-Driven SKU Differentiation Framework for Supply Chain Management 127
Alexander Kharlamov, Luís Miguel D.F. Ferreira and Janet Godsell

Deploying "Packaging Logistics" in Paper Napkins 139
Jesús García-Arca, A. Trinidad González-Portela Garrido and José Carlos Prado-Prado

Differentiation of the Difficulty Level of Supply Chain Management Integration Actions 149
Fernanda Arantes, Maria Leite and Antonio Cezar Bornia

Establishing a Link Between Lean Practices and Corporate Sustainability 157
Cristóvão Silva and Paulo Vaz

Explaining Alliance Success Factors in Spanish Food and Beverage Supply Chain: Case Analysis 167
Jesús Morcillo Bellido and Alfonso Durán Heras

How to Design an Efficient and Sustainable Box? 175
Jesús García-Arca, A.T. González-Portela Garrido,
J.A. Comesaña-Benavides and J.C. Prado-Prado

Interoperability Frameworks in Public Administration Domain: Focus on Enterprise Assessment 185
Alexandre da Costa Castro, José Marcelo Almeida Cestari,
Eduardo Rocha Loures, Edson Pinheiro de Lima
and Eduardo Alves Portela Santos

Main Factors Affecting the Development of Interorganizational Partnerships in Biodiesel Supply Chain in Brazil 197
Eliene Cristina Barros Ribeiro, António Carrizo Moreira,
Luís Miguel Domingues Fernandes Ferreira, Leonardo Leocádio Coelho de
Souza and Aldara da Silva César

Productivity Improvement, Considering Legal Conditions and Just in Time Principles in the Mixed-Model Sequencing Problem 205
Joaquín Bautista-Valhondo, Rocío Alfaro-Pozo and Cristina Batalla-García

Proposal of a Framework for Assessing Environmental Performance of Supply Chains 215
C. Silva, L.M.D.F. Ferreira and S.G. Azevedo

Root Cause Identification of Existing Barriers Detected by People with Disabilities in Air Transport 225
I. Garcia-Miranda and A. Duran Heras

Spare Parts Inventory Management Using Quantitative and Qualitative Classification 233
Fernanda Oliveira and Clara Bento Vaz

Sustainable Supply Chain Management: A Case Study 243
Cláudia Silva and Joaquim Borges Gouveia

Transport KPIs for Supply Chain Improvement. A Literature Analysis 251
Pablo Domínguez-Caamaño, Jesús García-Arca,
Arturo J. Fernandez-González and José Carlos Prado-Prado

Using Big Data for Competitive Dimensions Improvement in a Telco Company 259
Rafael Fernandes Novo and José Manoel Souza das Neves

Reduction of Drying Process Time of Natural Cork Stoppers Process in Lean Improvement Efforts 267
Tatiana M. Pinho, Daniel Campos, J. Boaventura-Cunha,
Américo Azevedo and A. Paulo Moreira

Waste Types in People Processing Services 277
José Dinis-Carvalho, Rui M. Lima, Andromeda Menezes
and Marlene Amorim

Part III Quality and Management in Engineering Systems and Networks

Characterising Knowledge Workers' Job Positions 289
Ana Moreno-Romero, Eva Ponce-Cueto and Ruth Carrasco-Gallego

Information Quality in Companies Committed to TQM 299
Marta Zárraga-Rodríguez, Manuel Francisco Suárez-Barraza, M. Jesús Álvarez, Elisabeth Viles and Carmen Jaca

Performance Measurement Systems for Designing and Managing Interoperability Performance Measures: A Literature Analysis 307
Alexandre da Costa Castro, José Marcelo Almeida Cestari, Eduardo Rocha Loures, Edson Pinheiro de Lima and Eduardo Alves Portela Santos

Systematic Analysis of Economic Viability with Stochastic Approach: A Proposal for Investment 317
José Donizetti de Lima, Marcelo Gonçalves Trentin, Gilson Adamczuk Oliveira, Dayse Regina Batistus and Dalmarino Setti

Thermal Comfort Field Study Based on Adaptive Comfort Theory in Non-residential Buildings 327
Elena Barbadilla, José Guadix, Pablo Aparicio and Pablo Cortés

Part IV Education in Engineering Systems and Networks

Agents Playing the Beer Distribution Game: Solving the Dilemma Through the Drum-Buffer-Rope Methodology 337
José Costas, Borja Ponte, David de la Fuente, Jesús Lozano and José Parreño

Organizational Engineering: The Emerging Stage of Industrial Engineering 347
Javier Carrasco, Carlos Mataix and Ruth Carrasco-Gallego

Perception of the Evolution of Industrial Engineering Areas Based on the Brazilian ENADE-INEP Assessment System 355
Eduardo Guilherme Satolo, Renato Luis Garrido Monaro, Milton Vieira Junior and Daniel Luis Garrido Monaro

The Skateboard Manufacturing Company: A Teaching Case on Production Planning and Control 365
Marco Aurélio de Mesquita, Leonel José Girotti and João Vitor Tomotani

Erratum to: The Skateboard Manufacturing Company: A Teaching Case on Production Planning and Control E1
Marco Aurélio de Mesquita, Leonel José Girotti and João Vitor Tomotani

Part I
OR, Modelling and Simulation

A Simulation-Based Analysis of a Cork Transformation System

J. Teles, R.B. Lopes and A.L. Ramos

Abstract This paper describes the development of a simulation study for the analysis of an important part of a factory which refers to the transformation of cork into stoppers that are 100 % natural. The aim of the simulation model is to assess the productive flow through the analysis of performance indicators such as the lead time, the rate of utilization of the simulated resources, as well as the respective queues. After the detailed analysis of the system, a plan of action is devised with suggestions of improvements for implementing in the near future for addressing the main problems.

Keywords Cork industry · Simulation · Arena software · Line balancing

1 Introduction

A system is defined as a set of distinct elements—entities (people or machines)—which are related with each other by means of an interaction or interdependence according to a particular purpose (Schmidt and Taylor 1970). In general, the systems are limited, with limits or boundaries that should be defined. According to Banks et al. (2005), a system is often affected by changes occurring outside of it.

J. Teles (✉)
Department of Economics, Management and Industrial Engineering, University of Aveiro, Campus Universitário de Santiago, 3810-193 Aveiro, Portugal
e-mail: jorgeteles@ua.pt

R.B. Lopes
CIDMA/Department of Economics, Management and Industrial Engineering, University of Aveiro, Campus Universitário de Santiago, 3810-193 Aveiro, Portugal
e-mail: rui.borges@ua.pt

A.L. Ramos
GOVCOPP/Department of Economics, Management and Industrial Engineering, University of Aveiro, Campus Universitário de Santiago, 3810-193 Aveiro, Portugal
e-mail: aramos@ua.pt

M. Amorim et al. (eds.), *Engineering Systems and Networks*, Lecture Notes in Management and Industrial Engineering, DOI 10.1007/978-3-319-45748-2_1

These changes occur in the system environment. In simulation models, it is necessary to decide and define a boundary between the system and its environment. This decision depends on the purpose of the study. Pedgen et al. (1995) describe simulation as a design of a process model that reflects a particular real system in which experiments are carried out in order to later understand the system behaviour and/or evaluate different strategies for the operation of the system. They consider that a simulation study includes the construction of a model and its use for experiments aimed to analyse the problem. Therefore it can be assumed that simulation models turned out to be a trial and analysis methodology in which the following is necessary:

- Describe the behaviour of the systems;
- Build theories or hypotheses;
- Use the model to predict future behaviours of the system, i.e. the effects caused by changes on the system or in its operating mode.

The simulation can be used before building a new system, when changing an existing one, or when creating a new integrated system, in order to reduce potential failures in the future or meet the specification, to eliminate unforeseen bottlenecks, to avoid underuse or overuse resources, and to optimize overall system performance (Anu 1997).

As can be observed in Fig. 1 the simulation is one possible way to study a specific system, being particularly advantageous when analytical solutions cannot

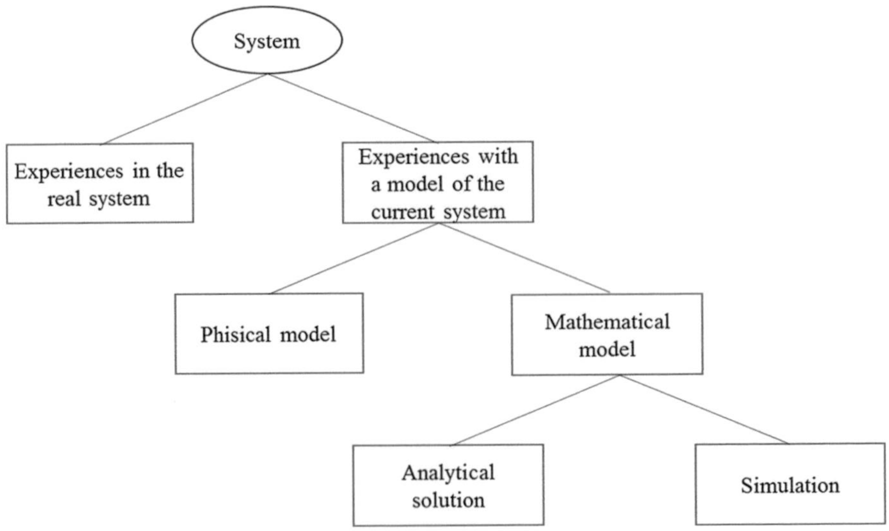

Fig. 1 How to study a system (adapted from Law and Kelton 2000)

be found (e.g. large and complex systems, marked stochastic behaviour, broad set of performance measures to analyse).

A simulation model can be of various types such as, static, dynamic, deterministic, stochastic, continuous or discrete. This classification is obtained though specific system characteristics, such as the probabilistic component, the time instant to be simulated and its impact on the behaviour of the system (Law and Kelton 2000).

2 Discrete Event Simulation—DES

Discrete Event Simulation (DES) is a technique widely used to study industrial processes, namely, for evaluating and analysing changes in industrial processes (Law and Kelton 2000). This technique consists in modelling a system that evolves over time through a representation of state variables that change instantaneously in distinct time sets. These time sets are those where there is a particular event, being defined as an instantaneous occurrence that can change the state of the system (Law and Kelton 2000).

Banks et al. (2005) add that DES is used for studying systems that change their state at discrete points in time, that is, a system is modelled with respect to its state at each point, where entities passing through the system, the affected resources, the activities and events cause a system status change at discrete points in time. Schriber et al. (2013) confirm that in DES a system is seen as a set of discrete traffic units moving (flow) point by point in the system while competing with each other to use the scarce resources (capacity constraints).

3 Steps in a Simulation Study

According to Law and Kelton (2000), Banks et al. (2005) and Sargent (2013) there are 8 steps to be followed for conducting a successful simulation study. In the first step (problem formulation and study planning) the most important aspects to remember are: the overall goals of the study; which questions the model has to address; which performance measures will be used in the evaluation of the system and its different settings; what is the model scope; which system settings will be modelled; the time required for the study; and, finally, what resources will be needed to develop the study. In the second step (data collection and definition of conceptual model) the modeller will have to collect the information regarding the system layout and operating procedures, the model parameters and probability distributions related to input data in the model, as well as the real system output data to validate the model. The conceptual model will also need to be defined which, according to Robinson (2013), is the abstraction of the real system under analysis. A modeller must understand the real system first, and only then he can model it, in order to be as representative as possible. The ideal would be to build a model as detailed as possible, but unfortunately there will not be enough time to do so, not even enough knowledge about the system for this purpose. Therefore, there are abstractions that should be done in modelling. This abstraction is called conceptual model. However, one must take into account the level of detail of the conceptual model. Figure 2 shows the relationship between model accuracy and its level of detail (complexity) according to Robinson (2008). It can be seen that the level of detail has to be weighted, always taking into consideration the purpose of the study. Therefore the author states that a model must be valid, credible, viable and useful. It is based on these requirements that the necessary and appropriate abstractions must be made. The steps of the verification and validation (V&V) of the simulation model are intended to determine whether the model and the results are correct, depending on the purpose of study, as explained by Sargent (2013). Also according to this author, V&V should be directed to the purpose of the model, i.e., depending on the purpose of the model it should be able to answer the most relevant questions.

Sargent (2012) argues that V&V can be done during the construction of the simulation model, or at the end of the design model. The same author states that the animation, comparison with other valid models, comparison of events of the model versus actual system, extreme condition testing, validation by experienced people, evaluation of the previous output data from the real system, graph analysis and running sensitivity tests are some of the techniques used for validating the simulation model. After V&V, in the experience of designing experiments, Sanchez and Wan (2011) stated that one of the first things that, as a rule, should be done is to set the experimental factors. The goal of a Design of Experiments (DoE) is to identify the important factors that will lead to the answers and those who do not. The latter can be removed from the analysis in order to reduce the experimental effort and simplify the task of interpreting the data. Finally, the simulation model needs to be executed, the output data analysed and the results of the study disseminated.

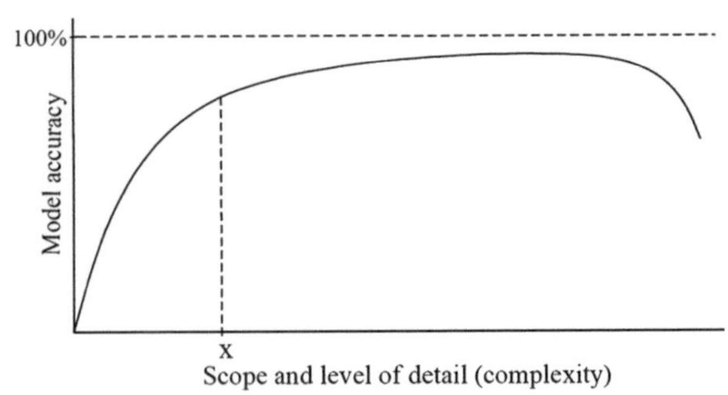

Fig. 2 Relationship between accuracy versus scope and level of detail of a model, adapted from Robinson (2008)

4 Case Study: Simulation of the Production Process of Cork Stoppers

This simulation study was carried out in a company operating in the business of natural cork stoppers and comprised a significant part of its production process (i.e. cutting of cork strips, drilling cork and stopper quality choice). The main goal of the study was to analyse and evaluate the current configuration of the process, which is relatively stable, and identify small improvements that will improve the operations of the existing system configuration. In order to achieve that, the following performance indicators were defined:

- Lead Time;
- Utilization rates of resources;
- Queues (dimension and waiting time);
- Input versus Output.

First the system that will be modelled needs to be understood. To meet the objectives of this simulation study it was defined to simulate a working day (1 shift), corresponding to a time span from 8 a.m. to 5 p.m.. Regarding the data collection that serves as input to the simulation model, it was conducted with the help of a statistical software tool part of the Arena® simulation software (Input Analyzer). The sample used as the basis for each time cycle had at least 20 records collected on the plant floor.

When defining the conceptual model some aspects were ignored, such as small conveyors and a logistics train carrying stoppers, since they did not influence the study objectives. As corks present various sizes, several cycle times were measured depending on the type of cork. Having the several sizes of cork into consideration, an ABC analysis was performed to find out where the major data collection efforts should be directed.

After this phase, the construction and verification of the logic model was conducted using the Arena software.

After the construction of the logical model and before its execution, a set of important parameters in the design of experiments was defined:

- Time length for the execution of the simulation model is 27 h;
- Warm-up time is 18 h;
- The number of replications of the model is 20.

Concerning the performance measures, as stipulated above, this study must answer the following questions:

- Which are the bottlenecks resources?
- On which processes there is a production accumulation and why?
- Where is a potential for improvement in the production flow?
- Is there a balance between sectors?

- How long does it take for processing cork pallets into stoppers?
- What is the system state in the shift change?

5 Results and Conclusions

Analysing the output given by the simulation model it was possible to extract several results (statistical estimates of the performance measures based on 95 % confidence intervals). Concerning lead times, the modelled system took about 1260 ± 28 min. In Fig. 3 data concerning cork stoppers production (output in) of each process is displayed. It can be seen that between the first and the second processes it occurs the biggest loss of capacity, and therefore these are bottlenecks resources. The same applies to the cork stoppers choice process. The first process is able to produce around 1250 thousand cork stoppers and by the end of the 1st selection 200 thousand less stoppers are produced. This difference is caused by the unbalanced operations between sectors and by the body rectifying machine, causing this bottleneck which affects the functioning of downstream resources.

Regarding the usage rate of the various processes/system resources, and looking at Fig. 4, it is becomes clear that some resources present very high utilization rates (e.g. stripping resources) while others are idle (e.g. supplier "RH").

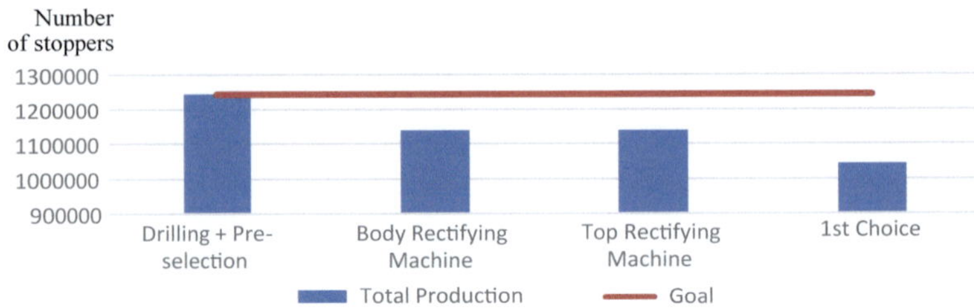

Fig. 3 Production of cork stoppers by sector

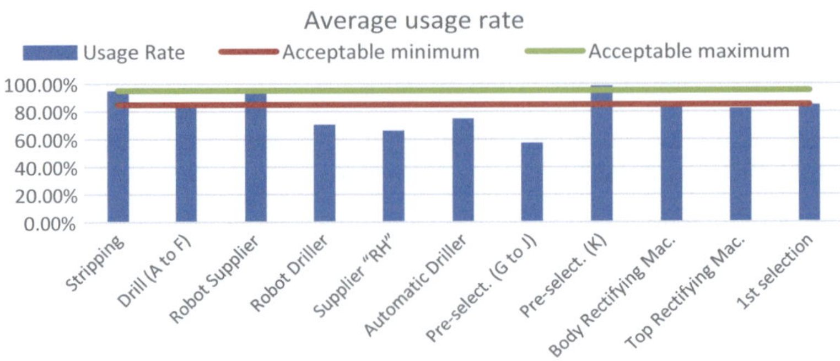

Fig. 4 Average usage rate of processes/system resources

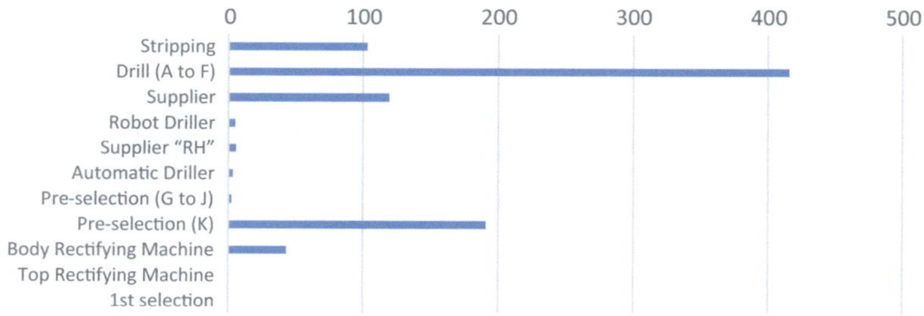

Fig. 5 Average queue time (minutes)

By analysing both Figs. 4 and 5 it can be concluded that there are resources that greatly affect the lead time of the system. It is on these resources that improvement actions should be addressed in order to make the production flow more fluid and thus reducing the lead time of the system. Unusual waiting time in queues also contributes to the poor performance of this indicator, as is the case of pre-selection (K) and body rectifying machine. The queues that have acceptable values are related to the resources of stripping, which correspond to the average half-pallet cork consumption and the resources of the drill, pedal and semiautomatic. For these, if one divides this line into six jobs placements, one comes to the conclusion that each operator takes about 65 min to consume his queue, which is consistent with the reality and considered acceptable.

Analysing the results of the developed simulation model, it can be concluded that the improvement actions should be focused on:

- The Robot supplier, is a bottleneck resource that causes the robot drill resource to be underused;
- The Supplier of K line, is an underutilized resource;
- The Pre-selection (G–J), is an underutilized resource;
- The Pre-selection (K), is a bottleneck resource;
- The Body Rectifying Machine, is a bottleneck resource that is causing Top Rectifying Machine and 1st Selection resources to be underutilized; it can be seen in Fig. 6 that there is no waiting queue on Top Rectifying Machine and 1st Selection resources while having only some entities on processing. The animation, in this case, hints that Body Rectifying Machine resource is a bottleneck to the production flow.

Briefly, it is noted that the current system configuration still contains some inefficiencies, created by the imbalance between processes, which must be minimized or eliminated. The type of industrial process in analysis also helps in creating intermediate stocks and significant lead times. Nevertheless, the resources to which the company should apply improvement actions to minimize the impact of the type of plant layout and existing system production flow were largely identified.

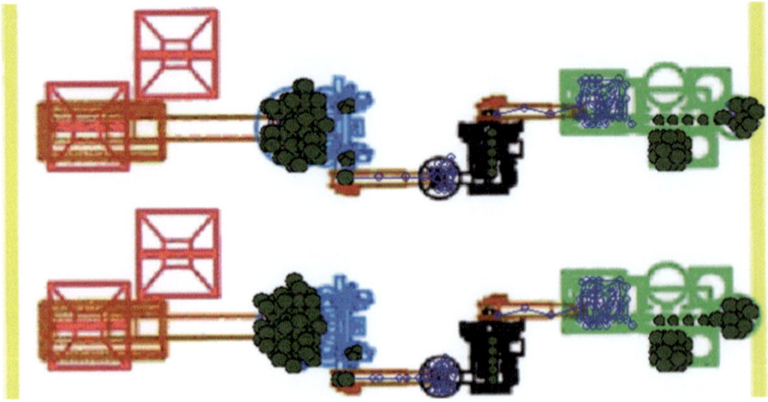

Fig. 6 Animation of the resources body rectifying machine, top rectifying machine and first selection

Concluding, this work describes the development of a simulation model for analysing the production of cork stoppers. Based on the analysis of the developed model several suggestions could be made for improving the system. The main suggestion concerned the balancing of the resources' cycle times, after increasing the processing time of the first resource. This suggestion was implemented in the real system, which led to a significant improvement in production (around 10 % production of cork stoppers).

Following this study, it can be concluded that simulation studies may be important for helping companies analysing and improving their real-life systems. In some cases these studies may have significant impact, thus increasing companies' competitiveness.

References

Anu M (1997) Introduction to modeling and simulation. Binghamton, pp 7–13

Banks J, Carson JS II, Nelson BL, Nicol DM (2005) Discrete-event system simulation. Pearson Education Inc, New Jersey

Law AM, Kelton WD (2000) Simulation modeling and analysis. McGraw-Hill Companies, United States

Pegden CD, Sahnnon RE, Sadowski RP (1995) Introduction to simulation using siman. McGraw-Hill Inc, Blacklick

Robinson S (2008) Conceptual modelling for simulation part I: definition and requirements. J Oper Res Soc 59:278–290

Robinson S (2013) Conceptual modeling for simulation. Loughborough, pp 377–387

Sanchez SM, Wan H (2011) Better than a petaflop: the power of efficient experimental design. pp 1441–1455

Sargent RG (2012) Verification and validation of simulation models. J Simul 7:12–24

Sargent GR (2013) An introduction to verification an validation of simulation models. Syracuse, New York, s.n., pp 321–327

Schmidt JW, Taylor RE (1970) Simulation and analysis of industrial systems. Richard D. Irwin, Homewood Illinois

Schriber TJ, Brunner DT, Smith JS (2013) Inside discrete-event simulation software: how it works and why it matters. pp 424–438

Using Overall Equipment Effectiveness (OEE) to Predict Shutdown Maintenance

Rolando Jacyr Kurscheidt Netto, Eduardo Alves Portela Santos, Eduardo de Freitas Rocha Loures and Rodrigo Pierezan

Abstract This research proposes an approach to predict equipment condition using OEE performance metrics. Statistical tools are used to correlate measurements of OEE factors and maintenance history from a real database. The results suggesting that there is a correlation between the Time Between Stoppages and the trend degree of the Mean and/or the Standard Deviation (SD) of cycle time. This approach intended to help the predictions of shutdowns for maintenance.

Keywords Condition-based maintenance · OEE · Trend analysis · Process mining

1 Introduction

Several maintenance strategies are proposed and applied to productive processes, for improving the operational capacity of processes, saving maintenance costs, and improving industrial competition (Kumar et al. 2013). An appropriate maintenance action, at the right moment, is necessary for reducing process failures and increasing equipment reliability (Wang 2012). In order to maximize preventive maintenance effectiveness, the Condition-Based Maintenance technique was proposed. The main motivation of this strategy is that equipment failures are preceded by signals, conditions or indications that the failure may occur (Ahmad and Kamaruddin 2012). According to Ahmad and Kamaruddin (2013), Overall Equipment Effectiveness (OEE) provides an estimation of the lifetime of the

R.J. Kurscheidt Netto (✉) · E.A.P. Santos · E. de Freitas Rocha Loures · R. Pierezan
Pontifical Catholic University of Parana, (PUC-PR), Curitiba, Paraná, Brazil
e-mail: rolando.k@pucpr.br

E.A.P. Santos
e-mail: eduardo.portela@pucpr.br

E. de Freitas Rocha Loures
e-mail: eduardo.loures@pucpr.br

R. Pierezan
e-mail: rodrigo.pierezan@pcupr.br

M. Amorim et al. (eds.), *Engineering Systems and Networks*, Lecture Notes in Management and Industrial Engineering, DOI 10.1007/978-3-319-45748-2_2

equipment where its intended use is influenced by its availability, performance and quality. This paper proposes a study to estimate degradation on an equipment operation using performance factor of OEE index, aiming to monitor the operating condition of equipment to find deviations from its normal condition. The present paper is organized as follows. In Sect. 2 the theoretical background are introduced. Section 3 describes the proposed methodology. A case study of methodology application in a real process is showed in Sect. 4, covering the found issues and discussions about results. Finally, a conclusion about methodology application and further results are presented in Sect. 5.

2 Theoretical Background

The approach proposed in this paper is based on the following key concepts: performance measurement based, Condition Based Maintenance and Trend Analysis with attention to the Double Exponential Smoothing trend prediction model. A theoretical foundation on manufacturing event logs is presented.

Van Der Aalst et al. (2011) highlights the growth of information systems as Process-Aware Information Systems (PAIS), which record process information in the form of structured and detailed records of events from activities performed by a process (Weber et al. 2011). Each event refers to an instance (case ID) and a set of activities (task ID), including information from the beginning, ending, scheduling of this activity, date and time (timestamp) and user who performed the task are possible in some of these (Rozinat et al. 2009). Factory Information Systems (FIS) are a subclass of PAIS, which are responsible for the management of data related to a manufacturing system, where the information is typically received in the form production events (Santos et al. 2008). The data acquired by FIS are stored in the form of sequential events into defined structures of data, entries from operators on the factory floor or from sensors installed on monitored machines.

CBM, which is a technique in the context of a predictive maintenance policy, is the most popular and modern technique discussed in the literature (Ahmad and Kamaruddin 2012), where its fundamental basis is condition monitoring, using signals measured by sensors continuously, allowing to identify if the operational condition of the equipment is deviated from a defined normal condition (Jardine et al. 2006). Frequent inspections when equipment is in good operation increases the costs of operation. On the other hand, deferring the inspection can increase the occurrence of unexpected failure and economic losses (Abeygunawardane et al. 2013). Many types of equipment failures present some kind of change in their condition that is detectable. When the equipment is in a good condition, it has performance or quality factors with known mean μ and variance σ^2 values (Venkatasubramanian et al. 2003). However, in a wear condition or in the eminence of a failure, the mean, the variance or both values might change.

The OEE, proposed by Nakajima (1988), is one of the most used indicators for performance measurement, and is used for measuring the progress made by the Total Productive Maintenance philosophy. It is a quantitative indicator, serving to measure performance, to identify areas for improvement, and to guide efforts to improve the areas that involve the equipment used in the process (Kumar et al. 2013). The basic formulation for the OEE is based on the product of three mutually exclusive factors: Availability, Performance, and Quality. Performance factor is the ratio of the actual cycle speed and theoretical cycle time. Its traditional formulation is represented by Eq. (1).

$$\text{Performance}\ (\%\ ) = \frac{\boldsymbol{Total\,Produced} * \text{Theoretical Cycle Time}}{\boldsymbol{Calendar\,Time} - \boldsymbol{Excluded\,Time} - (\sum \boldsymbol{DT} + \sum \boldsymbol{ST})} * 100 \tag{1}$$

It indicates the current diversion of in-time production with respect to an ideal cycle (Almeanazel 2010). The denominator in Eq. 1 represents the time which the machine is really producing. The term $\sum DT$ (Downtime) is the sum of all no planned stoppages. The term $\sum ST$ (Stoptime) represents the sum of planned stoppages. In this paper we make use of a direct measure from the equipment, provided by FIS data, and obtained from activity "Machine Working". Using the time difference between its Start and Complete events, it can be computed the measured time cycle.

Trend Analysis (TA) is a statistical tool used to fit a general trend model based on time series data. The application of TA is appropriate because the nature of the component deterioration trend is presented according to a time series form (Maurya et al. 2010). The Double Exponential Smoothing (DES) model is one of the practical forecasting models that can be used to predict trending, being specifically used to forecast the future point based on the non-linear trend model (Ahmad and Kamaruddin 2013). The DES model uses an equation for estimating the forecasting value and another equation to estimate the trending, as shown in Eqs. (2–4).

$$C_t = \alpha Y_t + (1 - \alpha)(C_{t-1} + T_{t-1}) \tag{2}$$

$$T_t = \beta(C_t - C_{t-1}) + (1 - \beta)T_{t-1} \tag{3}$$

$$F_{t+1} = C_t + T_t, \tag{4}$$

where Y_t is the actual value in time t, α is the value-smoothing constant, β the trend-smoothing constant, C_t is the smoothing value for time t, T_t is the smoothing trend value for time t, and F_{t+1} is the forecast value for time $t + 1$.

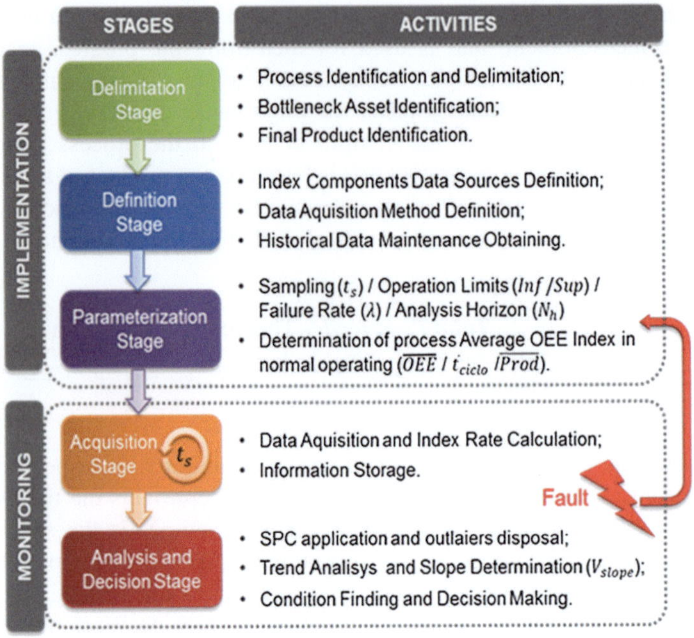

Fig. 1 Proposed methodology (Kurscheidt Netto et al. 2014)

3 Proposed Methodology

A methodology was proposed in Kurscheidt Netto et al. (2014) to use the performance measures obtained from an equipment to monitor its condition and therefore, define the optimal moment to preventive maintenance. Figure 1 illustrates the proposed methodology.

Analyzing the variation trend of these measures, it is possible to estimate the equipment condition and the moment for the preventive maintenance action. The phases are defined as Implementation phase, which aims to structure the methodology based on equipment information to be monitored and Monitoring phase, where the equipment condition is monitored and the decision for the maintenance action is taken.

4 Case Study

To validate the proposed methodology, correlation analyses were applied to a database collected from a lathe machine (CNC Turning) through a FIS, installed in an automobile industry. A period of eight months was considered for the event logs of the productive process, totalizing 85,830 events, structured in 85 cases. Each

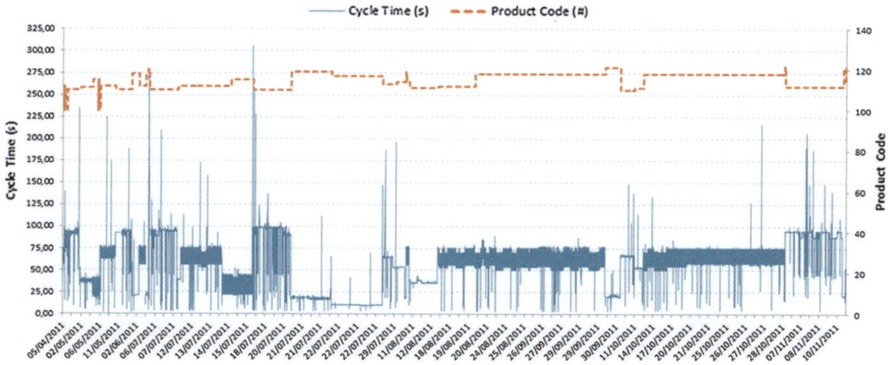

Fig. 2 Normal structure for production activities

case represents one production day. The equipment under analysis is used for the manufacturing of 13 types of products, with different cycle times and number of operations. Figure 2 graphically summarizes the cycle time variation according to the product type during the period under analysis. The curve represents the cycle time filtered, to eliminate incorrect events, e.g. production times higher than 1000 s. Nevertheless, a noise level was observed in the measures obtained, requiring its treatment for the conduction of the study object of this paper (Fig. 2).

5 Results and Discussion

Events registered by the FIS (taskID) were classified into Activities and Events, aiming to extract the process main structure represented by the Activities and identify the occurring deviations, represented by the Events. Events, in their turn, were classified into Production Events, Quality Events, and Maintenance Events. Table 1 summarizes the results obtained with the mining of the event log under analysis, identifying the four activities that compose the process and events classified as non-scheduled stoppages.

Four types of products were selected for the analysis, representing around 80 % of the production in the period considered, as pointed out in Table 2. Based on the execution time for activity “Machine Working”, the cycle times were obtained for each day of production and products selected in Table 2. In order to conduct the analyses, the outliers values were removed in order to keep the standard deviation of the cycle time acquisitions at about ±2 s for products #110 and #111, and about ±10 s for products #112 and #118. After that, days with 10 or less pieces produced were removed.

The DES method was employed, with mean smoothing factors and trend defined in 0.6 for the analysis of the respective cycle time trend (Mean and Standard Deviation). The compiled results are shown in Fig. 4, in which both the mean trend

Table 1 Structure activities and maintenance events

Activity description	Freq	Relative freq. (%)	Classification
Machine working	25,142	29.2928	Activity
Finished part	20,601	24.0161	Quality event
Product removal	20,437	23.8110	Activity
Short stoppage	18,662	21.7430	Activity
Adjustments	298	0.3472	Maintenance event
Tool wear	54	0.0629	Maintenance event
Autonomous maintenance	6	0.0070	Maintenance event
Electrical maintenance	2	0.0023	Maintenance event

Table 2 Selection of relevant products for analysis

Prod code	Quantity	Relative freq (%)	Cumulative (%)	Days of prod
118	9141	36.36	36.36	21
112	5494	21.85	58.21	20
110	3614	14.37	72.58	20
111	1697	6.75	79.33	9

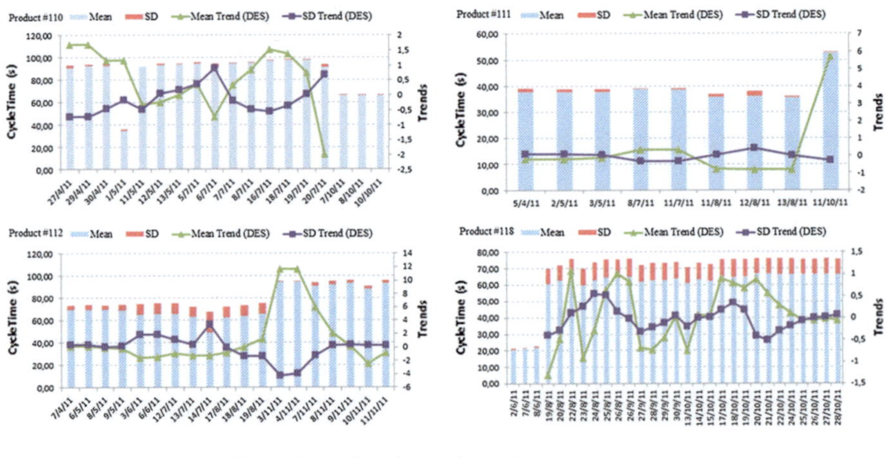

Fig. 3 Results of mean, SD and trending for each product

and the standard deviation follow normal distributions, at a reliability criterion of 95 %. It is possible to observe the occurrence of variations in the cycle time mean values of some products. Operation errors are regarded as factory-floor related, e.g. incorrect definition of the product type and/or acquisition errors of the events timestamp. The analysis is validated through the removal of those values for verification of trends and variations. As attested in the results shown in Fig. 3, the cycle times mean and/or standard deviation tend to increase in certain periods. Such

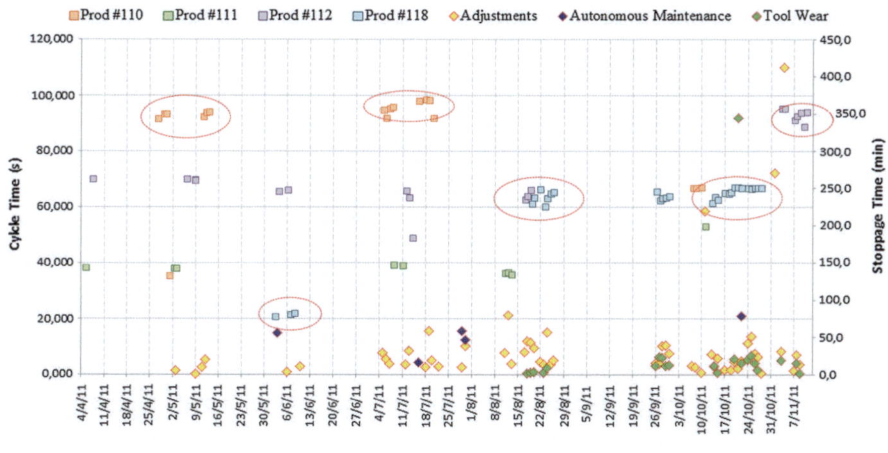

Fig. 4 Graphical correlation between mean cycle time and moment of stoppage

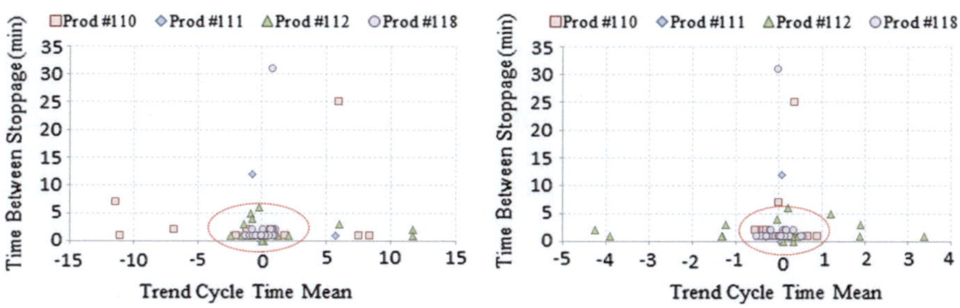

Fig. 5 Correlation between TBS and mean/SD trends for each product

trends were related to non-scheduled stoppage events (Downtimes) of the machine under analysis, specifically those related to Maintenance Events, as presented in Table 1.

Figure 4 brings the results of the stoppage total time measurement for each day of production analyzed and related to the cycle time mean for each product. The dashed circles highlight the periods in which an increase of the cycle time mean occurred. Considering that the database does not include additional information on the root of such deviations, it is assumed that they are caused by the equipment degradation. So as to found such hypothesis, the mean trend value and cycle time SD were correlated with the Time Between Stoppages (TBS).

Figure 5 presents the results obtained and is possible to verify the grouping of the mean and the SD measurements, represented by the points inside the highlighted circles. Regarding the correlation with the mean trend, the analysis of the results shows that product #110 data are more dispersed, while regarding SD trend product #2 is more dispersed. Values for other products are grouped in a certain region, suggesting that there is a correlation between the Time Between Stoppages and the trend degree of the mean and/or the SD. In this analysis, the correlation with all Maintenance Events is considered, although the Electrical Maintenance event has almost no occurrence.

6 Conclusions

In this paper, we consider the hypothesis that any reduction in the equipment speed is due to factors such as degradation, aging or fatigue-related failures. With the results obtained, it is possible to attest the correlation between the mean and standard deviation trend of cycle time with the occurrence of maintenance events. Although measures present relative dispersion, most values obtained are grouped.

Indeed, the measurement quality of the cycle time, contributing for the accuracy of the methodology proposed by Kurscheidt Netto et al. (2014), can be emphasized. The existence of noise in the values acquired must be observed, increasing the factor variance and influencing the operational limits, thus impacting on the correct identification of the slope value. The main cause determined after the database analysis refers to event registration errors by the operator. In addition to the main objective of this paper, the need for the correct use of events for the explicit and well defined delimitation of the start and end working shifts, machine operation state and start/end of in-progress production. These events help employ mining algorithms, reducing the noise and contributing to the application of the proposed methodology.

References

Abeygunawardane SK, Jirutitijaroen P, Xu H (2013) Adaptive maintenance policies for aging devices using a Markov decision process. IEEE 28(3):3194–3203

Ahmad R, Kamaruddin S (2012) An overview of time-based and condition-based maintenance in industrial application. Comput Ind Eng 63:35–149

Ahmad R, Kamaruddin S (2013) Maintenance decision-making process for a multi-component production unit using output-based maintenance technique: a case study for non-repairable two serial components. Unit Int J Performability Eng 9(3):305–319

Almeanazel OTR (2010) Total productive maintenance review and overall equipment effectiveness measurement. Jordan J Mech Ind Eng 4(4):517–522

Jardine AKS, Lin D, Banjevic D (2006) A review on machinery diagnostics and prognostics implementing condition-based maintenance. Mech Syst Signal Process 20(7):1483–1510

Kumar U, Galar D, Parida A, Stenstrom C, Berges L (2013) Maintenance performance metrics: a state-of-the-art review. J Qual Maintenance Eng 19(3):233–277

Kurscheidt Netto RJ, Santos EAP, Loures ER, Pierezan R (2014) Condition-based maintenance using OEE: an approach to failure probability estimation. In: Proceedings of 7th international conference on production research—Americas 2014, Lima

Maurya MR, Paritosh PK, Rengaswamy R, Venkatasubramanian V (2010) A framework for on-line trend extraction and fault diagnosis. Eng Appl Artif Intell 23(6):950–960

Nakajima S (1988) An introduction to TPM: total productive maintenance. Productivity Press, Portland

Rozinat A, Mans RS, Song M, Van Der Aalst WMP (2009) Discovering simulation models. Inf Syst 34(3):305–327

Santos EAP, De Freitas RL, Deschamps F, De Paula MAB (2008) Proposal of an industrial information system model for automatic performance evaluation. In: IEEE international conference on emerging technologies and factory automation, pp 436–439

Van Der Aalst WMP, Schonenberg MH, Song M (2011) Time prediction based on process mining. Inf Syst 47(2):237–267

Venkatasubramanian V, Rengaswamy R, Kavuri SN, Yin K (2003) A review of process fault detection and diagnosis: part III: process history based methods. Comput Chem Eng 27(3):327–346. Elsevier

Wang W (2012) An overview of the recent advances in delay-time-based maintenance modeling. Reliab Eng Syst Safe 106:165–178

Weber P, Bordbar B, Tino P, Majeed B (2011) A framework for comparing process mining algorithms. In: GCC conference and exhibition (GCC), 2011 IEEE, pp 625–628

Forecasting Cloud Computing: Producing a Technological Profile

Iñaki Bildosola, Rosa Rio-Bélver, Ernesto Cilleruelo and Javier Gavilanes

Abstract Migrating to cloud computing is one of the current enterprise challenges. In this sense, the small and medium enterprise should be the most interested, given that initial investments are avoided and the technology offers gradual implementation. However, 54.9 % of SMEs confess that they have no knowledge of cloud technology. Accordingly, this paper aims at generating a relevant profile of cloud computing technology, as the first part of a novel approach based on four families of technological forecasting methods to gather and structure information concerning an emerging technology, generating a relevant profile, identifying its past development, forecasting the short and medium-term evolution and integrating all of the elements graphically into a hybrid roadmap. The outcome of the approach will raise the awareness of such technology as well as facilitate its implementation.

Keywords Cloud computing · Forecasting · Technological information · Mapping the knowledge · Text mining

I. Bildosola (✉) · E. Cilleruelo
Foresight, Technology and Management (FTM) Group, Department of Industrial Engineering, University of the Basque Country UPV/EHU, alameda Urquijo s/n, 48030 Bilbao, Spain
e-mail: inaki.bildosola@ehu.es

E. Cilleruelo
e-mail: ernesto.cilleruelo@ehu.es

R. Rio-Bélver · J. Gavilanes
Foresight, Technology and Management (FTM) Group, Department of Industrial Engineering, University of the Basque Country UPV/EHU, calle Nieves Cano 12, 01006 Vitoria, Spain
e-mail: rosamaria.rio@ehu.es

J. Gavilanes
e-mail: javier.gavilanes@ehu.es

M. Amorim et al. (eds.), *Engineering Systems and Networks*, Lecture Notes in Management and Industrial Engineering, DOI 10.1007/978-3-319-45748-2_3

1 Introduction

Many attempts have been made at mapping the knowledge of a given technology to determine its possible evolution. To this effect, foresight techniques are continuously growing based on certain primary science characterization methods. Scientometrics is the quantitative study of scientific communication, which applies bibliometrics, among other techniques, to scientific literature (Börner et al. 2003). The conjunction of scientometrics and visualization techniques is the basis of science mapping. Maps of science are two- or three-dimensional representations of a scientific field in which the items on the map refer to themes in the mapped field (Noyons 2001). There have been attempts to summarize and structure the field, however, since initial studies, technology forecasting experts agree that models should be used in combination. On that basis, a previous work (Bildosola et al. 2015) was presented which proposed a novel approach to gathering and structuring information concerning an emerging technology, generating a relevant profile, identifying its past evolution, forecasting the short and medium-term evolution and integrating all of the elements graphically into a hybrid roadmap.

One of the most used definitions for cloud computing is from the NIST (Mell and Grance 2011) which states: "cloud computing is a model for enabling ubiquitous, convenient, on-demand network access to a shared pool of configurable computing resources that can be rapidly provisioned and released with minimal management effort or service provider interaction". It is expected to be the main driving force behind IT sector development with SMEs as its strategic market. Moreover, it is expected that 13.8 million jobs will be created over the 2011–2015 period (Gantz et al. 2012). However, 54.9 % of SMEs confess that they have no knowledge of cloud technology. This circumstance results in a loss of SME competitiveness, contrasted with the fact that this technology model is particularly beneficial for them (Rio-Belver et al. 2012). Thus, the need to generate a helpful profile and a technology roadmap for cloud computing in order to raise the awareness of such and to facilitate its implementation can be appreciated. Accordingly, this paper aims at generating a relevant profile of cloud computing technology based on the aforementioned approach. This information will provide a comprehensive basis to generate the hybrid roadmap. The paper is divided as follows: in section two the research approach is described based on three steps. The third section shows the results of its application to cloud computing. Finally, the most significant conclusions and the future work intentions are presented.

2 Research Approach, Generating the Profile

The research approach is that proposed by Bildosola et al. (2015) which was structured in eight steps containing four families of technological forecasting methods, namely: Statistical Methods in terms of Bibliometrics and Data Mining;

Trend Analysis; Technological Roadmapping; and Expertise. In the present work the first three steps are developed to obtain the cloud computing technology profile, which also initiate the remaining steps. In order to provide as much information as possible, the steps are clearly disaggregated and for those tasks where the text mining tool VantagePoint has been used, its utilization is clearly described.

Step 1. Retrieving Data and Refining the Search
Peer-reviewed (PR) journal articles stored in scientific databases are good information sources for tracking scientific activity (Porter and Cunningham 2005). Accordingly, the Science Citation Index (SCI) and the Scopus database were selected due to their broad coverage of several scientific disciplines and the availability of full citation data. More specific databases were discarded given that the main goal of the present work is to generate a profile centered on the technology itself, thus research into technology applications was of little interest to us. Even if a complete refinement is not feasible, in order to reduce the noise influence in the profile characterizations and its conclusions, such reduction was sought through refinement in terms of excluding articles not directly related to cloud computing technology field research. The whole process is performed directly in each database web tool, based on Boolean conditions. The data range is 2008–2013, from the very first cloud computing related publication to the last complete data year. Finally, a refined dataset of 4033 publications was obtained as shown in Table 1.

Step 2. Cleaning Up the Refined Dataset
The records containing the bibliographic information were downloaded and imported into VantagePoint software. Due to the limitations of the web tools when it comes to import the data, all the importation was done divided in time intervals and merging it all with the *Data Fusion* tool. Afterwards, the data set was cleaned

Table 1 Boolean refinement of the search

Searches	Search description	SCI + SCOPUS
1st search	Search *cloud computing*, in title, abstract or keywords fields	25,229 + 11,152
1st refinement	Search "*cloud computing*", in title, abstract or keywords fields	18,129 + 6300
2nd refinement	Limit the search to the title field	5054 + 2163
3rd refinement	Limit the search to the keywords field (only available in SCOPUS)	4720 + 2163
4th refinement	Exclude *overview*, *review*, *based on cloud computing*, *cloud computing application*, etc. terms	4287 + 1631
5th refinement	Refine to *computer science* and *telecommunications* related fields	3146 + 1216
6th refinement	Limit to *conference paper*, *article* and *book chapter*	2904 + 1129

in order to eliminate duplications; the title field was selected for that purpose and using the *Remove Duplications* tool the dataset was reduced to 3236 records. In addition to this, taking into account that the same article can be referenced twice as conference paper and as a publication, the same removal process was applied but based on the abstract field, producing 3232 elements. Lastly, those invalid records with the majority of the fields empty or corrupted were excluded, resulting in a cleaned database of 3194 records. Before mining the data, some fields had to be treated using fuzzy logic and *Thesaurus* tool with VantagePoint, in order to group elements of equal meaning. Furthermore, a manual post-refinement step had to be performed. However, a very conservative attitude was taken in this phase.

Step 3. Generating the Profile

Literature Profile and Research Community Profile
First of all, the publication itself is profiled, for which fields such as affiliation or authors are analyzed and mined within the records and presented by text summarization process. The aim of the sub-step is to obtain clear information about cloud computing research publication features such as the number of researchers and practitioners involved in it; the number of the publications regarding its academic or corporation origin or the publication mode such as journals and proceedings.

Secondly, the cloud computing technology research community is described. The community can be portrayed by identifying prominent authors along with their contributing institutions and professional networks (Gerdsri et al. 2013), which can be completed with reference country, conference and journal identification through text summarization. Text summarization is performed based on a diverse kind of summary list, which are easy to create and use with VantagePoint. Such lists can provide the first order of useful information. Moreover, two such lists can be combined to create a co-occurrence matrix to show the concentration of activity (Kongthon 2004). As a result, in order to identify professional networks and collaborations, the author–keyword map, namely a cross-correlation map, is generated as a visualization element based on the text summarization process.

State of the Research and Its Evolution
Once the research community has been identified, the technology itself is profiled. Based on text summarization, the main concepts dominating the technology field can be detected by looking at the top keywords present in each year under study. In addition to which usage of keywords throughout publications over time is performed, based on the study of trends to detect changes in keywords. This information makes it possible to present certain insights into emerging issues for the technology development over time (Gerdsri et al. 2013). The trend in the number of publications related to a technology helps to comprehend its position in the technology life-cycle, indicating if it is in an embryonic, growing or declining phase (Yoon et al. 2013). It

should be noted that keyword analysis has been performed specifically based purely on author's keywords for its availability (86 %) and homogeneity across the selected databases. Furthermore, this field contains the keywords that, in the opinion of the author, best characterize the research performed, creating a powerful tool with which to characterize the research (Garechana et al. 2014). In addition to this, reduction of the data set to the most frequent items is not strange in coword analysis, given the usually high number of infrequent keywords and for that purpose only the top 10 % of the most frequently used keywords were selected to discover trends and professional networks.

3 Results and Discussion, Cloud Computing Technology Profile

The following figures are derived by analyzing the obtained dataset in order to perform the literature review. The community holds 7023 active authors from 89 countries and the collaboration degree is high given that more than 90 % of the publications have more than one author; specifically the average author number is 3.16. There are 4558 organizations generating publications, an average of 1.61 per paper. Insofar as the organizations are concerned, the leading contributors are from academia with 77 % of the publications, followed by corporations with 7 %; public sectors with 2 %; and 14 % of the publications are collaborations. The type of publications can be separated into two main groups: journal papers which represent 22 % and conferences papers with 78 %. The research community is primarily described identifying those conferences and journals which strongly contribute to the body of knowledge. Table 2 shows that three of the top five journals are

Table 2 Leading conferences and journals

Journals publications	Articles	Leading conferences	Articles
Journal of supercomputing	22	IEEE international conference on cloud computing technology and science (CloudCom)	80
Tongxin Xuebao/journal on communications	18	IEEE 5th international conference on cloud computing (CLOUD)	49
Jisuanji Xuebao/Chinese journal of computers	16	International conference on cloud computing and services science (CLOSER)	41
Computer	15	International conference on computer and information science (ICCIS)	35
Journal of Huazhong University of Science and Technology	15	IEEE international conference on cloud computing and intelligence systems (CCIS)	35

Chinese journals, the country spearheading cloud computing research. It is significant that all the top conferences, save one, have been initiated in the 2010s as specific cloud computing conferences, with an emphasis on CloudCom as the first (2009) and most prolific conference, initiated in Beijing Jiatong University.

Table 3 presents the leading countries, contributing organizations and principal authors in each one. This information provides a starting point for researchers and practitioners to identify the appropriate institution and authors in their respective country with which to collaborate. Asian authors dominate the field and it is worth noting that IBM is the only business company among the country leaders; this is partly justified by the fact that academic institutions usually initiate the embryonic study of cutting edge technologies, whereas private companies (those with sufficiently large enough research oriented departments) follow behind focusing more on business oriented research.

Identifying shared research interests gives rise to future collaborations. Figure 1 shows the obtained author-keyword map for year 2013. For instance, Yan C and Yang Z from Beijing University of Posts and Telecom (China), share interest in "Mobile Cloud Computing" with Cho J from Kyungspook National University (South Korea), although they have never collaborated. Different shared interest clusters are identified through the map which can generate new research networks.

Analyzing the state of the research and its evolution, as shown in Table 4, ensure (data) security is the cornerstone of cloud computing technology. Thus, "security" has been the top issue from the very outset. "Virtualization" technology as one of the cloud computing enablers is also a common topic within top keywords. In

Table 3 Top countries, organizations and authors leading cloud computing research

Country (articles)	Top organizations (articles)	Principal authors (affiliation) (articles)
China (944)	Beijing University of Posts and Telecom (58)	Wang X (Northeastern University) (14)
	Tsinghua University (51)	Jin H (Huazhong University of Science and Technology) (12)
	Wuhan University (49)	Sun D (Northeastern University) (11)
USA (554)	IBM (41)	Ren K (Illinois Institute of Technology) (11)
	Arizona State University (27)	Fu S (University of North Texas) (11)
	University of North Texas (18)	Lou W (Worcester Polytechnic Institute) (11)
India (261)	Anna University (17)	Kundu A (Netaji Subhash Engineering College) (5)
	Pondicherry University (13)	Banerjee C(Netaji Subhash Engineering College) (5)
South Korea (169)	KyungHee University (25)	Huh EN (Kyunghee University) (10)
	Hannam University (16)	Lee H (Konkuk University) (10)
Taiwan (160)	Tunghai University (19)	Yang CT (Tunghai Univeristy) (12)
	National Taiwan university (17)	Chang BR (National University of Kaohsiung) (10)

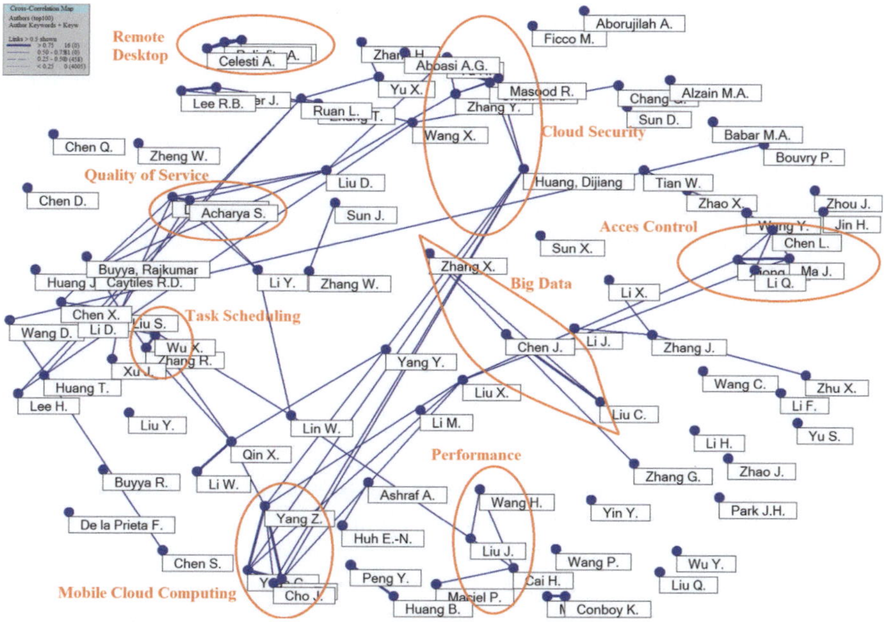

Fig. 1 Research network connections: shared interests (author-keyword map)

Table 4 Top most frequently used keywords and most increased keywords

2009 (265)	2010 (729)	2011(1262)	2012 (1514)	2013 (1626)
Grid computing (6.1) Service oriented architecture (6.1) Virtualization (5.6) Software as a service (5.0) Distributed computing (5.0)	Cloud computing security (6.4) Virtualization (6,4) Privacy (5.1) Distributed computing (4.9) Software as a service (4.0)	Cloud computing security (9.2) Virtualization (6.1) Software as a service (4.6) Infrastructure as a service (3.2) Service level agreement (3.1)	Cloud computing security (8.7) Virtualization (5.5) Virtual machine (4.4) Privacy (4.4) Service level agreement (4.0)	Cloud computing security (8.8) Virtual machine (4.9) Mobile cloud computing (4.7) Privacy (4.5) Distributed computing (4.3)
Grid computing (1000) Service oriented architecture (1000) Software as a service (800) Cloud computing security (600)	Service level agreement (700) Risk management (700) Trust computing (700) Infrastructure as a service (650)	e-Governance (900) Eucalyptus (900) Mobile cloud computing (850) Secure cloud computing (700) Dependability analysis (700)	Reliability (1200) Adoption of cloud computing (600) Game theory (600) Digital forensic analysis (500) Middleware (500)	Attributed based access control (500) Log analysis (500) Outsourced data (500) CRM (400) Trusted cloud computing platform (400)

The values in brackets for the years represent the number of keywords within the publications; for "top keywords" their frequency and for "most increased" their growth rate in percentage values addition to this, research is continuously focused on cloud computing delivery models, a huge, diverse field of study, represented by “software as a service” or “infrastructure as a service” keywords. Emerging keywords hint at possible future topics. For instance, “mobile cloud computing” first appeared within keywords increasingly used and then moved to the top position. In this sense, specific fields such as “CRM” or “attributed based access control” and other similar ones will be hot research topics from now on. The technology has almost finished its growth phase and, consequently, those top increasing themes at their embryonic phase (2008/09) which possessed a more generalized viewpoint, are being replaced by more specific ones.

4 Conclusions and Future Work

The cloud computing technology profile provides us with valuable information. The technology has surpassed its embryonic phase and it is already starting its application-focused or even business-focused research and innovation phase. In this regard, rising topics show us the future paths, represented among others by mobile cloud computing; different access control techniques and specific analytic methods such as log analysis. The technology is currently attracting a huge amount of research and practitioners and although the collaboration of different kind of organizations is high (14 %), it has further significant potential to create networks, with special mention to academic-enterprise collaborations. Asian universities are leading the field and it remains to be seen who will lead the next phase, even if some, such as IBM and Amazon (Eucalyptus), are already filling that space.

This paper has been focused on presenting a comprehensive profile of cloud computing technology. Several elements, such as the factor analysis map to identify clusters within the technology, have not been included due to size limitations and by the fact that they will be specially focused on the second phase of this work. This second phase will be centered on generating a technology roadmap as has been mentioned previously and will include elements such as more application-focused bibliometric data; a clustering based ontology-generation process and expertise based final reconstruction. In addition to this, “social mention” and “web scraping” techniques for extracting information from websites will be included in order to establish the latest evolution of the technology.

References

Bildosola I, Rio-Bélver R, Cilleruelo E (2015) Forecasting the big services era: novel approach combining statistical methods, expertise and technology roadmapping. In: Enhancing synergies in a collaborative environment, pp 371–379. Springer International Publishing, Cham

Börner K, Chen C, Boyack KW (2003) Visualizing knowledge domains. Ann Rev Inf Sci Technol 37(1):179–255

Gantz JF, Minton S, Toncheva A (2012) Cloud computing's role in job creation. IDC. Available: http://news.microsoft.com/download/features/2012/IDC_Cloud_jobs_White_Paper.pdf

Garechana G, Rio-Belver R, Cilleruelo E, Larruscain Sarasola J (2014) Clusterization and mapping of waste recycling science. Evolution of research from 2002 to 2012. J Assoc Inf Sci Technol 66(7):1431–1446

Gerdsri N, Kongthon A, Vatananan RS (2013) Mapping the knowledge evolution and professional network in the field of technology roadmapping: a bibliometric analysis. Technol Anal Strateg Manag 25(4):403–422

Kongthon A (2004) A text mining framework for discovering technological intelligence to support science and technology management. Georgia Institute of Technology, Georgia

Mell P, Grance T (2011) The NIST definition of cloud computing. NIST, Special Publication 800–145, Gaithersburg, MD

Noyons E (2001) Bibliometric mapping of science in a policy context. Scientometrics 50:83–98

Porter AL, Cunningham SW (2005) Tech mining: Exploiting new technologies for competitive advantage. Wiley-Interscience, Hoboken

Rio-Belver R, Cilleruelo E, Garechana G, Gavilanes J, Zabalza (2012) New management models based in cloud-computing. Paper presented at 7th international scientific conference business and management, Vilnius, Lithuania, 10–11 May

Yoon B, Phaal R, Probert DR (2013) Structuring technological information for technology roadmapping: data mining approach. Technol Anal Strateg Manag 25(9):1119–1137

Modelling the Strategies Alignment Process in the Collaborative Network Context

Beatriz Andres and Raúl Poler

Abstract A model to deal with the strategies alignment process, in the collaborative network (CN) context, is proposed. The main aim is to support the decision making of identifying which strategies to activate within the network in order to be aligned. This model provides a global view of all the strategies formulated in the CN, in order to identify those ones that have higher levels of alignment. The strategies alignment model is applied in an illustrative example.

Keywords Mathematical model · Strategies alignment · Collaborative networks

1 Statement of the Problem

Collaborative networks (CN) are characterised by the enterprises heterogeneity (Camarinha-Matos and Afsarmanesh 2005), each one defining its own objectives, heterogeneous too. Consequently, a high diversity of strategies is formulated, in order to reach the objectives defined. The strategies diversity may result in conflict situations among enterprises of the same CN. The strategies misalignment may, ultimately, lead to the failure of the collaborative partnership if the conflicts that arise are not tackled. In order to deal with these conflicts, the strategies alignment process is addressed, among enterprises of the same CN.

Two or more strategies are aligned when each activated strategy has positive influences on (i) the objectives defined in the same enterprise in which the strategy is formulated, and (ii) the objectives defined by the rest of the network partners. Therefore, the strategies are aligned when apart from being able to exist together

B. Andres (✉) · R. Poler
Research Centre on Production Management and Engineering (CIGIP),
Centre d'Innovació i Investigació, Universitat Politècnica
de València (UPV), Calle Alarcón, 03801 Alcoy, Spain
e-mail: bandres@cigip.es

R. Poler
e-mail: rpoler@cigip.upv.es

M. Amorim et al. (eds.), *Engineering Systems and Networks*, Lecture Notes
in Management and Industrial Engineering, DOI 10.1007/978-3-319-45748-2_4

harmoniously, if simultaneously activated, enhance the objectives' performance level (Andres and Poler 2014a). The collaborative strategies alignment enables to adequately coordinate the network objectives and appropriately identify those strategies that have positive impacts on the objectives defined by all the networked partners.

1.1 Motivation

The strategies alignment process is considered a relevant process to deal with form the CN context (Andres and Poler 2015). Considering a CN, the enterprises define the objectives to be reached and formulate a set of strategies. Amongst all of the formulated strategies, the enterprises can activate some of them in order to attain the defined objectives at the minimum cost. Considering this, the enterprises' interest lies in knowing what is the contribution of each activated strategy to achieve its objectives, taking into account the strategies costs. In order to measure this achievement the KPIs are used. When an enterprise has to deal with the decision of what strategies activate, in order to achieve its objectives, can opt to make the decision from an isolate perspective (non-collaborative scenario), or, on the contrary, from a common perspective (collaborative scenario). The non-collaborative decision may lead to a situation in which the strategies are misaligned deriving on the partnership failure, due to the enterprises only consider how the strategies affect on its objectives without taking into account how these strategies affect the other network enterprises' objectives. On the other hand, in the collaborative alternative, each enterprise decides which set of strategies activate, by not only taking into account the extent into which its own objectives are achieved, but also considering how the activated strategies influence the objectives defined by other network enterprises.

1.2 Objective and Research Question

Despite the importance of aligning strategies, in terms of avoiding partnership conflicts, to the best of our knowledge, it can be stated that there is a gap in the literature as regards contributions providing a holistic approach that allows considering all the strategies formulated by all the partners and model the influence that these strategies exert on the wide diversity of objectives defined. In order to fill this gap, the aim of this research is to propose an integrated approach to identify the aligned strategies form a holistic perspective, regardless of their nature and type, taking into account the CN context.

The following research question is raised to support the strategies alignment process, in order to solve them as the objective of this research.

What would be an appropriate model, method, guideline and tool to adequately support enterprises on the modelling, assessment and resolution of the strategies alignment process from a collaborative perspective?

2 Research Method

The research method used in this thesis is the *Constructive Research Approach* (CRA) proposed by Kasanen et al. (1993). CRA is more focused on providing solutions to relevant problems, these solutions are proposed in a novel way considering previous existing knowledge. In the CRA, the practical application of the proposed solution has to be demonstrated (Crnkovic 2010).

3 Contributions of the Thesis

3.1 Strategies Alignment Model

In order to model the strategies alignment process, representing the influences and relations between the KPIs and the strategies, a mathematical model is proposed. The strategies alignment model allows assessing possible misalignments among the strategies formulated within the enterprises of a CN. Supporting the decision making to identify which strategies to activate in order to attain higher levels of alignment not only in the same enterprise but also with the rest of enterprises of the CN. First of all, the set of parameters and decision variables, used to model the strategies alignment process, are defined in Table 1.

The main aim of the model is to identify amongst all the formulated strategies those that are aligned. That is, the strategies that positively influence all the objectives defined by the networked partners, increasing the performance at enterprise level and subsequently at network level. Thus, the developed model supports on the decision making as regards the number of units of strategy (u_str_{is}) to be activated and the time in which the strategies have to be activated (ti_str_{is}) with the main aim of maximising the network performance.

$$\text{Objective Function: max.} \quad \Delta kpi_{net} \tag{1}$$

$$\Delta kpi'_{ixk} = \frac{\Delta kpi_{ixk}}{\Delta kpi^{max}_{ixk}} \tag{2}$$

$$H' = \frac{H}{H} = 1 \tag{3}$$

$$d'_1\_str_{is} = \frac{d_1\_str_{is}}{H}; \quad d'_2\_str_{is} = \frac{d_2\_str}{H}; \quad d'_4\_str_{is} = \frac{d_4\_str}{H} \tag{4}$$

$$d'_3\_str_{is} = d'_4\_str_{is} - 2 \cdot d'_{2str_{is}} - d'_1\_str_{is} \tag{5}$$

Table 1 Model index and parameters

Index	
net set of networks, *net* = (1,...,*N*)	*x* set of objectives, *x* = (1, *X*)
i set of enterprises, *i* = (1, ..., *I*)	*k* set of key performance indicators, *k* = (1, ..., *K*)
	s set of strategies where *s* = (1, ..., *S*)
Model parameters	
n	Number of enterprises belonging to the network
o_{ix}	Objective *x* defined in enterprise *i*
bi	Budget owned by the enterprise *i* to invest in the activation of the strategies in monetary units [m.u.]
str_{is}	Strategy *s* defined by enterprise *i*
kpi_{ixk}	Key performance indicator (KPI) *k* used to measure the objective O_{ix}
Δkpi_{ixk}	Increase observed in the kpi_{ixk} when the str_{is} is activated
Δkpi_{ixk}^{intra}	Increase of the kpi_{ixk} when the $str_{is}$of the same enterprise *i* (e_i) is activated
Δkpi_{ixk}^{inter}	Increase of the kpi_{jxk} when the str_{js} of a different enterprise *j* (e_j) is activated
Δkpi_{ixk}^{max}	Maximum increase of kpi_{ixk} estimated by the *enterprise i*
$Threshold\_kpi_{ixk}$	Value from which the associated kpi_{ixk} is affected by the activation of a strategy str_{is}
Δkpi_{ixk}_T	Increase experienced by the kpi_{ixk} once the $Threshold\_kpi_{ixk}$ is computed
Δkpi_{ixk}_min	Minimum increase that the enterprise determines for the kpi_{ixk}, once the $Threshold\_kpi_{ixk}$ is computed
W_{ixk}	Relevance that the kpi_{ixk} has for enterprise *i* (weight of kpi_{ixk})
Δkpi_i	Increase of KPI at enterprise *i* level
Δkpi_{net}	Increase of KPI at network *net* level
$f\_inf\_str_{is}\_kpi_{ixk}(t)$	Function that models the behaviour of the kpi_{ixk} when str_{is} is activated
$f\_kpi_{ixk}(t)$	Function that models the overall behaviour of the kpi_{ixk} considering all the activated strategies
$f\_kpi'_{ixk}T(t)$	Function that models the behaviour of the kpi_{ixk} when the $Threshold\_kpi_{ixk}$ value is computed
c_str_{is}	Cost of activating one unit of strategy str_{is} [m.u.]
str_{is}_mu	Monetary units invested in the activation of str_{is}[m.u.]
$val\_str_{is}\_kpi_{ixk}$	Increase or decrease of the kpi_{ixk} when one unit of str_{is} is activated(u_str_{is} = 1)
$inf\_str_{is}\_kpi_{ixk}$	Maximum level of influence on the kpi_{ixk} when certain number of units of strategies (u_str_{is}) are activated
$slope\_str_{is}\_kpi_{ixk}$	Slope of the ramp in represented in $f\_inf\_str_{is}\_kpi_{ixk}(t)$
H	Horizon, time units [t.u.], period of time in which the set of strategies are to beactivated (H = 1)
$d_1\_str_{is}$	

(continued)

Table 1 (continued)

	Delay, time period between the initial time of activation of str_{is}.(ti_str_{is}) and the time when the kpi_{ixk} is started to be influenced by the activated str_{is} [t.u.]
$d_2\_str_{is}$	Time period between the str_{is} starts to influence the kpi_{ixk} until the maximum level of influence in is achieved ($inf\_str_{is}\_kpi_{ixk}$), [t.u.]
$d_3\_str_{is}$	Time period in which str_{is} is exerting the highest influence ($inf\_str_{is}\_kpi_{ixk}$) on the kpi_{ixk} [t.u.]
$d_4\_str_{is}$	Total duration of str_{is} [t.u.]
tf_str_{is}	Unit of time when str_{is} is finished [t.u.]
Decision variables	
u_str_{is}	Units of strategy str_{is} to be activated [u.s]
ti_str_{is}	Initial time of activation of str_{is} [t.u.]

$$t'_i\_str_{is} = \frac{t'_i\_str_{is}}{H}; \quad t'_f\_str_{is} = \frac{t'_f\_str_{is}}{H}; \tag{6}$$

$$t'_f\_str_{is} \geq H' \tag{7}$$

$$str_{is}\_mu = u\_str_{is} \cdot c\_str_{is} \tag{8}$$

$$b_j \geq \sum_s str_{is}\_mu \quad \forall\ s \tag{9}$$

$$inf\_str_{is}\_kpi'_{ixk} = u\_str_{is} \cdot val\_str_{is}\_kpi'_{ixk} \tag{10}$$

Depending whether $val\_str_{is}\_kpi'_{ixk}$ is positive or negative, the $\Delta kpi'_{ixk}$, will be increased or decreased when strategy str_{is} is activated. The function, $f\_inf\_str_{is}\_kpi'_{ixk}$, is a piecewise function that depends on the time (Fig. 1). The decision variable $t'_i\_str_{is}$ identifies the starting point of the $f\_inf\_str_{is}\_kpi'_{ixk}$, when str_{is} is activated. Besides $t'_i\_str_{is}$ allows modelling that not all the strategies are activated at the same time.

$$f\_\inf\_str_{is\_kpi'_{ixk}}(t) = \begin{cases} 0 \rightarrow t \leq t'_i\_str_{is} + d'_1\_str_{is} \Lambda t \geq t'_i\_str_{is} + d'_4\_str_{is} \\ slope\_str_{is}\_kpi'_{ixk} \rightarrow t'_i\_str_{is} + d'_1\_str_{is} < t < t'_i\_str_{is} + d'_1\_str_{is} + d'_2\_str_{is} \\ inf\_str_{is}\_kpi'_{ixk} \rightarrow t'_i\_str_{is} + d'_1\_str_{is} + d'_2\_str_{is} \leq t \leq t'_i\_str_{is} + d'_1\_str_{is} + d'_2\_str_{is} + d'_3\_str_{is} \\ -slope\_str_{is}\_kpi'_{ixk} \rightarrow t'_i\_str_{is} + d'_1\_str_{is} + d'_2\_str_{is} + d'_3\_str_{is} < t < t'_f\_str_{is} \end{cases} \tag{11}$$

$$slope\_str_{is}\_kpi'_{ixk} = \frac{val\_str_{is}\_kpi'_{ixk}}{d'_2\_str_{is}} \tag{12}$$

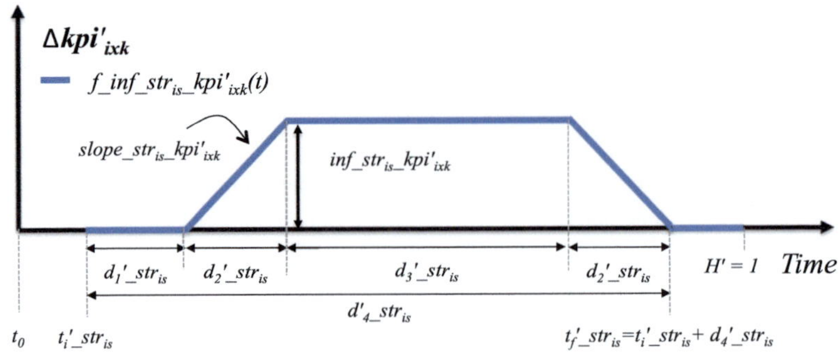

Fig. 1 Representation of $f\_inf\_str_{is}\_kpi'_{ixk}$

The $\Delta kpi'_{ixk}$ is caused by both intra-enterprise influence, $\Delta^{intra} kpi'_{ixk}$ and inter-enterprise influences, $\Delta^{inter} kpi'_{ixk}$, mathematically modelled through:

$$\Delta kpi'_{ixk} = \Delta^{intra} kpi'_{ixk} + \Delta^{inter} kpi'_{ixk} \tag{13}$$

$$\Delta^{intra} kpi'_{ixk} = \int_{t'_i\_str_{is} + d'_1\_str_{is}}^{H'} f\_inf\_str_{is}\_kpi'_{ixk}(t) \cdot dt \tag{14}$$

$$\Delta^{inter} kpi'_{ixk} = \int_{t'_i\_str_{js} + d'_1\_str_{js}}^{H'} f\_inf\_str_{js}\_kpi'_{ixk}(t) \cdot dt \tag{15}$$

The function representing the curve of the total increase of the kpi'_{ixk} is also a piecewise function that depends on time given by:

$$f\_kpi'_{ixk} = f\_inf\_str_{is}\_kpi'_{ixk}(t) + f\_inf\_str_{js}\_kpi'_{ixk}(t) \tag{16}$$

After being depicted the function $f\_kpi'_{ixk}$ the value estimated by the threshold ($Threshold\_kpi'_{ixk}$) must be considered.

$$f\_kpi'_{ixk}\_T(t) = f\_kpi'_{ixk}(t) - Th\_kpi_{ixk} \tag{17}$$

Accordingly

$$\Delta kpi'_{ixk}\_T = \int_a^b f\_kpi'_{ixk}(t) \cdot dt - \int_a^b Th\_kpi_{ixk} \cdot dt \tag{18}$$

where, $f\_kpi'_{ixk}(a) = Threshold\_kpi_{ixk}$ and $f\_kpi'_{ixk}(b) = Threshold\_kpi_{ixk}$

$$\Delta kpi'_{ixk}\_T \geq \Delta kpi'_{ixk}\_min \tag{19}$$

To finally characterise the model, two KPIs are defined. (i) at enterprise level,

$$\Delta kpi'_i = \frac{\sum_{x,k} \Delta kpi'_{ixk}\_T \cdot w_{ixk}}{\sum_{x,k} w_{ixk}} \tag{20}$$

and (ii) at network level the $\Delta kpi'_{net}$ gives a whole perspective on how the performance is behaving at network level.

$$\Delta kpi'_{net} = \frac{\sum_i \Delta kpi'_i}{n} \tag{21}$$

The proposed mathematical optimisation model has its main aim on maximising the network KPI ($\Delta kpi'_{net}$). Through identifying the number of units of strategies to activate (u_str_{is}) and the time when activate them ($t'_i\_str_{is}$), so that make positive impacts on the objectives defined by the nodes of the network.

3.2 Illustrative Example

In order to show how to implement the strategies alignment model, an illustrative example is proposed. Two enterprises are considered, each one defining two objectives, e_1: o_{11} and o_{12}, e_2: o_{21} and o_{22}. The achievement of the objectives is measured through the KPIs each one with its corresponding weights (w_{ixk}), e_1: kpi_{111} (w_{111}) and kpi_{121} (w_{121}), e_2: kpi_{211} (w_{211}) and kpi_{221} (w_{221}). In order to achieve the objectives each enterprise formulates two strategies (e_1: str_{11} and str_{12}, e_2: str_{21} and str_{22}) and defines its related data as regards the durations and costs. The enterprises have a certain budget to carry out these strategies. Finally the values of influence $val\_str_{is}\_kpi'_{xk}$ are estimated by each enterprise (Table 2).

According to the work developed by Andres and Poler (2014b) and Andres et al. (2014), the resolution of the strategies alignment model is done considering the System Dynamics method. Besides, the model is simulated in AnyLogic (2015) simulation software. An optimisation experiment is carried out considering the data of the illustrative example, obtaining that only the str_{11} of enterprise 1 and str_{22} of enterprise 2 are aligned and must be activated in a specific time [t.u.]: $ti\_str_{11} = 0.28$ and $ti\_str_{22} = 0.36$ in order to obtain the maximum network performance, $\Delta kpi'_{net} = 0.267$.

Table 2 Illustrative example data

Enterprise 1 $b_1 = 3$																
													kpi'_{111}		kpi'_{1121}	
													W_{111}	0.5	w_{121}	0.5
													Threshold_kpi'$_{111}$	0.2	Threshold_kpi'$_{121}$	0.1
str$_{11}$	u_Str$_{11}$	?	ti_str$_{11}$	?	c_str$_{11}$	2	d'$_1$_str$_{11}$	0.05	d'_2_str_{11}	0.01	d'$_4$_str$_{21}$	0.6	val_str$_{11}$_kpi'$_{111}$	0.8	val_str$_{11}$_kpi'$_{121}$	0.01
str$_{12}$	u-Str$_{12}$	?	ti_str$_{12}$	?	c_str$_{12}$	3	d'_1_str_{12}	0.2	d'_2_str_{12}	0.03	d'$_4$_str$_{21}$	0.5	val_str$_{12}$_kpi'$_{111}$	0.3	val_str$_{12}$_kpi'$_{121}$	0.7
													val_str$_{21}$_kpi'$_{111}$	0.1	val_str$_{21}$_kpi'$_{121}$	0.3
													val_str$_{22}$_kpi'$_{111}$	0.3	val_str$_{22}$_kpi'$_{121}$	0.2
Enterprise 2 $b_2 = 6$																
													kpi'_{211}		kpi'_{221}	
													W_{211}	0.5	w_{221}	0.5
													Threshold_kpi'$_{211}$	0.3	Threshold_kpi'$_{221}$	0.15
str$_{21}$	u_Str$_{21}$	?	ti_str$_{21}$	?	c_str$_{21}$	6	d'$_1$_str$_{21}$	0.1	d'_2_str_{21}	0.02	d'$_4$_str$_{21}$	0.75	val_str$_{21}$_kpi'$_{211}$	0.7	val_str$_{21}$_kpi'_{221}	0
str$_{22}$	u_Str$_{22}$	?	ti_str$_{22}$	?	c_str$_{22}$	5	d'_1_str_{21}	0.05	d'_2_str_{21}	0.01	d'$_4$_str$_{21}$	0.5	val_str$_{22}$_kpi'$_{211}$	0.2	val_str$_{11}$_kpi'$_{221}$	0.8
													val_str$_{11}$_kpi'$_{211}$	0.3	val_str$_{11}$_kpi'$_{221}$	0.4
													val_str$_{12}$_kpi'$_{211}$	0.2	val_str$_{12}$_kpi'$_{221}$	0.3

4 Stage of the Research and Expected Contributions of the Thesis

The research conducted in this thesis has contributed with the proposal of a model, a method (SD) and a tool to automatically support the decision making as regards the strategies alignment, from a collaborative perspective. Misalignments among the strategies defined by the CN partners lead to the reduction of performance and ultimately to the collaboration failure. The strategies alignment process is addressed to support the enterprise decision makers of a CN on the assessment of the strategies influences in the objectives attainment, achieving higher levels of performance in both enterprise and network levels. Despite the advantages of the model application, the main drawback is related with the data gathering as regards the value $val\_str_{is}\_kpi'_{ixk}$. Future research lines are devoted to improve the mechanisms provided so far and propose a guideline to deal with the data collection process that will allow to feed the proposed model and solve it. Another line is related with the verification of the proposed model with a set o experiments modelling networks with high number of nodes. Besides the validation of the proposed model, method, guideline and tool in a real supply chain will be also carried out.

References

Andres B, Macedo P, Camarinha-Matos LM, Poler R (2014) Achieving coherence between strategies and value systems in collaborative networks. IFIP Adv Inf Commun Technol 434:261–272

Andres B, Poler R (2014a) Computing the strategies alignment in collaborative networks, pp 29–40. In: Mertins K, Bénaben F, Poler R, Bourrières J-P (eds) Enterprise interoperability VI. Springer International Publishing, Cham. Retrieved 16 Dec 2014 (http://link.springer.com/10.1007/978-3-319-04948-9)

Andres B, Poler R (2014b) Identifying the strategies to be activated for optimising the performance in collaborative networks. In: The global reach of industrial engineering. Enhancing synergies in a collaborative environment. 8th international conference on industrial engineering and industrial management XX, international conference on industrial engineering and operations management, p 25

Andres B, Poler R (2015) Models, guidelines and tools for the integration of collaborative processes in non-hierarchical manufacturing networks: a review. Int J Comput Integr Manuf 29 (2):166–201

AnyLogic (2015) AnyLogic. Retrieved (http://www.anylogic.com/)

Camarinha-Matos LM, Afsarmanesh H (2005) Collaborative networks: a new scientific discipline. J Intell Manuf 16(4–5):439–452

Crnkovic GD (2010) Constructive research and info-computational knowledge generation. In Magnani L, Carnielli W, Pizzi C (eds) Studies in computational intelligence. Model-based reasoning in science and technology. Abduction, logic, and computational discovery. Springer, Berlin, pp 359–380

Kasanen E, Lukka K, Siitonen A (1993) The constructive approach in management accounting research. J Manage Account Res 5:243–264

The Role of Complexity and Flexibility of the Instance in the Joint Solution Approach

Raul Pulido, Álvaro García-Sánchez, Alessandro Brun and Miguel Ortega-Mier

Abstract Many pieces of research address the development of new algorithms and new solution techniques for decision-making; however, most of them do not consider the characteristics of instance in their analysis, such as the complexity and flexibility of the instance. Building a complex model, such as a joint model, requires a huge amount of time and effort while the resulting solution of such joint models may or may not be the best solution for all the actors involved in the process. Therefore, it is important to make an in-depth analysis of the instance before investing the time and effort to build a joint model. In this regard, this paper provides an instance evaluation procedure to help decision-makers decide whether to use joint decision or not for a particular instance.

Keywords Joint decision · Flexibility · MILP · Flexibility · Complexity

1 Introduction

The traditional decision-making process is usually sequential where the best decision is taken for the first stage of the process and then this output of the first stage is used as a basis for the next stage decisions and so on. However, by using

R. Pulido (✉) · Á. García-Sánchez · M. Ortega-Mier
Escuela Superior de Ing.Industriales, UPM, José Gutiérrez Abascal 2, 28006 Madrid, Spain
e-mail: raul.pulido@upm.es; raul.pulido@polimi.it

Á. García-Sánchez
e-mail: alvaro.garcia@upm.es

M. Ortega-Mier
e-mail: miguel.ortega.mier@upm.es

R. Pulido · A. Brun
Politecnico di Milano, Dipartimento di Ingegneria Gestionale, via Lambruschini 4/b, Milan, Spain
e-mail: alessandro.brun@polimi.it

M. Amorim et al. (eds.), *Engineering Systems and Networks*, Lecture Notes in Management and Industrial Engineering, DOI 10.1007/978-3-319-45748-2_5

a sequential decision-making process it is difficult to reach an overall optimal solution as the final decisions completely depend on the first-stage decision. To overcome this limitation a joint decision approach offers a great opportunity to reach an overall optimal solution by enlarging the search space. Joint decision-making can have many implications since, besides the intrinsic cost and time to develop the joint model, it may involve a possible change in the organizations in order to allow different actors to share information and to persuade global goals instead of local goals. This requires a close collaboration and coordination among the different actors involved in the overall process. Interestingly, many operations management researchers assume that "integration is a must" and that cross-functional coordination and integration are necessary (Ketokivi et al. 2006). However, in later research, Turkulainen and Ketokivi (2012) argued that the benefits of integration and cross-functional coordination are context-dependent and sometimes disaggregation is beneficial.

Joint decisions usually result in a paradox since the different actors may not achieve the optimal solution for their sub-operations in order to achieve an overall optimal solution. Therefore, a joint decision-making process could be attractive for some circumstances and unappealing in other situations. Considering this paradoxical nature of joint decisions, this research attempts to explore "When is it advisable to use a joint decision-making process and when is it better not to use it? The contribution of this paper is to create an "instance evaluation procedure" based on the complexity and flexibility of the instance to help the decision-makers to decide before investing their time and effort in the preparation of a joint decision model. To this end, the authors argue that it is highly important to consider instance characteristics before setting out on a joint decision model.

1.1 Characteristics of the Instance

The instance is the complete set of data that defines the problem space e.g. in the case of a scheduling problem the number of days, workers, job shops, production lines, units to produce and so on. One of the most important characteristics of an instance is the size of the problem, which is determined by the number of continuous and binary variables that represent all the relationships among variables and parameters. The problem size is considered as a major contributor to the complexity of the instance. However, there are many other factors that need to be considered when analyzing the instance complexity. The complexity and flexibility of the instance plays a crucial role in the decision-making process. For dynamical systems theory the complexity measures are usually computational complexities that are a measure of the interactions (Adami 2002). Similarly, Heylighen (2008) highlighted that a fundamental part of any complex system is the connected parts via interactions. These parts can be distinct and/or connected as well as autonomous and/or to some degree mutually dependent. This interdependence can create conflicting goals since the improvement of one part could lead to the decrement of the other part. Therefore, just considering the total number of variables present in a problem space as the only parameter/measure of complexity is not the right approach. Many other factors need be considered when analyzing the complexity of an instance. An important work in this regard is by Vanhoucke and Maenhout (2009) where they characterize the Nurse scheduling problem. In their work, they highlight four factors to analyze the complexity of the indicators: (a) problem size, (b) the preference distribution measures, (c) the coverage distribution measures, and (d) the time related constraints.

Similarly, flexibility of an instance is another key characteristic that needs to be considered when analyzing a joint decision. Flexibility is "the ability to change or react with little penalty in time, effort, cost, or performance" (Upton 1994). Thus, a flexible instance of workers means the extent to which the employees can perform different tasks. In this research, we propose considering three new factors while analyzing the instance characteristics. These factors include preference distribution, coverage distribution and cost dispersion.

2 Factors for Instance Analysis

In this section we propose and define three indices that need to be considered when analysing an instance for a joint decision. These factors are discussed briefly.

2.1 Preference Distribution (PD)

The preference distribution measures the dispersion among the needs or requirements of resources by the different entities over the scheduling horizon. If all the requirements are similar the preference of distribution will be low, on the other hand if all the requirements are different this index will be high. It can be measured using Eq. 1. Where $Entity_i$ is the requirement of resource i of the entity.

$$PD = \frac{\sqrt{\frac{\sum_{i=1}^{N} (Entity_i - \overline{Entity})^2}{Number\ Of\ Entities}}}{\overline{Entity}} \tag{1}$$

2.2 Extra Coverage Constrainedness (ECC) Rigidity/Flexibility

The coverage requirements are expressed by the average of the extra capacity (availability) of all machines (resources). When this number is close to 0, we could say that it is a rigid instance that there is no extra resource; when it is close to 1 its means that it is a flexible instance meaning we have some extra resources. This factor can be measured with the help of Eq. 2.

$$ECC = \frac{\sum_{Machine}\left(1 - \frac{requirement_{machine}}{capacity_{machine}}\right)}{number\ of\ resources} \tag{2}$$

2.3 Cost Dispersion (CD)

The cost dispersion is a measure that is used to quantify the variation of cost among the different areas, in which the decision will be made together. We will refer to the total cost of each part, for example in the case of the inventory it will be the total cost of the inventory not the cost of each unit of inventory. It can be measured using Eq. 3.

$$CD = \frac{\sqrt{\frac{\sum_{i=1}^{N}(Cost_i - \overline{Cost})^2}{Number\ Of\ Costs}}}{\overline{Cost}} \tag{3}$$

In the next section, these factors are studied using 2 case studies where different combinations of the preference distribution, extra coverage constrainedness and cost dispersion are tested. The size of the instance is constant and the amount of resources available helps to characterize the instance. The three indexes vary between low (close to 0) and high (close to 1).

3 Case Studies and Results

The computational experience was performed in a Windows-PC with an Intel Core 7, 8 GB of RAM, running Windows 7, with the AIMMS 3.14 mathematical modeler and Gurobi 6.0. A maximum stop criterion of 3600 s was set for all instances.

3.1 Case Study 1

In a car assembly line, the production sequence has to be decided for the planning period. Each workstation could deal with a production rate, which means that a workstation could install X high trim components each Y cars. In the event that the number of high trim components is higher, an extra utility worker has to come to help, with a penalty cost. Each station installs a different type of component that needs to be next to the assembly line before it is needed. The transportation vehicles carry these components from the warehouse to the workstations where it is assumed that all the components exist. Each model has a set of characteristics, such as engine, rims, tires, steering, etc. These components could have different trims (Low or High). All the models are different from the other models in at least one type of component. The components required at each workstation are delivered as a kit. The model was implemented using mixed integer linear programming. A detailed description can be found in Pulido et al. (2014a). There are 3 main decisions that have to be taken and are usually taken sequentially.

1. The production sequence that minimizes the use of extra utility workers.
2. The distribution cost of components that minimizes transportation cost.
3. The inventory level that minimizes the inventory cost.

The first index will be calculated as the deviation of the number of high trim elements that each car requires, and the average entity will be the one for which the assembly line was designed. For the second index, the machines will be the workstation and transportation vehicle. The requirement of workstations of the car assembly line will be the requirement of each car model for this workstation, while between the production ratios and for the transportation vehicles, the requirement will be the demand for the use of a vehicle, and its capacity will be the transportation capacity.

The results of the experimentation are presented in Table 1. Where the first column of the instance defines the instance with respect to three indexes, and the left part of the table is the result of the traditional sequential approach while in the right

Table 1 Result of case study 1

PD, ECC, CD	Sch	Transp	Invent	SeqD	Sch	Transp	Invent	JointD	Savings (%)
L, L, L	594	1432	928	2954	638	1424	796	2858	3
L, L, H	594	716	9283	10,593	638	715	7921	9274	14
L, H, L	0	1439	902	2341	0	1426	813	2239	5
L, H, H	0	720	9023	9743	22	716	7979	8717	12
H, L, L	308	1435	944	2687	308	1434	834	2576	4
H, L, H	308	718	9436	10,462	396	714	8124	9234	13
H, H, L	0	1436	947	2383	0	1424	795	2219	7
H, H, H	0	718	9469	10,187	0	712	7952	8664	18

part is the result of the joint approach. Promising results appear when the preference distribution is high, and there is diversity among the tasks that have to be done. Also when there is extra coverage of resources since there is flexibility of the allocation. And finally when there is diversity of the cost there are promising results, especially when the biggest cost contributor is the last of the sequential model.

3.2 *Case Study 2*

The teaching hospital plays a key role in the health care system. Inside the hospital, the main part of this structure is the operating rooms, since the majority of the patients go through the operating room.

The scheduling of the surgeries is important since the vacant time of the operating room, the idle time of a surgeon, and the extra time cost of the operating room impact directly on how the hospital functions. A detailed explanation of the model can be found in (Pulido et al. 2014b). There are 2 main decisions that have to be taken and are usually taken sequentially.

1. The doctor who perform the surgery that minimizes the extra time. Each surgeon has a different expertise and could perform a surgery faster or slower.
2. The operating room schedule that minimizes the vacant time of the surgeons and idle operating rooms.

The first index will be calculated as the deviation of the length of the surgery against the average surgery duration. In order to calculate the second index the machine will be the surgeons and operating rooms. The requirement of surgeons will be the total length of the duration of the surgeries that could be performed by a surgeon between the shift length (capacity) while the requirement of the operating room will be the total length of surgeries that can be performed in this operating room between the shift lengths.

Table 2 presents the results. First the indexes used, then the overtime, vacant OR time and surgeon waiting time cost for the sequential decision and the same three

Table 2 Result of case study 2

PD, ECC, CD	OverT	VacT	WaitT	SeqD	OverT	VacT	WaitT	JointD	Savings (%)
L, L, L	1307	885	362	2554	1387	885	210	2482	3
L, L, H	980	885	620	2485	1040	905	360	2305	8
L, H, L	0	1020	350	1370	13	885	210	1108	24
L, H, H	0	1020	600	1620	0	1135	120	1255	29
H,L,L	1027	1545	58	2630	1307	1065	58	2430	8
H, L, H	770	1545	100	2415	980	1065	100	2145	13
H, H, L	40	990	362	1392	40	1065	58	1163	20
H, H, H	0	1050	540	1590	0	1035	100	1135	40

cost for the joint decision, and the savings. The Joint Decision is advisable with promising results when the preference dispersion is high, because when it is low the results are negligible. The role of the ECC is minor, since it plays a complicated role, as there is a penalty for the extra resources (vacant time cost). The dispersion of the cost is also important, especially when the cost of vacant time is high.

4 Prescriptive Framework

A prescriptive framework is developed based on results, and acquired experience is presented. With a more detailed analysis of the input data, we can assess the preference distribution of the instance, the extra coverage (flexibility/rigidity) of instance, and the homogeneity of the cost. If the results of this preprocessing of the data are promising, we can decide to take the next step and start to build a joint model.

When the preference distribution is low, which means similar products or tasks need to be produced/performed, the benefits of the Joint Decision decrease. On the other hand, if we have bigger diversity the results could be promising.

The extra coverage constrainedness plays a key role and will depend on whether there is a penalty or not for having extra resources. However, if there is no extra coverage the possible savings will decrease.

The major influence that we found is cost dispersion since when it is low the results are not so promising, but when there is a high dispersion then it is necessary

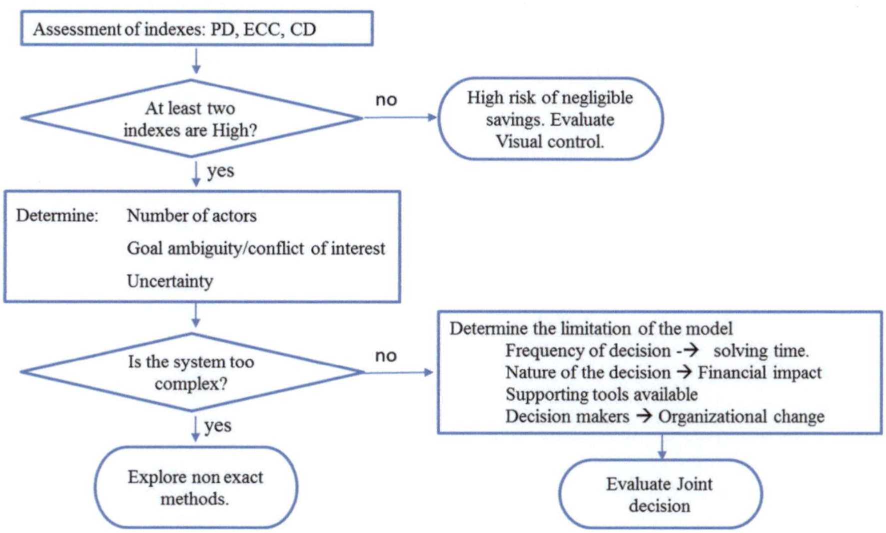

Fig. 1 Instance evaluation process

know the position of the most expensive cost as this plays a key role. When the highest cost is the final decision, the possible savings increase considerably.

However, for choosing the right type of model, it is important to take into consideration the number of actors or the size of the problem since this will determine if we use exact or non-exact methods. The goal ambiguity, frequency of decision and uncertainty should be taken into consideration in advance in order to decide on exact or non-exact methods.

For the reason mentioned previously, we suggest analyzing the likelihood of savings based on the preference of distribution, coverage of the instance and homogeneity of cost. Figure 1 shows the proposed evaluation process/procedure.

5 Conclusion

The major benefit of a joint decision is the possible cost savings, thanks to the better utilization of key resources. The decrease in the cost is accompanied by an improvement in the key performance indicator such as the use of resources or the decrease in overtime. As expected, the use of a joint model increases the size of the model, the complexity and the solving time. When the computational time is high other non-exact solution methods should be evaluated with the risk of a decrease in saving for non-achievement of the global optimum.

The range of possible savings for the same problems using a joint model depends on the data is quite large, but the bottom line is close to zero. Then, before deciding the type of model, we suggest pre-evaluating the instance as in some cases the implementation cost can be higher than the savings. Therefore, in some cases it is better to use the traditional sequential approach since the preparation of a complex model does not guarantee enough savings to justify the development of the joint model. Hence, this research concludes that any possible savings of a joint decision are case-dependent and every case should be evaluated before investing time and effort in a joint decision approach. The case/instance should be evaluated from its complexity and flexibility perspective.

As a further work we intend to test the proposed framework using more case studies, which can be generalized to other areas to try to help other researchers and practitioners to decide when or not it is better to use a joint decision.

References

Adami C (2002) What is complexity? BioEssays 24(12):1085–1094

Heylighen F (2008) Encyclopedia of library and information sciences, pp 1215–1224

Ketokivi M, Schroeder RG, Turkulainen V (2006) Organizational differentiation and integration-A new look at an old theory. Helsinki University of Technology, Department of Industrial Engineering and Management Working Paper, (2006/2), Espoo, Finland

Pulido R, Garcia-Sánchez Á, Ortega-Mier M, Brun A (2014a) MILP for the inventory and routing for replenishment problem in the car assembly line. Int J Prod Manage Eng 2(1):37–45
Pulido R, Aguirre AM, Ortega-Mier M, García-Sánchez Á, Méndez CA (2014b) Managing daily surgery schedules in a teaching hospital: a mixed-integer optimization approach. BMC Health Serv Res 14(1):464
Turkulainen V, Ketokivi M (2012) Cross-functional integration and performance: what are the real benefits? Int J Oper Prod Manage 32(4):447–467
Upton D (1994) The management of manufacturing flexibility. Calif Manag Rev 36(2):72–89
Vanhoucke M, Maenhout B (2009) On the characterization and generation of nurse scheduling problem instances. Eur J Oper Res 196(2):457–467

Towards Increasing Sustainability in Large Urban Mobility Attractors

Jesús Muñuzuri, Luis Onieva, José Guadix and Elena Barbadilla

Abstract Private-car based mobility typically constitutes the main option in the case of large urban mobility attractors, generating congestion and parking problems, as well as increased levels of emissions, related both to the displacement itself but also to the search for a parking space. We have analyzed the case of a large work-related mobility attractor and proposed specific policies aimed at increasing the sustainability of the mobility patterns, both environmentally and in terms of the livability of the area. The proposed policies are tested with a specifically built simulation model, enabling us to draw conclusions with respect to the applicability of the new system and its acceptance and "willingness to use" on the side of the general public.

Keywords Simulation · Mobility · Sustainability · User acceptance

1 Introduction

The increase of sustainability in urban transport constitutes an ongoing process where small, locally applied initiatives are embedded in the overall long-term strategy (Roseland 2008). Despite the long-term goal being perfectly defined and feasible, it is usually the transitory evolution from the current base scenario to that long-term objective what hinders progress during the intermediate stages. The

J. Muñuzuri (✉) · L. Onieva · J. Guadix · E. Barbadilla
School of Engineering, University of Seville, CM Descubrimientos,
41092 Seville, Spain
e-mail: munuzuri@us.es

L. Onieva
e-mail: onieva@us.es

J. Guadix
e-mail: guadix@us.es

E. Barbadilla
e-mail: ebarbadilla@us.es

M. Amorim et al. (eds.), *Engineering Systems and Networks*, Lecture Notes
in Management and Industrial Engineering, DOI 10.1007/978-3-319-45748-2_6

appropriate design of small, incremental steps is what ultimately causes the overall system to move in the right direction.

We focus here on the analysis of one of those incremental steps, aimed at reducing the emissions and energy consumption linked to parking search. In congested areas, the search for parking can lead to a significant increase in the overall mileage (Thompson and Richardson 1998; Weinberger et al. 2008), causing large inefficiencies in the transportation system, and also resulting in a non-negligible waste of time for drivers.

This work looks at the introduction of a specific policy to reduce parking search mileage in a large mobility-attracting area. We tested the expected effects of this policy using a specifically built microsimulation model, in order to determine the degree of acceptance for the potential users of the new system. It is important to keep in mind that this degree of acceptance is normally one of the main factors that determine the success of a sustainability-oriented incremental action.

2 Area Description

The Cartuja Technological Park is located at the northwest end of the city of Seville, on the old grounds of the 1992 Universal Expo. After growing continuously for two decades, and despite the economic crisis, it nowadays represents the main concentration of companies in the city, containing also two large university buildings and several offices of the local and regional administration. Besides, and even though it has almost reached its saturation point, there are still several developments planned for the near future.

The main problem to be faced by the Technological Park is mobility. The companies located within the Park represent a daily mobility of more than 15,000 persons, including permanent staff and visitors, and the administration personnel and university students represent another 10,000. However, the public transport coverage of the Park is very poor, with only two bus lines crossing it and no underground connection. Therefore, most of the daily trips with destination in the area are car-based, with the subsequent saturation of the parking space in the area. The Park is considering building an underground parking lot, but the cost-effectiveness of such an investment is not very clear, and in any case it would require several years before starting operations. Also, two large (and currently unused) areas located north and west of the Park may be used as parking lots. Figure 1 shows the location of the different current and future mobility infrastructures in the Park.

The mobility issue in the Park is also strongly influenced by the marked peak periods of entry (between 7.30 and 8.30) and exit (between 13.30 and 15.30) from the Park, which result in a series of bottlenecks for vehicles trying to pass through one of the few gates connecting the Park with the outside network. Figure 2 shows a typical distribution of these mobility patterns. Also, the degree of saturation of the

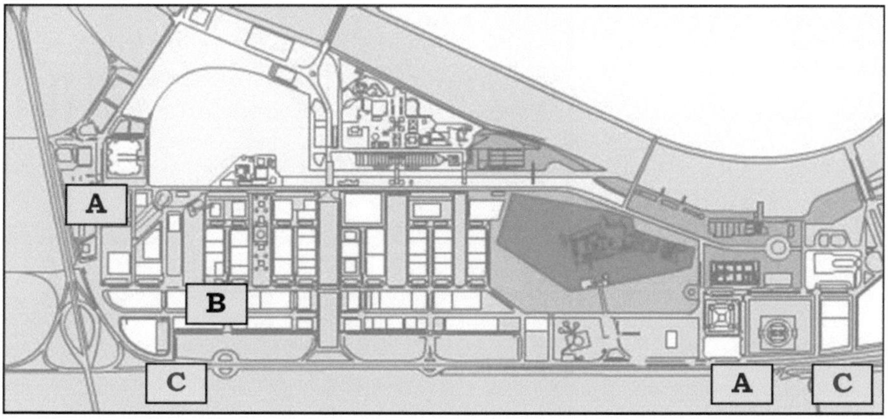

Fig. 1 Location of the main short and medium-term transport-related developments in the Park (*A* areas available for open-air parking lots; *B* location of the planned underground parking lot; *C* train stations)

Fig. 2 Typical distribution of entries and exits in the Cartuja Technology Park on an average workday

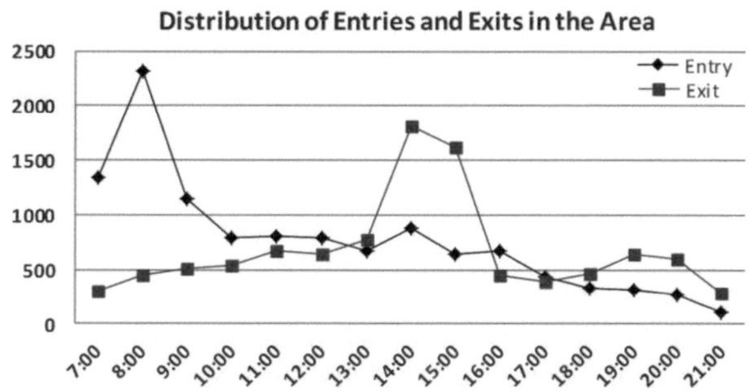

parking spaces inside the Park makes it virtually impossible for business-related visitors to access the area by car during normal business hours.

3 Proposed Solution

In view of the current scenario, and given that the Park does not have room for more cars and that further developments have already been started, the only possibility to reach sustainable mobility seems to be the introduction of a park-and-ride system (Bon and Steiner 2008; Meek et al. 2011). The two large areas located north and west of the Park may be used as parking lots with sufficient capacity to fulfil the Park's needs, and buses could connect these areas with the Park's buildings. Besides, the system could also provide access from two train stations which are expected to start operating in the medium term, and it could also facilitate inner mobility within the Park.

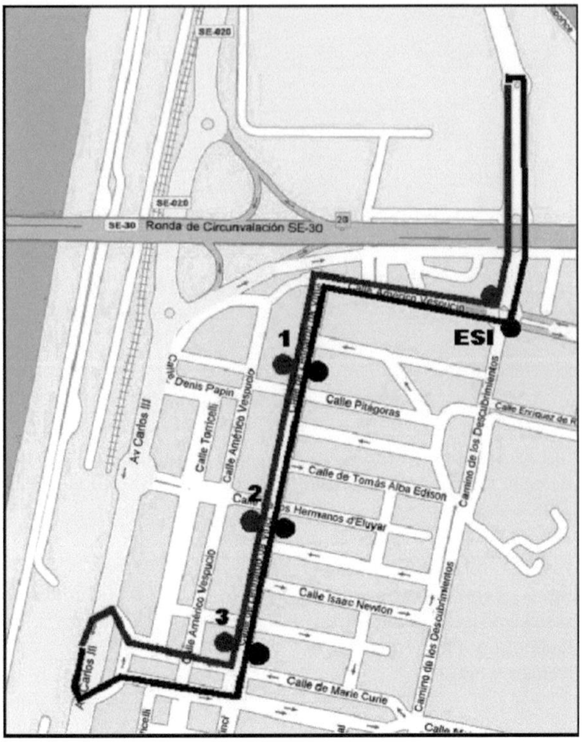

Fig. 3 Route of the shuttle buses operating in the park-and-ride system

The existence of this park-and-ride scheme would also indirectly benefit the park visitors, since they could also use the facilities. Another option would be to free some of the currently used parking spaces in the Park to be used by visitors on a parking fee basis.

Figure 3 shows the chosen route for the shuttle buses servicing the park-and-ride scheme. Their termini are located next to the two external parking areas, and the stops cover the main worker destination areas within the Park, and also the Seville School of Engineering (ESI in the figure).

4 Simulation

Despite its limitations, normally inherent to any representation of a complex reality by means of a simplified model (Lawson 2006), simulation is widely used for ex-ante assessments of production, service or transportation systems (Moeckel et al. 2002). In order to determine the requirements of our park-and-ride system in terms of number of buses, frequency and routes, we used microscopic simulation, using an approach similar to Lawson et al. (2011). However, instead of using a dedicated transportation simulation tool, we decided to favor a specifically built microsimulation model using the commercial package Arena®. This allowed us to gather detailed information on certain aspects of the simulation which are neglected by transportation simulation packages, like users getting on and off buses, waiting times at bus stops, etc.

The data used for the simulation building included aspects like:

- Total number of workers accessing the Park every day
- Location and size of the companies
- Park layout and distances
- Location of the future parking lots outside the Park
- Cycle time of the traffic lights inside the Park.

We built different scenarios with different system configurations and gathered information on the expected time required for transfer between the parking lots and the final destinations of the Park workers. The main parameter defining the scenarios was the number of shuttle buses incorporated to the system, which has a direct impact on waiting times and level of service.

We modeled the circulation of vehicles using conveyors, which represents traffic flows appropriately except for the fact that all the vehicles on the same conveyor have the same speed, and no lane changes or overtaking are possible. Nevertheless, this approach still simulates traffic flows with a high degree of reality (Muñuzuri et al. 2013).

Figure 4 shows the module sequences used to simulate the operation of park-and-ride zones. Also, we modelled the crossings between streets using the "Resource" tool in Arena®, whereby each resource is occupied when a vehicle enters the crossing, and liberated after it exists the crossing, so that a new vehicle may have access to it (Fig. 5). The precedence rules imposed by traffic lights were also included in the simulations. Finally, Fig. 6 shows the simulation modules corresponding to the getting on and off of users at shuttle bus stops, and Fig. 7 shows a view of the simulation in a small area of the network.

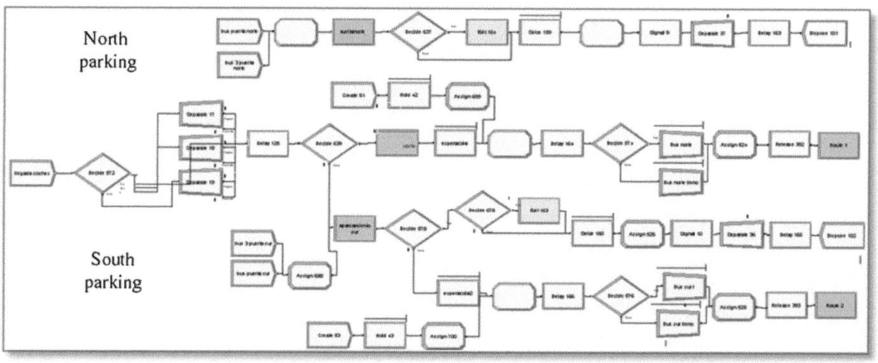

Fig. 4 Simulation module for the park-and-ride zone operation

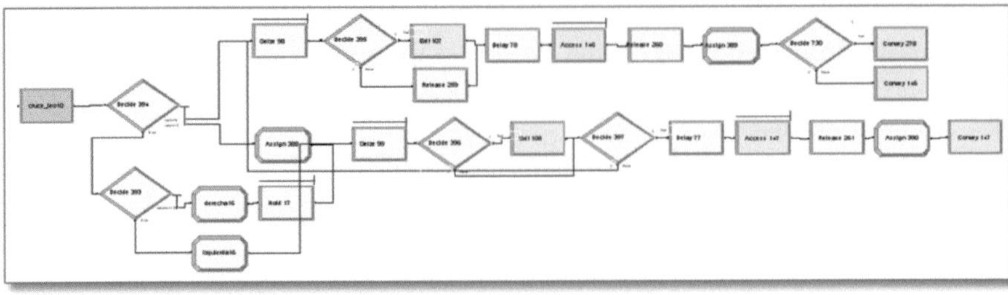

Fig. 5 Simulation module for traffic-light regulated crossings

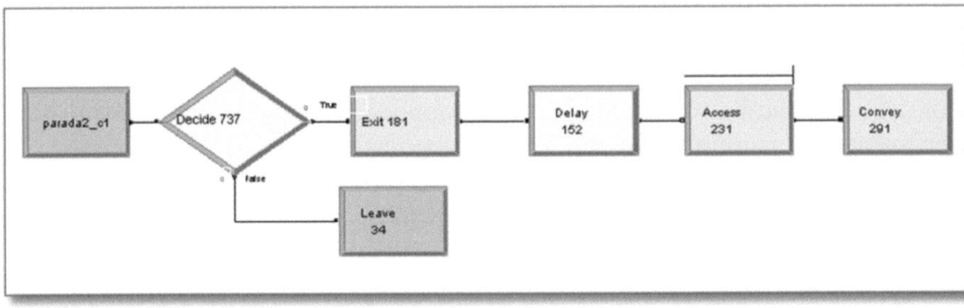

Fig. 6 Simulation module for bus stops

Fig. 7 View of the simulation process in a small section of the network, containing both private cars and shuttle buses

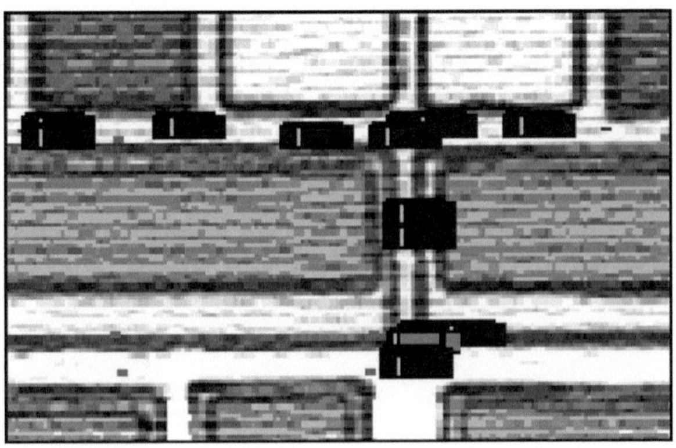

5 Results

The use of Arena® for the simulations enabled us to generate a series of reports using the Category Overview tool. Table 1 contains the results we obtained for the different simulations carried out over three different scenarios, built depending on the number of shuttle buses operating in the park-and-ride system, which ranges between 4 and 8. The results show the average and maximum waiting times for

Table 1 Simulation results for the 8 stops of the shuttle buses, including average and maximum waiting times (in minutes) both for the morning and the lunchtime rush hours

Stop	4 shuttle buses				6 shuttle buses				8 shuttle buses			
	Morning		Lunch time		Morning		Lunch time		Morning		Lunch time	
	Avg	Max	Avg	Max	Avg	Max	Avg	Max	Avg	Max	Avg	Max
1	3.53	14.46	4.06	16.33	2.12	7.73	2.96	13.38	2.25	6.91	1.87	8.82
2	3.68	10.99	4.12	17.91	2.50	9.74	2.74	12.78	2.54	8.47	1.92	9.74
3	4.05	12.79	4.41	19.48	2.76	10.64	2.63	13.55	2.29	10.64	1.90	10.64
4	4.11	11.63	4.29	16.25	3.09	8.99	2.80	14.29	2.60	6.70	1.91	8.99
5	4.47	12.48	3.96	19.97	2.58	7.68	2.68	13.36	2.27	7.27	1.79	8.16
6	4.00	12.23	3.84	18.31	2.47	10.04	2.77	11.58	2.48	7.83	1.89	10.04
7	3.88	12.81	4.02	17.88	2.58	9.92	2.78	11.56	2.47	8.51	1.92	9.92
8	4.49	14.06	3.81	15.09	2.50	9.59	3.01	12.73	2.11	6.34	1.69	9.59

users in the bus stops, both during the morning (entry flows) and lunchtime (exit flows) rush periods, which constitute the demand peaks of each working day, according to Fig. 2. Each figure was obtained as the average result over a 100 simulation runs.

The results depicted in the Table show a clear descent in the user waiting times, linked to the increase in the number of vehicles operating in the system. However, whereas the cost can be expected to grow somewhat linearly with the number of buses, the level of service does not increase likewise. Even though the average waiting times are kept low in the scenarios with 6 and 8 buses, the maximum waiting times (normally due to the fact that users are sometimes forced to wait because shuttle buses arrive completely full at their stop) are clearly unacceptable when it comes to consider willingness to use.

6 Conclusions

In view of the results shown in the previous section, the park-and-ride scheme can be considered unacceptable given the existing mobility patterns and regulations in the Cartuja Park area. The lack of enforcement of parking regulations have resulted over the recent years in a steep increase of illegal parking, with private cars occupying sidewalks, pedestrian areas and even street lanes. Even if the number of workers in the Parks increases as expected over the next years, the expectations are that these illegal behaviors will further increase along with it. After all, the alternative is to leave the car in one of the external parking areas and suffer the queues, waiting times and en-route times until reaching the final destination for hundreds of workers.

The only option for this park-and-ride scheme is then to combine it with other measures focused on restricting the accessibility of private vehicles to the Park.

This includes aspects like the monitoring of license plates in the accesses, prosecuting illegal parking and increasing pedestrian areas. The introduction of park-and-ride can only be contemplated as embedded in a more extensive mobility policy for the park, given also that its implementation does not require important investment costs, so even pilot implementations could be easily put into practice.

On the other hand, our work has only contemplated the operational part of the service, but the operational costs are also likely to have a large influence on its degree of acceptance. The calculation of the operational costs (depending on the number of shuttle buses) and the decision of who should cover those expenses (the Park, the companies located in it or the users themselves) would have a strong impact on the final perception and demand of the system.

Acknowledgments The authors wish to express their gratitude for the financial support of the Andalusia Region (project SIFMOV, ref no GGI3003IDIT) for the completion of this work.

References

Bon A, Steiner R (2008) Sustainable campus transportation through transit partnership and transportation demand management: a case study from the University of Florida. Berkeley Plann J 19(1):125–142

Lawson CT (2006) Microsimulation for urban transportation planning: miracle or mirage? J Urban Technol 13(1):55–80

Lawson S, Chamberlin R, Choi J, Swanson B, Kiser B, Newman P, Monz C, Pettebone D, Gamble L (2011) Modeling the effects of shuttle service on transportation system performance and quality of visitor experience in rocky mountain national park. Transp Res Rec 2244:97–106

Meek S, Ison S, Enoch M (2011) Evaluating alternative concepts of bus-based park and ride. Transp Policy 18:456–467

Moeckel R, Schürmann C, Wegener M (2002) Microsimulation of urban land use. 42nd European Congress of the Regional Science Association, Dortmund, pp 27–31

Muñuzuri J, Cortés P, Onieva L, Guadix J (2013) Simulating the effects of pedestrianisation on urban freight deliveries. Eur Transp/TrasportiEuropei 54:1–19

Roseland M (2008) Toward sustainable communities. New Society Publishers, Canada

Thompson RG, Richardson AJ (1998) A parking search model. Transp Res A 32(3):159–170

Weinberger R, Seaman M, Johnson C (2008) Suburbanizing the city: how new york city parking requirements lead to more driving. Transportation Alternatives, New York

Stock Market Firm Value Effects of Research and Development Expenditures in the Oil and Gas Industry

Eduardo Pontual Ribeiro, Willian Freitas de Almeida and Rosemarie Bröker Bone

Abstract This paper evaluates the market firm value response to research and development (R&D) expenditures in the Oil and Gas Industry (O&G). R&D projects are seen as key to long term firm survival in the Oil and Gas industry. But the uncertainty associated with R&D success may reduce firm value market when a firm expands R&D. Using a panel of O&G firms traded in the New York Stock Exchange, we econometrically estimate the market value effect of R&D. We conclude that R&D expenditures increase firm market value, but with a one year lag, even after controlling for firm differences in book value or earnings and oil and gas reserves.

Keywords Firm valuation · R&D expenditures · Oil and Gas industry · Panel data

1 Introduction

Research and Development (R&D) expenditures are key to the long term survival of technology intensive firms, such as integrated Oil and Gas firms (Besson 2006). Technological advances are required to increase effective oil and gas extraction rates from known reserves, at economically feasible costs. Bret-Rouzaut and Thom

The opinions expressed here do not represent the official view of Petrobras.

E.P. Ribeiro (✉)
Economics Institute, Federal University of Rio de Janeiro, Rio de Janeiro, Brazil
e-mail: eribeiro@ie.ufrj.br

W.F. de Almeida
Petrobras, Petróleo Brasileiro, S.A., Rio de Janeiro, Brazil
e-mail: wawalmeida@gmail.com

R.B. Bone
Department of Industrial Engineering, Polytechnic School of the Federal University of Rio de Janeiro, Rio de Janeiro, Brazil
e-mail: rosebone@poli.ufrj.br

M. Amorim et al. (eds.), *Engineering Systems and Networks*, Lecture Notes in Management and Industrial Engineering, DOI 10.1007/978-3-319-45748-2_7

(2005) argue that O&G firms must treat technology as an explicit business strategy, fostering research centres. These expenditures are required to generate innovations and patents that allow firms to maintain costs under control over time in the production and exploration processes and allow them to reach new reserves, such as deep-sea, pre-salt resources or unconventional hydrocarbons.

Market value effects of R&D announcements have attracted research interest (e.g. Chan et al. 2001) with mixed results. Nevertheless, recent evaluations of these effects for the O&G industry are lacking.

We collected detailed information from the largest O&G firms traded in the New York Stock Exchange, considering also regional coverage. Firm data was obtained from firm 20-F and 10-K fillings at the SEC for the 2000–2011 period for British Petroleum P.L.C, ConocoPhillips Holding CO, Chevron Corporation, Eni SPA, Exxon Mobil Corp., Repsol YPF, Petróleo Brasileiro S.A., Statoil ASA, Royal Dutch Shell PLC and Total S/A.

R&D expenditures present a particularity regarding its accounting rules. In many situations and countries, they are considered expenditures, and not investments. Thus, their effect on firm value cannot be readily measured from balance sheet results. Alternative methods are required to extract their contribution to firm value.

We explore an empirical model proposed by Ohlson (1995) and used by many researchers. Based on the well-known Divided Discount Model (DDM), the model proves an association between market firm value and firm value drivers, such as O&G firm reserves, conditional of firm book value and earnings in an econometrically convenient form. We use this model to test the hypothesis of whether higher R&D expenditures are perceived by the market as a sign of future profitability. That is, we evaluate whether the possibility of technological advances within a firm are valued positively by stock market investors. A positive answer to our question is relevant to firms seeking to finance its R&D expenditures, as investors would be, in principle, willing to subscribe to such expenditures in search of higher profitability.

2 R&D Expenditures at the Oil Majors and Very Large Producers

Technological innovation is seen as a key factor in an integrated O&G firm. It is used as a tool to improve operational procedures, reduce costs, increase reserves and foster profitability. Nevertheless O&G corporations invest relatively less than other sectors. Using data obtained from our sample firms, the ratio of R&D expenditures to revenues (an indicator known as R&D intensity) for the O&G industry is about 0.3 % of revenues, while other R&D intense sectors are pharmaceuticals (15 %), software and computers (16 %) and airspace and defence (14 %), according to Helfat (2004).

R&D expenditures cover, in general, all areas of an integrated O&G firm activities, namely, exploration, production, transportation, refining and distribution,

Table 1 R&D investments across O&G areas

Areas	BP	Conoco	Chevron	Eni	Exxon	Petrobras	Shell	Statoil	Total	YPF
Sismic mapping	•			•			•	•	•	
Reserve management	•	•	•	•	•	•	•	•	•	•
Drilling	•		•	•	•	•	•	•		
Deep sea exploration	•	•	•		•	•	•	•		
Refining	•	•	•	•	•	•	•		•	•
Gas liquification		•			•	•	•			•
Bio-fuels	•	•	•	•	•	•	•		•	•
Renewable energy	•	•	•	•	•	•	•	•	•	•
Unconventional reserves		•	•		•			•	•	•
Carbon capture	•	•					•	•	•	•

Source Firms' 20-F and 10-K SEC files, 2000–2011

as well as renewable resources. Table 1 presents a comparable description of R&D activities across firms. We see that the firms develop R&D projects in a wide range of areas.

The R&D investments generate technological breakthroughs that lead new to patents. Exxon, the largest listed O&G firm has approximately 10,000 active patents. According to the 20-F and 10 K forms, Total, Petrobras, Conoco and Eni, registered 250, 180, 122 and 79 patents, respectively, in 2011 alone.

The literature has mixed results regarding the market value effect of R&D expenditures. Ali et al. (2012) conclude that the market under-prices R&D expenditure benefits, also because of its accounting status (not clear whether expenditures or investment). However, there are papers that point to a positive effect of R&D. Sougiannis (1994) showed that the stock market values R&D expenditures with a more than proportional effect, but after such investments mature, i.e., with a lag. With older data (from 1987–1998), under different industry profile and oil price levels and volatility, Vijay et al. (2002) conclude on positive R&D effects of O&G market value, but with a small effect. Our paper revisits the O&G industry, given the significant changes in the 2000, with the rise of the 'new seven sisters' and unprecedent oil price levels and volatility.

3 Empirical Model and Results

Firm valuation models differ in their value drivers and underlying hypotheses. Most follow a discounted cash flow paradigm. Damodaran (2002) highlights the dividend discount model (DDM), the free cash flow approach (FCF) and the residual income model (RIM).

Ohlson (1995) introduced an empirically tractable version of the DDM model, that allows for econometric estimation of the role of equity, earnings and other firm relevant profit shifters on firm value, without the need to forecast future dividends. It is a generalization of the popular multiples valuation method. In short, the listed firm market value (P_t) is related to firm book value (B_t), earnings (E_t) and profit shifters (X_t), in addition to random errors that deviate market value from fundamentals.

$$P_t = \alpha + \beta_1 B_t + \beta_2 E_t + \beta_3 X_t + \varepsilon_t \tag{1}$$

A key point in Ohlson's model is that the value shifters are off balance information, so that it is not priced in the firm book value. In the O&G industry, such value drivers may be proven reserves, as measured by the SMOG principles (Ribeiro et al. 2011). We include R&D expenditures (RD_t) as one of the value drivers to evaluate a direct effect on firm value. It should be noted that if the market perceives R&D investments as part of a firm equity, no direct effect of R&D on market value should be found. Thus, we use a conservative model, in the sense of bias going *against* our hypothesis, to provide sound empirical evidence.

We use a longitudinal data set of O&G firms with available comparable R&D expenditures for the 2000–2010 period. Econometric panel data set techniques are used to control for unobserved heterogeneity and avoid confounding effects. Special care is dedicated to the econometric evaluation of the model. This required modelling the autocorrelation of the error term (Stata 2014; Wooldridge 2010). Hence, our empirical model, using logs, is based on

$$P_{it} = \alpha + \beta_1 B_{it} + \beta_2 E_{it} + \beta_3 RD_{it} + \beta_3 SMOG_{it} + u_i + \varepsilon_{it} \tag{2}$$

where u_i is a firm fixed effect, that incorporates time invariant firm specific market value effects. We allow for serial correlation in the time-varying error term, namely, $\varepsilon_{it} = \rho\varepsilon_{it-1} + \varpi_{it}$, where the last term is a white noise random variable.

There is significant heterogeneity across firms. That requires care and the use of the above mentioned fixed effects. Descriptive statistics are presented on Table 2. Average firm market value in a year is about US$132 billion, but ranging from US $14 billion to $448 billion. R&D expenditures vary significantly too, ranging from US$43 million to US$1.4 billion, with an yearly average of US$459 million.

Our estimates are presented on Table 3. Specification tests pointed to the presence of autocorrelation in the regression errors. The estimates below incorporate this adjustment. Failing to correct for the presence of autocorrelation generally

Table 2 Descriptive statistics—large O&G firms listed in the NYSE

Variables	Mean	Median	Standard deviation	Minimum	Maximum
Market value	132,050	111,740	96,847	14,760	446,840
R&D expend	459	379	325	43	1454
Book value	63,053	62,211	40,993	5771	186,640
Smog	61,264	45,076	47,162	8813	238,800
Earnings	12,641	11,490	10,148	−16,998	45,220

Note In US$ current million
Original source Firms 20-F and 10-K files, 2000–2011. Data from British Petroleum P.L.C, ConocoPhillips Holding CO, Chevron Corporation, Enni SPA, Exxon Mobil Corp., Repsol YPF, Petróleo Brasileiro S.A., Statoil ASA, Royal Dutch Shell PLC and Total S/A

Table 3 Econometric estimates: dependent variable—firm market value

Variables	(1)	(2)	(3)	(4)	(5)
Current R&D expenditures	−0.037	0.052			
	(0.094)	(0.101)			
Lagges R&D expenditures		**0.248**	**0.228**	**0.376**	**0.221**
		(0.093)	**(0.083)**	**(0.053)**	**(0.072)**
Earnings				**0.226**	**0.143**
				(0.051)	**(0.057)**
Book value					**0.323**
					(0.108)
SMOG			**0.275**	**0.265**	**0.220**
			(0.058)	**(0.050)**	**(0.051)**
No. of observations	110	100	100	97	97
R^2	0.156	0.817	0.814	0.873	0.896
Autoregressive coef. (ρ)	0.725	0.542	0.560	0.458	0.424

Note Panel data model with fixed effects with autocorrelation correction (xtregar in Stata)
Bold Coefficients significant at the 1 % confidence level. Coefficients standard deviation in parenthesis. Sample size n = 10 firms, T = 11 years. All variables in logs

leads to coefficient standard errors underestimation (Wooldridge 2010) biasing inference.

The first column shows that current (same year) R&D expenditures have no statistically significant effect on firm market value. Only when we use one year lagged R&D expenditures we find a positive, significant correlation with firm market value.

This significant effect may be spurious in the sense that R&D expenditures are correlated to other firm value drivers, such as reserves in the O&G industry. Column (3) shows a positive, robust effect of R&D expenditures on market value, even after controlling for the firm proven reserves (measured by the SMOG concept).

The model in column 4 also controls for firm earnings. Interestingly the R&D expenditure effect is now numerically larger and estimated more precisely. As expected both earnings and reserves have a positive impact on firm value.

Last, column 5 presents the full model, with earnings, book value and reserves as explanatory variables, as well as lagged R&D expenditures. All coefficients are significant and with expected signs. Book value, reserves and earnings increase have a positive effect on market valuation of integrated O&G firms, based on our sample. The effect of R&D expenditures is significant and positive. A 10 % increase in firm R&D expenditures leads to an expected 2.2 % firm market value increase, an year later.

4 Conclusion

Strategic positioning in the Oil and Gas (O&G) industry requires constant innovation to foster competitiveness, as in other cost driven industries. Innovation comes about as a result of research and development (R&D) efforts. At the same time, R&D efforts carry high risk and include long term projects and heavy sunk costs outlays. These factors may limit the market view that such R&D expenditures are lucrative or blur the perspective that such expenditures can improve a firm financial position, to investors' benefits.

The O&G industry is not one of the most R&D intense industries, with R&D expenditures at about 0.3 % of revenues. Yet the sheer size of these firms point to large R&D outlays, reaching US$1.4 billion annually.

The evidence for market value effects of R&D expenditures is mixed in the literature, as pointed above. This paper contributes to the literature using a sample of very large, worldwide, integrated O&G firms to investigate whether the market values R&D expenditures in this industry. That is we evaluate if investors perceive R&D expenditure changes as related to technological advances that will affect firm profitability.

The empirical strategy explores the Ohlson (1995) empirical model, based on the well-known Dividend Discount Model (DDM) used for firm valuation. We apply advanced econometric methods to avoid common biases such as the comparisons across firms, size, location, governance and reputation effects. This is brought about with a fixed effect specification for panel data (Wooldridge 2010). In addition, we care to evaluate the model implicit time series assumptions in the error term and conclude for the need to correct for autocorrelation in the model residuals.

In the many specifications evaluated, to avoid spurious results due to omitted variables, R&D expenditures appear to positively influence firm market value, with a lag. This result is valid even controlling for the different firm profitability, book value or reserves levels in our sample.

Acknowledgement Support from CNPq is acknowledged.

References

Ali A, Ciftci M, Cready WM (2012) Market underestimation of the implications of R&D increases for future earnings: the US evidence. J Bus Finan Acc 39(3):289–314

Besson C (2006) Scientific challenges in oil and gas exploration and production. In: OECD global science forum paris. OECD, Paris

Bret-Rouzaut N, Thom M (2005) Technology strategy in the upstream petroleum supply chain. Les cahiers de l'économie—no 57 Série Analyses et synthèses. Institut Français du Pétrole

Chan LKC, Lakonishok J, Sougiannis T (2001) The stock market valuation of research and development expenditures. J Finan 56(6):2431–2456

Damodaran A (2002) Investment valuation, 2nd ed. Willey, New York

Helfat CE (2004) Evolutionary trajectories in petroleum firm R&D. Manage Sci 40(12):170–1747

Ohlson JA (1995) Earnings, book values, and dividends in equity valuation. Contemp Acc Res 11 (2):661–687

Ribeiro EP, Neto LTM, Bone RB (2011) Reservas de Óleo e Gás em Modelos de Avaliacão para Empresas Petrolíferas. Revista Brasileira de Finanças 9(4):549–569

Sougiannis T (1994) The accounting based valuation of corporate R&D. Acc Rev 69(1):44–68

Stata (2014) Stata v.13. Stata Corp., College Station

Vijay G, Yatin B, DeBruine M (2002) Evaluation of the impact of R&D on EPS in the oil and das industry. Acad Acc Stud J 6(2):66–77

Wooldridge J (2010) Econometrics of cross section and panel data, 2nd edn. MIT Press, Cambridge

Part II
Logistics, Production and Information Systems

A Decision Support Framework for Production Flow Coordination Using Supply Chain Management Practices, Ordering Systems and Modeling Techniques

Wagner de Barros Neto, Laisa Caroline de Paiva Gomes, Maico Roris Severino and Moacir Godinho Filho

Abstract One of the greatest challenges faced by companies is flow coordination in its supply chain (SC). For such coordination, it can be found in the literature and practice the use of modeling techniques, ordering systems (OS) and practices in supply chain management (SCM). These, however, are not jointly used for coordinated operations in SC, admitting then a theoretic gap to be explored. In that way, this work has the objective to present a decision support framework for production flow coordination by combining the use of those mechanisms. For the development of this framework, a literature review about methodologies in production flow coordination in SC was made. The developed framework has purpose to assist managers in the decision making process about the practices and systems that are more suitable for each situation, thus obtaining the best coordination in its supply chain. We highlight that the project is still in progress and the framework is in phase of evaluation.

Keywords Supply chain coordination · Supply chain management · Ordering systems · Modeling · Process analysis

---

W. de Barros Neto (✉) · L.C. de Paiva Gomes · M.R. Severino (✉)
Unidade Acadêmica Especial de Engenharias, Universidade Federal de Goiás (UFG), Av. Dr. Lamartine Pinto de Avelar, 1120, Universidade de Goiás, Catalão-GO, Brazil
e-mail: w_barros@hotmail.com

M.R. Severino
e-mail: maicororis@gmail.com

L.C. de Paiva Gomes
e-mail: laisacaroline06@hotmail.com

M. Godinho Filho
Departamento de Engenharia de Produção, Universidade Federal de São Carlos (UFSCar), Rod. Washington Luís—Km 235, São Carlos-SP, Brazil
e-mail: moacir@dep.ufscar.br

M. Amorim et al. (eds.), *Engineering Systems and Networks*, Lecture Notes in Management and Industrial Engineering, DOI 10.1007/978-3-319-45748-2_8

1 Introduction

A major difficulty faced by companies is to manage a supply chain in an effective way and build relationships between suppliers and clients, allowing a production system to obtain competitive advantages and achieve its desired performance objectives related to cost, quality, speed, flexibility and reliability (Slack et al. 2007) or even timeliness, custom ability or suitability (Godinho Filho and Fernandes 2005).

In the last decades, there has been an exponential growth of studies and researches related to supply chain management, both in academia and business world, with many advances in the development of new practices, whether they are new methodologies, tools or techniques. These set of practices are found in the current literature on studies related to Supply Chain Management (SCM).

According to Mentzer et al. (2001), SCM can be defined as systemic and strategic coordination of the functions and traditional business tactics, through negotiations between functional areas within a particular company and through negotiations in the supply chain. The authors explain that the purpose of this coordination is to improve the performance of an individual company and, in the long run, the chain as a whole. Lambert (2004) describes eight essential processes for the SCM. They are: Customer Relationship Management (CRM), Supplier Relationship Management (SRM), Customer Service Management (CSM), Demand Management (DM), Order Fulfillment (OF), Manufacturing Flow Management (MFM), Product Development and Commercialization (PD&C) and Returns Management (RM).

The SCM's aim is to improve operational efficiency, responsiveness and profitability of the firms and partners in the supply chain. This goal is achieved through the development and implementation of models, which must adapt to each different scenario and consequently there is not a single model that captures all aspects of the processes of a supply chain (Badole et al. 2013). Such models can be developed using different techniques of Operations Research and can be classified as a deterministic model, stochastic or hybrid. According to Taha (2010), in Operations Research there is not a single technique to solve all mathematical models that may arise; it is the type and complexity that determine the nature of the solution method.

Problems relating to the production flow coordination are not restricted to relationships between organizations in the supply chain; they may also occur in industrial plants during the production process. However, when it comes to problems in this scope, we find in the literature a variety of practices for planning, programming and production control known as Ordering Systems (OS), which can be production, services and/or purchase orders.

These OS, according to Fernandes and Godinho Filho (2011), can be defined as systems that schedule or organize/explode the requirements in terms of components and materials, and/or control the emission/release of production, purchase and services orders, and/or program/sequence tasks on the machines. Among these OS, we can cite the following systems: kanban, drum-buffer-rope (DBR), constant work in process (CONWIP), materials requirements planning (MRP), period batch control (PBC), among others.

Despite the widespread use of the OS for coordination within companies, there is strict implementation of these systems for orders' coordination between suppliers and customers, and most of the time they do not get equivalent performance when compared to coordinated orders between machines within a plant. It is noteworthy that depending on the characteristics of products and markets and the specifics of the strategies of each company or supply chain, an OS is more suitable than another, or may even support a combined use.

Flow coordination mechanisms are designed to manage products and information flows in supply chains, among which we can mention the Vendor Managed Inventory (VMI), Quick Response (QR), Collaborative Planning, Forecasting and Replenishment (CPFR) among others. Sahin and Robinson (2002) provide an extensive review of the literature on flow coordination of goods and information sharing in supply chains, classifying the literature based on the degree of information sharing and coordination.

The objective of this study is to develop a process for analysis and support in the decision making process about implementing mechanisms for production flow coordination in supply chain using modeling techniques, SCM practices and OS.

To reach this objective, a literature review was made on the conditions under which use of each SCM practice, each OS and each modeling technique is more suitable. Subsequently, a decision support framework to choose the best combined use of coordination mechanisms was elaborated.

2 Process Analysis Framework

From the used methodology, was developed a framework that presents a detailed characterization process of a relationship between a supplier and a customer for later decision making (supported by a set of criteria) for suitable production flow coordination mechanisms in supply chains. Subsequently, a qualitative and quantitative assessment would be made of this decision. Thus, through the use of this elaborated process, any supply chain may implement the methodology to meet their expected performance objectives.

One of the challenges of production managers is to decide which practices to use to achieve production and materials flow coordination in the supply chain. The developed process is presented as a useful tool to assist in decision making, impacting positively on companies' competitiveness which make use of it. Thus, it was created a flowchart through a conceptual framework that consists of 11 steps to assist managers in coordinating the production flow of relationships between supplier and customer in a supply, as shown in Fig. 1.

Step 1:
The first stage of this framework deals with the characterization of the relationships between agents in a supply chain (e.g. subcontractor, supplier, manufacturing,

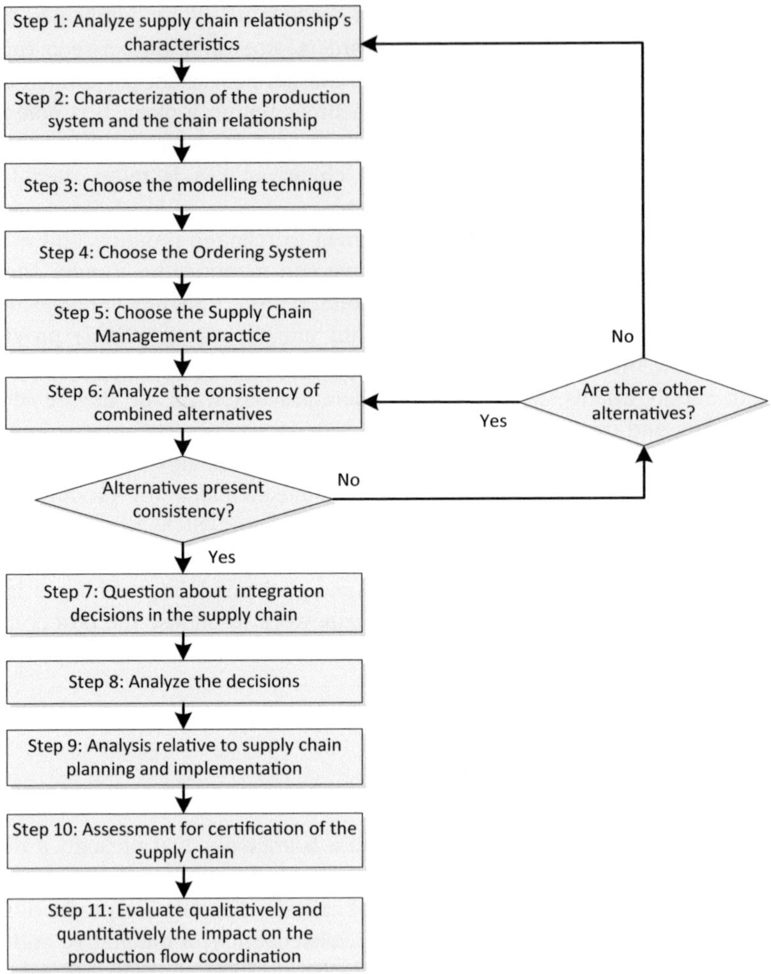

Fig. 1 Decision support framework

customers) and their characteristics with the rest of the chain. For more details, see Galasso et al. (2009).

Step 2:

This step consists on the characterization of the production system, impacting the choice of which modeling technique, OS and SCM practices to be used. Examples of characterizations may be: item class (A, B or C), volume ratio range, degree of repeatability, stability demand level, demand predictability and flexibility, setup time, etc.

Step 3:

The modeling technique that is most appropriated for the company's scenario and strategy to achieve its optimization goal should be chosen in this step.

Badole et al. (2013) show several existing techniques and their various applications found in the literature.
Step 4:
You must select the most appropriate OS to the case. To this end, Fernandes and Godinho Filho (2011) present the set of conditions under which each OS are recommended.
Step 5:
In this step, the most appropriate SCM practice(s) to the case is (are) selected. Thus, Severino (2012) presents several practices used in supply chain management and their contribution to production flow coordination.
Step 6:
The analysis consist in verifying whether the modeling techniques, the OS and the SCM practices complement each other or if the use of one prevents the use of another in the proposal for the combined use. In this sense, this complementarity is evaluated and a planning of the implementation of the combined use is done, defining the role of each and how the operating mechanism establishes the production flow coordination. It is noteworthy that when doing this planning it should be verified the impacts on infrastructure, resources and in the current work routines situations. Such evaluations allow detecting possible resistance in terms of the use of the possibilities.
In the event of incompatibility between the chosen alternatives, two situations can happen: if there are other possible alternative, its consistency will be analyzed; if there are no other alternative, the previous steps must be repeated to verify if the characterization was done correctly, in case there were any misunderstanding in the terms of the choice.
Step 7:
This step is to examine the types of waste in the supply chain, as shown by Liu et al. (2013), identifying the behavior of integration decisions.
Step 8:
The analysis of the decisions is relevant to specify which situations will be evaluated in relation to production, transportation, distribution, demand and control of the product, analyzing how much they will influence the quality, quantity and adequacy of transport, point of sale and product demand scenarios. For deeper understanding, see van der Vorst et al. (2009).
Step 9:
Evaluation about the ideal (planned) state and the actual (executed) state in the supply chain, describing its different levels (strategy, design, planning and operations) so that the applied tools are updated, making a planning and detailing on the functional level of the operations, as seen in Ivanov (2010).
Step 10:
For the evaluation and certification of the supply chain, as shown in Acar et al. (2010), is necessary to run simulation models relating to transport, product type, quality inspections, use of optimization techniques for production, evaluate the influence of the location on the delivery of orders and if the forecast will influence product demand.

Step 11:
It is recommended to test or simulate the chosen formulated proposal that presented theoretical consistency. Among the simulation goals highlights the verification of the proposal applicability in the real case or if such proposal impacts in other relationships in the supply chain in case a single item was analyzed.

After the simulation, it should be evaluated the impact of the elaborated proposal in the production flow coordination in the studied supply chain. Therefore, it is suggested to carry out qualitative and quantitative evaluations. For the qualitative assessment, it is recommended the use as reference the activities of the sub processes of Manufacturing Flow Management Process (MFMP) as suggested by Goldsby and García-Dastugue (2003). The authors state that to have a coordinated production flow in the supply chain some activities should be developed in the relationship between supplier and customer. In this way, it is analyzed if the activities of MFMP proposed by the authors are being accomplished by implementing the proposal suggested by this framework.

In terms of quantitative evaluation, it is suggested the use of performance indicators that are suitable for the case in question as, for example, the approaches of Beamon (1999), Gunasekaran et al. (2001) and the SCOR model, with the approaches of Stewart (1995) and Dreyer (2000).

3 Conclusions

Nowadays the production managers develop solutions for production flow coordination in the relationships between suppliers and customers through their professional experience, using tools they had the opportunity to meet in their academic or professional career. However, this decision making is performed by trial and error and does not obtain optimum results. It is in this sense that a decision support framework is developed, providing a tool that helps achieving better production flow coordination in the supply chain with the combined use of modeling techniques, supply chain management practices and ordering systems. With such process, the manager knows the relationship between supplier and customer with more details, allowing through a set of criteria to define the mechanisms of production flow coordination that are more suitable for the case, performing qualitative and quantitative evaluations before implementing it. Thus, using such a framework, the decision making process increases the probability of successful with the choice.

References

Acar Y, Kadipasaoglu S, Schipperijn P (2010) A decision support framework for global supply chain modelling: an assessment of the impact of demand, supply and lead-time uncertainties on performance. Int J Prod Res. doi:10.1080/00207540902791769

Badole CM, Jain DR, Rathore DA et al (2013) Research and opportunities in supply chain modeling: a review. Int J Supply Chain Manage. Available via Exceling Tech. http://ojs.excelingtech.co.uk/index.php/IJSCM/article/view/662. Cited 29 Jan 2015

Beamon B (1999) Measuring supply chain performance. Int J Oper Prod Manage. doi:10.1108/01443579910249714

Dreyer D (2000) Performance measurement: a practitioner's perspective. Supply Chain Manage Rev. Available via HighBeam Research. http://www.highbeam.com/doc/1G1-64972565.html. Cited 30 Jan 2015

Fernandes FCF, Godinho Filho M (2011) Production control systems: literature review, classification, and insights regarding practical application. Afr J Bus Manage. doi:10.5897/AJBM11.184

Galasso F, Mercé C, Grabot B (2009) Decision support framework for supply chain planning with flexible demand. Int J Prod Res. doi:10.1080/00207540802426508

Godinho Filho M, Fernandes FCF (2005) Paradigmas estratégicos de gestão da manufatura (PEGEMs): elementos-chave e modelo conceitual, 3rd edn, vol 12. Gestão & Produção, São Carlos, pp 333–345, set./dez. Available via SCIELO. http://www.scielo.br/pdf/gp/v12n3/28023.pdf. Cited 29 Jan 2015

Goldsby TJ, García-Dastugue SJ (2003) The manufacturing flow management process. Int J Logistics Manage. doi:10.1111/j.1540-5915.2002.tb01654.x

Gunasekaran A, Patel C, Tirtiroglu E (2001) Performance measures and metrics in a supply chain environment. Int J Oper Prod Manage. doi:10.1108/01443570110358468

Ivanov D (2010) An adaptive framework for aligning (re)planning decisions on supply chain strategy, design, tactics, and operations. Int J Prod Res. doi:10.1080/00207540902893417

Lambert DM (2004) The eight essential supply chain management processes. Supply Chain Manage Rev 8(6)

Liu S, Leat M, Moizer J et al (2013) A decision-focused knowledge management framework to support collaborative decision making for lean supply chain management. Int J Prod Res. doi:10.1080/00207543.2012.709646

Mentzer JT, DeWitt W, Keebler JS, Min S et al (2001) Defining supply chain management. J Bus Logistics. doi:10.1002/j.2158-1592.2001.tb00001.x

Sahin F, Robinson EP (2002) Flow coordination and information sharing in supply chains: review, implications, and directions for future research. Decis Sci. doi:10.1111/j.1540-5915.2002.tb01654.x

Severino MR (2012) Coordenação do fluxo de produção por meio do uso combinado de práticas utilizadas na gestão da cadeia de suprimentos e de sistemas de coordenação de ordens puxados. 2012. 171 f. Tese (Doutorado em Engenharia de Produção)—Universidade Federal de São Carlos (UFSCar), São Carlos, 2012. Available via BDTD UFSCar. http://www.bdtd.ufscar.br/htdocs/tedeSimplificado//tde_busca/arquivo.php?codArquivo=5666. Cited 29 Jan 2015

Slack N, Chambers S, Johnston R (2007) Administração da produção. Atlas, São Paulo

Stewart G (1995) Supply chain performance benchmarking study reveals keys to supply chain excellence. Logistics Inf Manage. doi:10.1108/09576059510085000

Taha HA (2010) Operations research: an introduction, 9th edn. Pearson Prentice Hall, Upper Saddle River

van der Vorst JGAJ, Tromp SO, van der Zee DJ (2009) Simulation modelling for food supply chain redesign; integrated decision making on product quality, sustainability and logistics. Int J Prod Res. doi:10.1080/00207540802356747

A Greedy Primal-Dual Type Heuristic to Select an Inventory Control Policy

Nazanin Esmaili, Bryan A. Norman and Jayant Rajgopal

Abstract We propose a greedy primal-dual type heuristic to jointly optimize the selection of an inventory control policy and the allocation of shelf space in order to minimize the expected counting and replenishment costs, while accounting for space limitations. The problem is motivated by an application in the healthcare sector. It addresses the limitations in designing an inventory control system for hospitals stockrooms and the drawbacks of the common approach of using a single policy such as a two-bin Kanban or a Periodic Automatic Replenishment (PAR) system for all items. In the proposed approach, we not only choose policies to use available storage space more efficiently but also consider changing the policies or their parameters to use the space within a selected storage bin more efficiently. On numerical examples where a mathematical programming formulation can be solved in a reasonable amount of time, our experiments indicate that the proposed algorithm is very efficient.

Keywords Greedy · Multi-item inventory system · Shelf allocation · Healthcare · Integer programming

1 Introduction

A contemporary challenge that all countries face is to be able to provide health care to citizens that is both high in quality and reasonable in cost. In many developed countries affordability is especially an issue given that the populations are older (Hall 2012). In the United States, where health care is largely managed by the

N. Esmaili · B.A. Norman · J. Rajgopal (✉)
Department of Industrial Engineering, University of Pittsburgh, Pittsburgh, PA 15261, USA
e-mail: rajgopal@pitt.edu

N. Esmaili
e-mail: nae22@pitt.edu

B.A. Norman
e-mail: banorman@pitt.edu

M. Amorim et al. (eds.), *Engineering Systems and Networks*, Lecture Notes in Management and Industrial Engineering, DOI 10.1007/978-3-319-45748-2_9

private sector, there is heightened competition between health care organizations and many of these organizations have initiated projects to improve quality and provide more cost-effective service. One area in particular is hospital logistics; studies have estimated inventory costs in the health care sector to be between 10 and 18 % of net revenues (de Vries 2011). Hospitals typically handle hundreds of different items in medical supplies and pharmaceuticals, and a sufficient inventory of these items needs to be on hand to enable hospital staff to perform their daily work and provide high-quality health care. Typically, these medical supplies are stored at numerous locations within a hospital and in sufficient quantities to ensure high levels of availability when required. This however, has to be traded off against the fact that hospitals often lack adequate storage space (Little and Coughlan 2008).

In this paper, we study a multi-item system with shelf space constraints, where any of several different inventory policies including PAR level, (S, s), (Q, s) and two-bin Kanban could be an option for an item. We assume periodic review and a lead time of zero (as is common in the replenishment of individual hospital stockrooms from a central storage location) and consider the same reorder point s for an item across all inventory control policies: s is selected for an item such that if its inventory is at or below s at the time of review, the probability of stocking out before the next review is unacceptably high. We then determine the best policy for each item based on the overall shelf space available and compute its order quantity for its policy. To our knowledge, no prior work has considered a model with these specific characteristics. The remainder of this paper is organized as follows. Section 2 reviews the different inventory policies considered here while Sect. 3 describes the model. Section 4 proposes a greedy primal-dual type algorithm to solve the model. Section 5 presents computational test results motivated by a real hospital. Finally, Sect. 6 provides concluding remarks and ideas for future research.

2 Trade-Offs Between Kanban and Other Policies

In an individual area or department stockroom within a hospital, traditional inventory holding costs are generally not relevant since these costs are incurred regardless of whether the items are stored in the stockroom or in the hospital's central warehouse or storage facility (from which the stockroom is replenished). The main tradeoffs to consider in inventory control operations here are space, replenishment costs and counting costs. We consider these in the context of four common policies in healthcare: (S, s), (Q, s), PAR and two-bin Kanban policies. In the (S, s) policy a replenishment order is placed whenever the inventory on hand at the time of review is less than or equal to the reorder point (s). The reorder point triggers a replenishment action for an amount equal to the difference between S and the current amount on-hand. Thus the order-up-to level (S) is the inventory on-hand at the beginning of each cycle following an order cycle; the actual amount ordered in each cycle varies. In our setting the value of s should be large enough to ensure that demand until the next review point is satisfied with a very high probability (e.g., 0.99 or higher). In the (Q, s) policy an order is triggered in a similar way as with the order-up-to policy but with the difference that we order the same amount each time (Q); however, the inventory on-hand at the beginning of each cycle varies. The PAR level for an item is the amount that should be available at the beginning of any replenishment cycle for that item; if at the time of review the inventory level for the item is below "PAR" it is replenished. Since we assume that the trigger point for a new order equals s, we set the PAR level in this paper to $s + 1$.

The two-bin Kanban policy used in many manufacturing settings has also been adopted in healthcare: units of an item are stored in two identical bins and when the first bin is empty, we place an order for a full bin while switching to the second one. Consistent with the service level requirement of the previous paragraph we set bin capacity to s. The main advantage of this policy is that it is much easier from an operational perspective. Unlike the previous policies, no counting is needed and this system eliminates counting errors, simplifies logistics and reduces costs.

Several studies like Rosales et al. (2014) and Esmaili et al. (2015) show that changing from a PAR system to a two-bin Kanban system can substantially reduce total cost of counting and replenishment per review cycle. However, these studies do not emphasize the fact that although a two-bin Kanban system is simpler and reduces operational costs, it also has several limitations. Generally speaking, it requires more shelf space, carries more inventory, and might be inefficient or unsuitable for some items. We examine each of these issues a little further.

First, unless we can use a card separator within a single bin, we are restricted to using two separate bins. With the other policies it might be possible to store everything needed in a single bin. Thus we generally need more shelf space. Moreover, we need to consider both bins as one item since the bins should be together. With this restriction, we are less flexible when optimizing the allocation of space. Second, we generally have more inventory; e.g., with the PAR system we have a maximum inventory of $s + 1$ but here we could have as many as $2s$ units (with the other two systems it would depend on the values of S and Q). Third, this policy may not be suitable for all items. For example, s_i units of a larger item i may not fit in any of the available bins. Also, we can only store $2s_i$ units of an item i with a two-bin Kanban system, and sometimes this is not optimal because we might want to store more. On the other hand, the two-bin Kanban policy is optimal for an item i for which the required amount of space to fit $2s_i$ units (using two bins) is the same as the amount of space to fit $s_i + 1$ units with a PAR level system.

3 Model Description

If space is unlimited it would be optimal (in terms of operational costs) to use a two-bin Kanban system for every item where it is feasible to do so (i.e., Esmaili et al. 2015). In this paper we assume that space restrictions preclude the assignment of a two bin Kanban policy to all items. On the other hand, we do assume overall feasibility, i.e., that available space is enough to store all items using a PAR level policy (which takes up the least amount of space among all the policies, but is expensive in terms of operational costs). We consider standard steel shelving with K identical shelves with dimensions $H \times W \times D$. There are a total of I items, where item i is a rectangular solid of dimensions $h_i \times w_i \times d_i$ and there are M different bin types where bin type m has dimensions $\tilde{h}_m \times \tilde{w}_m \times \tilde{d}_m$. We assume that the amount of time to count an item (t_c) is the same for all items, and the amount of time to replenish an item (t_r) is also the same across all items. Item demands (D_i) are independent random variables. If required, we can use a vertical card to divide the space within a bin into two equal-sized subspaces.

Because of the nature of a two-bin Kanban system, we only use (1) a single bin with a "divide" card or (2) two bins. Thus, a bin is feasible for an item i if either half or the entirety of its space is big enough to hold s_i units of that item. We assume there is at least one bin that is large enough to accommodate a unit of each item and that an item is stored using only one bin type on all shelves. The depths and heights of all bins are assumed to be such that we cannot stack them on top of each other or have more than one bin back to back on a shelf. We also assume that an item is managed using only one type of policy, there is no lot-size requirement or quantity discount, the review interval (time between two successive reviews) is constant and backorders are not permitted. Finally, we assume that lead time is negligible and items retain their quality over time. Note that the only reason that we might want to hold more than $2s_i$ units of an item i using an (S, s) or (Q, s) policy is because we might wish to fill a bin to its capacity.

We assume that we do not have enough space to fit all items on the shelves using a two-bin Kanban policy for every item, but that we do have enough space to use a PAR policy for all items. This assumption guarantees that the problem is feasible while also eliminating the trivial solution of a two-bin Kanban policy for all items.

For each pair of item i and policy j, S_{ij} is the value of the stock level of item i immediately after the replenishment, using policy j, s_i is the 99th percentile of demand for item i. Let $x_{ij} = 1$ (decision variable) if policy j is assigned to item i; 0 otherwise. Let c_{ij} denote the expected cost per cycle of assigning policy j to item i. We use simulation to calculate this parameter. The simulation uses the individual probability distribution of demand for each item, policy characteristics, and its parameters, and simulates each policy over a suitable length of time to estimate counting and reorder effort. In fact, these expected counts for a policy only depend on the expected value of demand (μ_i) and maximum inventory level (S_{ij}). Then, we multiply the expected number counted and the expected number of reorders by t_c and t_r respectively to calculate the objective function coefficients for item i with policy j; clearly, c_{ij} is sensitive to t_c and t_r. The objective function thus represents the total expected replenishment and counting effort over each review period (i.e., $Z = \sum_i \sum_j c_{ij}x_{ij}$).

4 Description of the Algorithm

Let T_i be the set of policies for item i that is defined as follows: $T_i = \{\text{PAR}, (S = s_i + 2, s_i), (S = s_i + 3, s_i)\ldots(S = 2s_i, s_i), (Q = s_i + 1, s_i), \text{Kanban}\}$. Note that we define the set in the same order of its elements as given above, with PAR being the first element and the two-bin Kanban being the last. The algorithm consists of four steps. First, a preprocessing step assigns the most space efficient bin to each combination of item i and policy j and then eliminates dominated policies for all items. Second, an initial policy is assigned to each item i based on cost and space. Third, a space allocation step allocates available space to items in descending order of their width. Finally, an iterative step uses the initial assignment and primal-dual characteristics and iterates to successively either (a) improve a feasible set of suboptimal policies until no further improvement is possible or (b) find a feasible set of policies from an infeasible but super-optimal one by trading off against cost increases. The details of our algorithm are as follows:

1. *Preprocessing Step*: Select the best bin to use for each item with each of the policies in T_i so as to minimize the total required width and let W_{ij} be width of the bin row assigned to item i with policy j and n_{ij} be the total number of bin rows required for item i using policy j. If there is a policy in T_i for the item that has smaller cost and also takes less or equal space than another one, then the latter policy is dominated and therefore we remove all such dominated policies from T_i. If T_i has only one element, assign this as the optimal policy for item $i(x^*_{ij} = 1)$ and never consider changing it in the algorithm.
2. *Initial Policy Assignment Step*: This step assigns initial policies to items. Start with the first item without an assigned policy.
   Calculate $j^* \in \text{argmin}_j\{c_{ij} | j \in \{PAR, [(S = s_i + 2, s_i), W_{iPAR} = W_{i(S=s_i+2,s_i)}], [(Q = s_i + 1, s_i), W_{iPAR} = W_{i(Q=s_i+1,s)}]\},\}$. If
   - $j^* = \text{PAR}$: check whether Kanban is feasible for that item; if so, reset j^* to *Kanban* and assign $x_{iKanban} = 1$; otherwise assign $x_{iPAR} = 1$,
   - $j^* = (S = s_i + 2, s)$: assign $x_{i(s_i+2,s)} = 1$; then increase S from its current value until (a) the last bin is filled or (b) the item's cost is minimized,
   - $j^* = (Q = s_i + 1, s)$: assign $x_{i(Q=s_i+1,s)} = 1$; then increase Q from its current value using the same procedure as with (S, s) above.

   Repeat until all items have an initial policy assigned. Then continue to the next step. Note that at this time an item can have the PAR policy assigned only if it is less expensive or take up less space than $(S = s_i + 2, s_i)$ or $(Q = s_i + 1, s_i)$, and also cannot be assigned a Kanban policy because of the item's size.

3. *Shelf Space Allocation Step*: Rank items in decreasing order of bin width. Consider all items with the same width to be one class of items. Sequentially assign items of the same class to shelves until the class is completed. Start the next class on a new shelf. Continue this procedure until there are no shelves

available. Then fill any remaining empty space with items starting with the current class and with the first shelf. If no items remain then the current set of policy assignments is feasible; otherwise, it is infeasible.

4. *Iterative Step*: This step sequentially updates the policy for each item until there is nothing left to change. If the set of policies from the previous step is infeasible, then execute step 4.1; otherwise, execute step 4.2.

   4.1. Start with $i \in \text{argmin}_{i \ni x_{iKanban}=1}\left\{\frac{c_{iPAR}-c_{iKanban}}{W_{iKanban}-W_{iPAR}}\right\}$. Then assign $x_{iPAR} = 1$ and go to the Shelf Space Allocation Step. If it is still infeasible, repeat step 4.1 using the next i with the same criterion and continue until the problem becomes feasible; then, stop and use the current set of policy assignments as the final ones.

   4.2. Start with $i \in \text{argmax}_{i \ni x_{iKanban} \neq 1}\left\{\frac{c_{ij^*}-c_{i,j^*+1}}{W_{i,j^*+1}-W_{ij^*}}\right\}$. Then assign $x_{ij^*} = 0$, assign $x_{i,j^*+1} = 1$ and update $j^* = j^* + 1$. Then go to the Shelf Space Allocation Step. If the updated set of assignments is still feasible, repeat step 4.2 using the next i and continue until the problem becomes infeasible; then, stop and use the last set of feasible policy assignments as the final ones.

In step 2 we start by assigning the lowest cost policy with space usage no higher than that required by the PAR policy. If we move to a Kanban policy, for items that use the $(S = s + 2, s)$ or $(Q = s + 1, s)$ policy our potential for cost improvement is not as high as with one that currently uses PAR. Therefore, whenever feasible we switch to a Kanban policy (ignoring space for now). For items that are assigned (S, s) or (Q, s) it must mean that replenishment costs are high and dominate counting costs; otherwise the PAR with its lower counting costs would dominate these two. To decrease replenishment effort we therefore try and increase S or Q as long as there is space left over in the (last) bin or the increase in expected counting costs exceed savings from reduction in expected replenishment costs.

Entering step 4 there are two possibilities. The current set of policies is (1) infeasible and require more space than we have available, or (2) feasible and could potentially be improved by making use of available space. In the first case, we trade off increase in cost against space savings: we choose an item that currently has a Kanban policy and switch to a PAR policy based on the increase in cost per unit of space saved and repeat until we attain feasibility (note that we are guaranteed to stop because of our feasibility assumption). In the second case we trade off increased space usage against lower costs: we choose an item that currently does *not* use a Kanban policy and switch to a less expensive policy but with a higher space usage, based on the reduction in cost per unit of space added. These two might be viewed as a dual/primal approach; in one case we are better than optimal and try to attain feasibility, in the other we are always feasible but try to obtain optimality.

As a point of comparison, we also formulated and solved an integer programming (IP) model to solve this problem optimally; however, this formulation can only be used to solve relatively small problems. Let y_{ijk} be the number of bin rows of item i located on shelf k using policy j the IP model is as follows:

$$\max \sum_i \sum_j c_{ij} x_{ij}$$
$$\sum_i \sum_j W_{ij} y_{ijk} \le W, \forall k$$
$$\sum_j \sum_k y_{ijk} = n_{ij} x_{ij}, \forall i$$
$$\sum_j x_{ij} = 1, \forall i$$
$$x_{ij} \in \{0,1\}, \forall i,j$$
$$y_{ijk} \in \mathbb{Z}^+, \forall i,j,k.$$

5 Numerical Results

In this section, we illustrate the proposed greedy algorithm using actual data from a hospital. We select a subset of 10 items with diverse characteristics, stored in a stockroom with multiple shelves, each with height 5 in., width 30 in. and depth 21 in. There are 3 different bin types with height 5 in., depth 21 in. and width 7, 10, 15 in. respectively. The review interval is assumed to be one week and s_i is computed for item i via $P(D_i \le s_i) = 0.99$.

Figure 1 shows relevant data for each item, along with s_i computed from demand over 20 weeks, and the current PAR levels in use at the hospital. It also shows the average weekly replenishment and counting costs for each item obtained by our simulation. We assume that the fixed cost per replenishment is ten times the unit counting cost ($t_C = 1; t_r = 10$). Figure 2 shows total expected replenishment and counting effort over each review period for each of the ten items arranged in increasing order of demand. As expected, we can expect to expend more effort on faster moving items. Recall that in general, the PAR policy needs more replenishment and less counting effort than the (S = 2s, s) and (Q, s) policies. When item demand is low the reduced replenishment effort with the two latter policies

Item	Items Dimensions			D	s	A subset of the objective function coefficients			
	Height	Width	Depth			PAR	(S=2s,s)	(Q=s,s)	Kanban
Tens Unit	1.75	7.75	11.25	4	9	15.82	14.64	14.73	4.99
Walker Handles	3.25	3.50	7.00	3	8	15.59	13.43	13.63	4.51
Ortho Wedge	2.50	3.00	5.00	11	19	18.96	25.13	24.20	5.73
Dolomite Handles	4.00	8.00	13.50	10	18	18.92	24.16	23.32	5.54
Gel Pack	2.00	6.25	10.25	2	6	14.08	11.06	11.30	4.35
Knee Brace	3.00	3.00	12.00	9	17	19.13	23.43	22.55	5.43
All Purpose Boot	3.50	5.00	12.50	7	14	17.90	20.16	19.70	5.29
Ankle Brace	2.25	5.25	8.00	6	12	16.97	17.91	17.86	5.26
CAM Walker	2.00	3.00	4.00	8	15	18.07	21.01	20.66	5.41
Post Op Shoe	4.50	4.50	12.00	1	4	11.56	8.24	8.57	3.70

Fig. 1 Parameters and simulation results

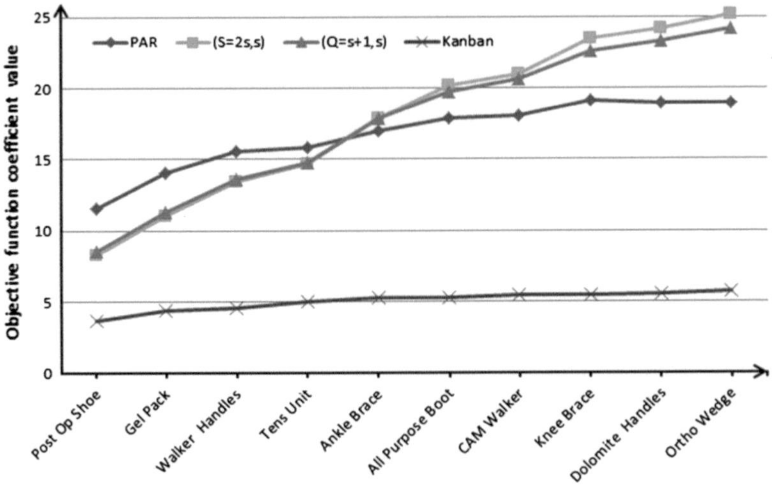

Fig. 2 Total expected replenishment and counting effort over each review period

Items	Example 1: K=15		Example 2: K= 17		Example 3: K=25	
	Algorithm	Optimization	Algorithm	Optimization	Algorithm	Optimization
Tens Unit	PAR	PAR	(s+6,s)	(s+6, s)	(2s,s)	(2s,s)
Walker Handles	(S,s)= (s+4,s)	(s+4,s)	Kanban	Kanban	Kanban	Kanban
Ortho Wedge	Kanban	Kanban	Kanban	Kanban	Kanban	Kanban
Dolomite Handles	PAR	PAR	PAR	PAR	PAR	PAR
Gel Pack	(S,s)= (s+2,s)	(2s,s)	(s+2,s)	(2s,s)	Kanban	Kanban
Knee Brace	PAR	PAR	PAR	PAR	PAR	PAR
All Purpose Boot	PAR	PAR	PAR	PAR	PAR	PAR
Ankle Brace	PAR	PAR	PAR	PAR	PAR	PAR
CAM Walker	Kanban	Kanban	Kanban	Kanban	Kanban	Kanban
Post Op Shoe	PAR	PAR	PAR	(2s,s)	(2s,s)	(2s,s)
Z*	139.27	137.04	128.18	123.68	115.79	115.79
Percentage Difference	0.016		0.036		0	
Avg. Difference	0.017					

Fig. 3 Greedy and IP results for three examples with different number of shelves

dominates the additional counting effort, but when demand increases, counting effort eventually dominates and the PAR policy becomes more effective. In our instance, we reach this point with the Ankle Brace, when demand is 6. Increasing t_r will move the point to the right while increasing t_c will shift it to the left.

Figure 3 shows the result of solving the example considering different numbers of shelves with the proposed greedy algorithm and the IP model. Note that while it is possible to solve the IP model for a small problem such as this, it rapidly becomes intractable as the number of items, number of bin sizes and the number and sizes of shelves increase. Therefore we restrict our comparisons with the IP model only to this smaller problem.

Finally, we point out that a FFDW heuristic can be also used for our space allocation step (Step 3) without starting a new shelf for a new class of items. We have shelves that are all of the same size and therefore only the width of an item is important in the space allocation. However, the FFDW might fail to find a feasible assignment even when one exists, unlike our simpler approach.

6 Conclusions

We present a greedy algorithm for optimizing the choice of which inventory policy to employ for each item based on characteristics of the items and storage space in the healthcare setting being evaluated. The objective is to minimize the average effort (cost of labor) to count and replenish all items, while providing an acceptably high level of service (avoiding stock outs) and taking into account the available space. We consider PAR level, (S, s), (Q, s), and two-bin Kanban policies. Each product can have any one of these policies. We illustrate the methodology using actual data from an actual hospital.

References

De Vries J (2011) The shaping of inventory systems in health services: a stakeholder analysis. Int J Prod Econ 133(1):60–69

Esmaili N, Norman BA, Rajgopal J (2015) A heuristic for selecting multi-item inventory review policies. In: Proceedings of the 2015 industrial and systems engineering research conference, Nashville, TN

Hall Randolph W (2012) Handbook of healthcare system scheduling. Springer Science + Business Media LLC, New York

Little J, Coughlan B (2008) Optimal inventory policy within hospital space constraints. Health Care Manage Sci 11(2):177–183

Rosales CR, Magazine M, Rao U (2014) The 2Bin system for controlling medical supplies at point-of-use. Eur J Oper Res 243(1):271–280

A Model that Integrates Direct and Reverse Flows in Omnichannel Logistics Networks

Javier Guerrero-Lorente, Eva Ponce-Cueto and Edgar E. Blanco

Abstract A more complex logistics network has to be managed by retailers that also offer online sales, since new shipping and drop off options are offered to consumers in order to satisfy their expectations. The main goal of this paper is to propose a mixed integer linear programming (MILP) model that integrates forward and reverse material flows in a retailer's omnichannel logistics network. The model proposed helps to determine the mix of orders and returns flows that minimizes costs, and also allows to quantify key trade offs associated to the different options offer in omnichannel models.

Keywords Commercial returns · Reverse logistics networks · Omnichannel

1 Introduction

Online shopping and electronic commerce (e-commerce) continues to grow in importance since the advent of the Internet in the 1990s. In 2011, it represented 4.7 % of the total retail trades (U.S. Census Bureau 2013) and it is expected to grow to 10 % by 2017 with an average annual growth rate of 9 % (Forrester et al. 2013). At the end of 2010 mobile e-commerce represents 3 % and just one year later, mobile e-commerce was 9 % of e-commerce (Kleinman 2012). Growth rates may even be higher as Internet access has shifted from traditional computers to

J. Guerrero-Lorente (✉) · E. Ponce-Cueto (✉)
Dpto. de Ingeniería de Organización, Administración de Empresas
Y Estadística. ETSII, Universidad Politécnica de Madrid, Madrid, Spain
e-mail: javier.guerrero@upm.es

E. Ponce-Cueto
e-mail: eva.ponce@upm.es

E.E. Blanco
MIT Center for Transportation and Logistics, Massachusetts Institute
of Technology, Cambridge, MA, USA

M. Amorim et al. (eds.), *Engineering Systems and Networks*, Lecture Notes
in Management and Industrial Engineering, DOI 10.1007/978-3-319-45748-2_10

smartphones, increasing multichannel capabilities and the adoption of web retailing by large brick and mortar retailers.

The surge of Business to Consumer (B2C) e-commerce not only has made direct shipments from manufacturers and retailers to individuals grow, but it has also increased the rate of commercial returns that retailers need to manage in this new context. According to Guide et al. (2006) the return rates vary widely by product category, by season, and across global markets. For example, large traditional retailers, such as Home Depot, can have return rates of 10 % of sales or even higher due to liberal returns policies (Guide and Van Wanssenhove 2009). In the fashion apparel industry, this rate could increase up to 35 %. Returns percentages are also typically much higher for catalog sales and online sales (Guide et al. 2006).

Commercial returns are products returned to the retailer (or other actor in the reverse supply chain) by consumers after purchase. Tibben-Lembke (2004) provides a more detailed explanation of different types of commercial returns. Different retailers have different return policies (30, 60, 90 days), other have more liberal policies, allowing 365 days or even more. For instance, the returning online orders policy in Lands' End, an online retailer in the USA, allows that online customers can return products at any time for an exchange or refund or tis purchase price (this policy only applies within the U.S.).

Reasons for returns also vary from products and industries. In the online retail fashion industry, where return rates are higher, color matching or sizing are the more typical reasons for returns. Many companies have identified the proper management of these commercial returns as a key aspect to achieve competitive advantage. According to Mollenkopf et al. (2007) online retailers should invest more in a more responsive and effective management of commercial returns.

There are also online retailers that are working on solutions that help them reduce online returns. For instance, Running Warehouse, an U.S. footwear retailer, has introduced an app which allows customers to find more accurate information about their shoe size and, as a result, this has reduced fit-related returns rates by 23 %.

In this new context more coordination between consumers, retailers and other actors in the direct and reverse supply chain is needed for both traditional and online sales. On the other hand, a more complex logistics network has to be managed by retailers and other actors, since new shipping and drop off options are offered to consumers in order to satisfy their expectation.

The management of a third party logistics (3PL's) or a retailer's network for the transportation and processing of commercial returns may require returns centers and collection points. Several authors, like Fleischmann et al. (2001), Fleischmann et al. (2003) and Jayaraman et al. (2003) have developed a zero-one mixed integer linear programming (MIP) models to define the right path for returns in reverse logistics systems. Such network design problems can be used to determine the flow between an intermediate transshipment sites, the refurbishment/recycling/return processing sites, and the collection points (origination sites) that minimize overall costs.

Due to the recent increasing development of omnichannel retailing, the newness of this paper rests on solving a network design problem that considers countless channels for both direct and reverse flows. In Sect. 2 a conceptual framework is introduced in order to identify the physical flows and associated costs of products in the logistics network of omnichannel retailers. Section 3 relates the trade offs in a MILP model to determine the optimal combination of flows that minimizes costs. Finally, conclusions and further research lines are proposed in Sect. 4.

2 Physical Flows of Products in Omnichannel Retailers

There are several options for both consumers and retailers to manage physical or online purchases and their associated commercial returns. The factors that affect the decisions for both depend on the costs and the lead times involved in the process.

Based on the available information found in the web sites of significant retailers that represented 79 % of online retail sales in 2012 in Spain (Euromonitor 2012), Table 1 shows the different options offered by retailers to online consumers for pickup and return their online purchased products.

The order fulfillment facilities are the facilities where online orders are processed:

- Distribution centers for store replenishment and online orders fulfillment.
- Fulfillment centers, exclusively for online order fulfillment.
- Stores, that allow traditional purchases and fulfillment of online orders.

Once the merchandise is ready for being delivered, it can be transported to any of these product exchange points (PEP):

- Stores. The customer picks up the online order in a traditional retailer.
- Convenience collection points for pickup online orders such as: post offices; kiosks; automated package stations (APS); gas stations; and convenience stores.
- Homes. Merchandise is delivered to consumers' homes.

There are also different options in omnichannel models where consumers can drop-off their commercial returns. We named these points as "return exchange points" (REP), and the choices identified are the same as the PEP.

Finally, commercial returns can be processed in the order fulfillment facilities or they can be sent to a dedicated returns processing center.

The red arrows in Fig. 1 represent the forward material flow, ε represents the demand (online purchased orders) and X_{ij} the ordered units prepared at order fulfillment facilities (*i*) and delivered through the Product Exchange Points (*j*). Customers are depicted by *l* index, and Z_{jl} represents the ordered units collected by customer *l* at PEP *j*. The blue arrows represents the reverse material flows associated with commercial returns (θ is the total commercial returns that enter into the system). Drop-off options for end customers that want to return their products are represented by REP at facilities *j*. Fixed and variables costs are associated to each facility represented in Fig. 1. In addition, customer costs associated to use each channel for pickup or drop-off options are also included.

Table 1 Retailers' pick up and returns options for e-commerce

Retailer	Spanish E-com sales share (%)	Type	Merchandise	Orders fulfillment place	Product exchange point	Return exchange point	Returns processing
Ebay.es	11.10	Pure play	General merchandise	Home, distribution center, fulfillment center, store	Home, collection point, store	Home, collection point, store	Home, collection point, store
Amazon.es	10.20	Pure play	General merchandise	Fulfillment center	Home, collection point	Home, collection point	Fulfillment center
El Corte Ingles	8.70	Brick and mortar	General merchandise, fast moving consumer	Distribution center, store	Home, collection point, store	Home, store	Distribution center, store
Apple	7.90	Brick and mortar	Electronics	Fulfillment center, store	Home, store	Home, store	Fulfillment center, store
Privalia	7.10	Pure play	General merchandise	Fulfillment center	Home, collection point	Home, collection point	Fulfillment center
Buy Vip	6.60	Pure play	General merchandise	Fulfillment center	Home	Home	Fulfillment
Carrefour	6.30	Brick and mortar	General merchandise, fast moving consumer	Store	Home, store	Home, store	Store
Mercadona	5.50	Brick and mortar	Fast moving consumer goods	Store	Home, store	Home, store	Store
Redcoon	5.40	Pure play	Electronics, home appliances	Fulfillment center	Home	Home	Fulfillment center
Vente-Privee	5.20	Pure play	General merchandise	Fulfillment center	Home, collection point	Home, collection point	Fulfillment center
Eroski	5	Brick and mortar	Fast moving consumer goods	Store	Home, store	Home, store	Store

Blackburn et al. (2004) identified the trade offs between responsive (short lead times) and efficient (low costs) supply chains and defined the product marginal value of time as a depreciation cost of the product affected by the total time of the recovery process and by the product characteristics. Depending on the combination of the locations in Fig. 1, costs and lead times will vary.

3 Model Proposed

Based on Fig. 1, a MILP model is proposed in this paper to determine the optimal combination of flows that minimizes overall costs in a forward/reverse logistic network that combines different channels for pickup online orders (different pickup collection options for customers) and different shipping returns options (drop-off options offer to customers for commercial returns).

Index sets

i orders fulfillment facilities
j product/return exchange points
k returns processing facilities
l customers

Decision variables

X_{ij} ordered units prepared at i and sent to j in a period of time T
$\tilde{X}_{jk}$ returned units collected at j and sent to k in a period of time T
$\tilde{X}_{ki}$ processed units at k and sent to i to be redistributed in time T
Z_{jl} ordered units pickup/received by customer l from PEP j
$\tilde{Z}_{lj}$ returned units by customer l to REP j
Y_i, Y_j, Y_k binary variables, 1 if facility/option index is open, 0 otherwise

Parameters

p_i order fulfilment cost (cost of preparing the order in facility i)
p_j processing cost (of ordered units) in j

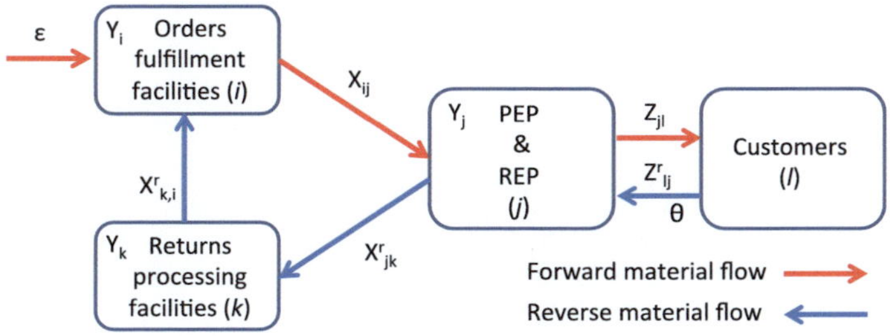

Fig. 1 Conceptual framework that describes the *forward* and *reverse* flows analyzed in this paper

r_j	returns management cost (of returns collected) in j
p_k	returns processing cost in k
f_i, f_j, f_k	fixed cost of opening a facility in the locations represented by the index sets
c_{ij}	transportation costs from i to j
c_{jk}	transportation costs from j to k
c_{ki}	transportation costs from k to i
d_{jl}	customer costs associated to pickup option from PEP j
r_{lj}	customer costs associated to drop-off returns through REP j
g_i, g_j, h_j, g_k	maximum capacity of processed units in i, j, k
ε_l	total units ordered by customers l
ε	total demand
θ_l	total units returned by customers l
θ	total units returned by customers (total commercial returns collected)
ρ_k	total units processed in k, but not sent to i in order to be re-distributed
M	very large number

The objective function and constraints of the model are:

Transportation variable cost

$$Min \sum_{i,j} c_{i,j} \cdot X_{i,j} + \sum_{j,k} c_{j,k} \cdot \tilde{X}_{j,k} + \sum_{k,i} c_{k,i} \cdot \tilde{X}_{k,i}$$

Fixed cost of facility operations

$$+ \sum_{i} f_i \cdot Y_i + \sum_{j} f_j \cdot Y_j + \sum_{k} f_k \cdot Y_k$$

Processing cost of facility operations

$$+ \sum_{i} \left[p_i \cdot \sum_{j} X_{ij} \right] + \sum_{j} \left[p_j \cdot \sum_{i} X_{ij} \right] + \sum_{j} \left[r_j \cdot \sum_{k} \tilde{X}_{jk} \right] + \sum_{k} \left[p_k \cdot \sum_{i} \tilde{X}_{ki} \right]$$

Consumer variable cost associated to pick up and drop/off choices

$$+ \sum_{j,l} d_{j,l} \cdot Z_{j,l} + \sum_{l,j} r_{l,j} \cdot \tilde{Z}_{l,j}$$

s.t.:

$$\sum_{j} X_{i,j} \leq g_i \quad \forall i \tag{1}$$

$$\sum_{i} X_{i,j} \leq g_j \quad \forall j \tag{2}$$

$$\sum_{k} \tilde{X}_{j,k} \leq h_j \quad \forall j \tag{3}$$

$$\sum_{l} Z_{j,l} \leq g_j \quad \forall j \tag{4}$$

$$\sum_{l} \tilde{Z}_{l,j} \leq h_j \quad \forall j \tag{5}$$

$$\sum_{j} \tilde{X}_{j,k} \leq g_k \quad \forall k \tag{6}$$

$$\sum_{i,j} X_{i,j} = \sum_{j,l} Z_{j,l} = \sum_{l} \varepsilon_l \tag{7}$$

$$\sum_{j,k} \tilde{X}_{j,k} = \sum_{l,j} \tilde{Z}_{l,j} = \sum_{l} \theta_l \tag{8}$$

$$\sum_{i,j} X_{i,j} \geq \sum_{j,k} \tilde{X}_{j,k} \tag{9}$$

$$\sum_{k,i} \tilde{X}_{k,i} \geq \sum_{j,k} \tilde{X}_{j,k} + \rho_k \tag{10}$$

$$\sum_{j} X_{i,j} \leq M \cdot Y_i \tag{11a}$$

$$\sum_{i} X_{i,j} \leq M \cdot Y_j \tag{11b}$$

$$\sum_{j} \tilde{X}_{j,k} \leq M \cdot Y_k \tag{11c}$$

$$\sum_{l} Z_{j,l} \leq M \cdot Y_j \tag{11d}$$

$$\sum_{l} \tilde{Z}_{l,j} \leq M \cdot Y_j \tag{11e}$$

$$\sum_{i} X_{i,j} = \sum_{l} Z_{j,l} \quad \forall j \tag{12}$$

$$\sum_{k} \tilde{X}_{j,k} = \sum_{l} \tilde{Z}_{l,j} \quad \forall j \tag{13}$$

$$\sum_j Z_{j,l} = \varepsilon_l \quad \forall l \tag{14}$$

$$\sum_j \tilde{Z}_{l,j} = \theta_l \quad \forall l \tag{15}$$

$$X_{i,j} \geq 0 \tag{16a}$$

$$\tilde{X}_{j,k} \geq 0 \tag{16b}$$

$$\tilde{X}_{k,i} \geq 0 \tag{16c}$$

$$Z_{j,l} \geq 0 \tag{16d}$$

$$\tilde{Z}_{l,j} \geq 0 \tag{16e}$$

$$Y_i, Y_j, Y_k \in \{0, 1\} \quad \forall i, j, k \tag{17}$$

Constraints (1)–(6) reinforce that any flow entering at each facility does not exceed its processing capacity. Constraint (3) and (5) refers to the capacity of location j to process returns (the capacity of each REP). Constraint (7) guarantees that all demand (ordered units) received at facilities i, are sent to customers l through j facilities (PEP options), and constraint (8) guarantees the balance of the reverse flows (all returned units collected through all REP j flow out to one of the k facilities in order to be processed). Constraint (9) means that returns can´t be bigger than the ordered units sent to customers, and (10) represents that the units of returns sent to i for being re-distributed can't be bigger than the amount of returns plus the unit processed at k, but not sent for re-distribution. Constraints (11a)–(11e) allow inflow of units only to open facilities. Constraint (12) represents that orders collected by customers l from a PEP (j) need to be prepared in i facilities. Same idea represents constraint (13) but for commercial returns (the reverse flow), where all the returns collected in REP j need to be sent to k facilities to be processed. Constraints (14) and (15) represent the balance between customers l and demand per each channel j (PEP), and among customers l and returns per each j (REP) respectively. Finally constraints (16a)–(16e) and (17) define non-negativity and binary variables respectively.

4 Conclusions and Further Research

The main contribution of this paper is the proposal of a MILP model to evaluate key trade-offs when different options for pick up online purchase and drop-off commercial returns arise in omnichannel models. In this research paper, a conceptual framework of a complex forward and reverse logistics problem is proposed and the formulation of the MILP model is developed.

As a further research we propose to apply the model to a real case study (any of the retailers included in Table 1).

For any retailer that offers online sales, the comparison between their current logistics network and the solution of the model proposed in this paper would provide an interested insight of the appropriateness of the different channels and options offered in an omnichannel network. In order to improve business performance, the model could help retailers to determine which channels should have to be limited and which ones should be boosted.

A sensitivity analysis could be also conducted to analyse different commercial return rates in product categories, and how these rates could affect network configurations. Commercial return rates for fast moving consumer goods are much lower than the rates for electronics or apparel products, and also require less collection and returns processing resources. As online sales have higher commercial return rates there is also a greater interest to analyse and quantify the costs related to these reverse flows.

References

Blackburn JD, Guide VDR Jr, Souza GC, Van Wanssenhove LN (2004) Reverse supply chains for commercial returns. Calif Manage Rev Sci 46(2):6–22

Euromonitor (2012) Internet retailing in Spain

Fleischmann M, Beullens P, Bloemhorf-Ruwaard JM, Van Wassenhove LN (2001) The impact of recovery on logistics network design. Prod Oper Manage 10(2)

Fleischmann M, Bloemhof-Ruwaard JM, Beullens P, Dekker R (2003) Reverse logistics network design. In: Dekker R et al (eds) Reverse logistics, quantitative models for closed-loop supply chains. Springer, Berlin, pp 65–94

Forrester, Mulpuru S, Johnson C, Roberge D (2013) U.S. Online retail forecast, 2012 to 2017. Forrester. Forrester Research

Guide VDR Jr, Van Wanssenhove L (2009) The evolution of closed-loop supply research. Oper Res 57:10–18

Guide VDR Jr, Souza LN, Van Wanssenhove L, Blackburn JD (2006) Time value of commercial product returns. Manage Sci 52:1200–1214

Jayaraman V, Patterson RA, Rolland E (2003) The design of reverse distribution networks: models and solution procedures. Eur J Oper Res 150(2003):128–149

Kleinman S (May 2012) Online shopping customer experience study. ComScore

Mollenkopf DA, Rabinovich E, Laseter TM, Boyer KK (2007) Managing internet product returns: a focus on effective service operations. Decis Sci 38(2)

Tibben-Lembke R (2004) Strategic use of the secondary market for retail consumer goods. Calif Manage Rev 46:90–104

U.S. Census Bureau (2013) U.S. retail trade sales total and E-commerce

A Nonlinear Integer Programming Model for Warehousing Sustainable Logistics

Francesco Boenzi, Salvatore Digiesi, Francesco Facchini, Giorgio Mossa and Giovanni Mummolo

Abstract The warehouse management problem is a critical issue in Operation Management. In modern competitive market, many firms are automating their basic warehouse activities in order to be cost effective. However, traditional mechanized warehousing systems (MWS) still represent the 75 % of the overall installations. In MWS, forklifts are adopted for the load handling. The adoption of efficient "internal logistic strategies" could help in reducing time required and costs of warehousing activities. In recent years, many firms adopted green supply chain practices (GSCP) in order to improve their environmental performances while also achieving economic goals (Wu et al. 2015). Furthermore, in planning Smart City logistics, warehousing in port or railway station storage areas is being receiving wide attention since they contribute effectively to a sustainable development of modern cities. Under this perspective, an optimal "internal logistic strategy" allowing to jointly minimizing jointly costs and environmental impacts of warehousing activities has to be adopted. The aim of this study is to develop a Nonlinear Integer Programming Model to solve a storage location assignment problem (SLAP) for optimizing the environmental performance of the internal logistic

---

F. Boenzi · S. Digiesi · F. Facchini · G. Mossa · G. Mummolo (✉)
Department of Mechanics, Mathematics and Management, Politecnico of Bari,
Viale Japigia, 182, Bari, Italy
e-mail: giovanni.mummolo@poliba.it

F. Boenzi
e-mail: francesco.boenzi@poliba.it

S. Digiesi
e-mail: salvatore.digiesi@poliba.it

F. Facchini
e-mail: francesco.facchini@poliba.it

G. Mossa
e-mail: giorgio.mossa@poliba.it

M. Amorim et al. (eds.), *Engineering Systems and Networks*, Lecture Notes in Management and Industrial Engineering, DOI 10.1007/978-3-319-45748-2_11

activities in a warehouse. Suitable storage strategies are identified on the basis of the type of the forklifts adopted (internal combustion or electric engine equipped) as well as the sizes and the weight of the loads to be handled.

Keywords Sustainable logistic · Material handling · Warehouse management · Optimized product allocation

1 Introduction

Warehousing can be defined as the process in which three main functions are accomplished: receiving products from a supplier, storing products as long as necessary until they are requested (internally or externally), and retrieving the products when they are demanded (Queirolo et al. 2002). Warehouses still play a key role in supply chains, since they allow mitigating variations in supply and demand, and provide value-added services (Sainathuni et al. 2014; Gnoni et al. 2003).

Nowadays 75 % of warehouses are still characterized by mechanized warehousing systems (MWS) technology (Burinskiene 2011), in which manual equipment (forklifts) are adopted. Order picking is the most labour—intensive and time—consuming task in conventional warehouses (Accorsi et al. 2014). Up to 55 % of total warehouse operating costs results from order picking operations (De Koster et al. 2007). Different methods have been developed in order to optimize order picking activities. These methods are often based on heuristics and integer programming approaches, and allow reducing time and costs of material handling activities. Nowadays, both public and private subjects are increasingly paying more attention to environmental effects of logistic; indeed the environmental performances of the logistics activities are one of the most important characteristics for a Smart City. An increasing request for more sustainable solutions of logistic issues able to minimize both logistic (forklift routing, pick orders, etc.) and external costs has been observed (Digiesi et al. 2012). Traditional optimization methods reveal themselves to be inadequate for the sustainable management of a warehouse in a green supply chain, since they focus on economic goals, neglecting the environmental performance of the system.

In this work, a model allowing optimizing the environmental performance of the internal logistic activities in a warehouse is proposed. Suitable storage strategies are identified on the basis of the type of the forklifts adopted and the sizes and the weight of the loads to be handled.

This paper is structured as follows: in Sect. 2 the model is presented and input and output parameters are detailed; in Sect. 3 the model is applied to a full-scale case study and the results obtained are presented and discussed; conclusions are in drawn Sect. 4.

2 The Model

The model developed is a non-linear integer programming model allowing the evaluation of the overall emission due to forklift operation in each phase of material handling cycle: transport, picking and retrieving of the stored items. Suitable storage strategies allowing minimizing overall emissions are identified. Strategies are identified on the basis of the type of the forklifts adopted (internal combustion or electric engine equipped) and the sizes and the weight of the loads to be handled. The external costs of the resulting impacts are evaluated.

Notations and assumptions adopted are listed below:

Q storage capacity of the warehouse: $Q= nx*ny*nz$

n_x, n_y, n_z number of items stored according to the x, y and z-axis respectively, defining the storage geometrical configuration (unit)

L_x, L_y distance between loading/unloading and stockpiles areas (m)

d_x, d_y, d_z uniform dimensions of a single item in the stock (m)

$n_{y,z}$ average number of forks movements required for storing or retrieving one item only, for a stockpile with dimensions n_y and n_z (#)

h_{t,d_z} average fork height required for retrieving one item from a stockpile with dimension t units in height and stocking units sized d_z (m)

h^{store}_{t,d_z} average fork height required for storing one item in a stockpile with dimension t units in height and stocking units sized d_z (m)

h_{n_z} average fork height required for picking up one item in a stockpile with dimension n_z (m)

$H_{y,z}$ average fork vertical movement required for picking up one item (m)

v_v, v_h fork lifting speed (v as subscript) and forklift travel speed (h) (m/s)

t_v, t_h average time spent by forklifts in fork vertical and horizontal shifting (s)

$\rho_{u,v}$, $\rho_{u,h}$ engine utilization factors in vertical and horizontal shifting (-)

P_n forklift engine power rating (kW)

E_h average energy required for horizontal shifting of one item (kWh/unit)

E_v average energy required for vertical lifting of one item (kWh/unit)

E average overall energy required for picking up one item (kWh/unit)

E_R^p emission factor of the p-th pollutant (g/kWh)

e_C^p external cost of the emission due to the p-th pollutant (€/g)

e_C average total external cost of emission per unit (€/unit)

Assumptions:

- the considered warehouse layout considered is depicted in Fig. 1;
- number of items to be handled is given;
- the material handling in the warehouse is operated by means of counterbalance forklifts carrying on one item only per each load/unload cycle;
- all the items stocked in the warehouse have prismatic form and are characterized by the same sizes (d_x, d_y, and d_z) and weight;
- the items are stocked in stockpiles of the same height;

Fig. 1 Layout of the warehouse considered in the model

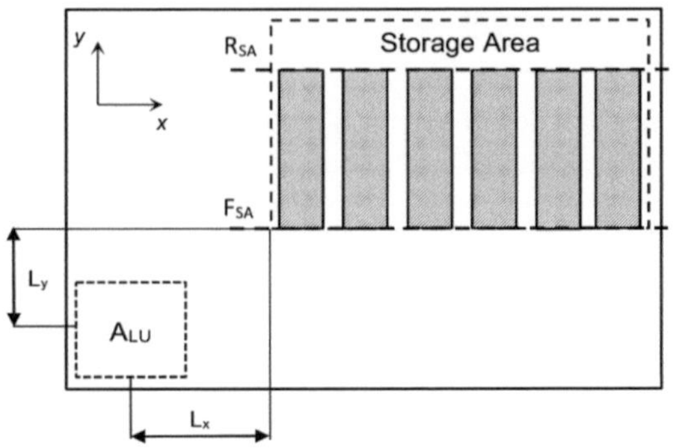

A_{LU}: Area for load/unload of the items

F_{SA}/R_{SA}: Frontal/Rear access to stockpile for material handling activities

- the distance between the loading/unloading area (A_{LU}) of the warehouse and the storage area is given;
- each stockpile can be accessed by both sides (F_{SA} and R_{SA} in Fig. 1);
- the distance between the stockpiles (aisles) according to the x-axis can be neglected;
- a storage configuration is univocally identified by two integer numbers (n_y, n_z), being $n_x = Q/(n_y \cdot n_z)$.

No constraints about the size and the geometrical characteristics of the warehouse are considered.

Theoretical formulation: The material handling activity considered in the model consists of the following main steps: forklift starts from the A_{LU} and reaches the storage area, picks up or stores one item and then goes back to A_{LU}. In the storage area, lifting and lowering phases required for storing or retrieving one item are considered. In order to evaluate the emissions due to forklift operation, the average overall energy (E) required by the forklift for the handling of one item is preliminary evaluated (2.1). E_v and E_h are obtained as the product of the power rate of the engine that equips the forklift, P_n, the engine utilization factors ($\rho_{u,v}$; $\rho_{u,h}$), and the time required for the completion of the task (t_v; t_v).

$$E = E_v + E_h \tag{2.1}$$

t_v and t_h values are obtained from the distance and the speed values in the lifting and in the shifting phases, respectively; as regards the lifting phase, the average forks vertical movements values are evaluated by means of Eq. 2.2.

$$H_{y,z} = n_{y,z} h_{n_z} \tag{2.2}$$

Equations (2.3) and (2.4) are adopted for evaluating the average number of forks movements ($n_{y,z}$) required in case n_y is an even or an odd number, respectively.

$$n_{y,z} = \frac{(n_z + 1)(n_y + 2)}{8} \tag{2.3}$$

$$n_{y,z} = \frac{(n_z + 1)}{2n_y}\left[\frac{\left(n_y^2 - 1\right)}{4} + \frac{(n_y - 1)}{2} + 1\right] \tag{2.4}$$

Equation (2.5) allows evaluating the average forks height required for picking up (retrieving and then storing) one item from a stockpile characterized by n_z.

$$h_{n_z} = \frac{1}{n_z}\sum_{t=1}^{n_z}\left(h_{t,d_z} + h_{t,d_z}^{store}\right) \tag{2.5}$$

The first term in Eq. (2.5) measures the average forks height required for retrieving one item from a stockpile of type (n_x, n_y, n_z), and the second term the analogous value for storing one item in the same stockpile. Parameter h_{t,d_z} measures the overall forks vertical movements required for retrieving one item with dimension d_z from a stockpile of type (n_x, n_y = 1, t), and must be computed for all the values of t ranging in (1; n_z) by means of Eq. (2.6).

$$h_{t,d_z} = \frac{1}{t}\sum_{k=1}^{t}\left[(t - 1)^2 - \left(k^2 - 3k + 2\right)\right]d_z \tag{2.6}$$

$$h_{t,d_z}^{store} = \frac{1}{t}\sum_{k=1}^{t}(k - 1)d_z \tag{2.7}$$

Parameter h_{t,d_z}^{store} is the average forks height required for storing one item in a stockpile of type $(n_x, n_y = 1, n_z^t)$ and is computed by means of Eq. (2.7).

Environmental impact assessment: Many models of forklifts are available on the market for fulfilling physical requirements of warehousing applications. Forklifts commercially available are powered with three different types of engine: diesel, electrical and Liquefied Petroleum Gas (LPG) engine. Each of them is characterized by different value of the nominal capacity, tailpipe emissions, management costs, etc. Starting from the average amount of energy required for picking up one item from a stockpile, the related average external costs are evaluated by means of Eqs. (2.8) and (2.9) in case of Diesel/LPG engine equipped forklifts and electrical powered forklifts, respectively. In both cases, the external costs are evaluated as the product of the average amount of energy required for picking up one item and the unitary external cost (€/kWh). In case of Diesel/LPG engine equipped forklifts, the unitary external cost is obtained as the sum of the products of the pollutants emission rates and the related unitary monetary cost (see Eq. 2.8). Emission rate values and monetary costs of emissions assumed in this work are from (Gaines et al. 2008; Spadaro and Rabl 1999) (see Table 1). In case of electrical powered forklifts

Table 1 Diesel/LPG engine emissions and external cost in Euro

	Emission rate (E_R) (g/kWh)		External cost (e_C) (€/kg)
Pollutant	Diesel	LPG	
NO_x/CO	10.8/45.0	15.6/10.9	16.00/0.02

(see Eq. 2.9), the unit external cost assumed is referred to the Italian mix of power generation: 0.05 (€/kWh) (ExternE Project-Series 2012).

$$e_C^{Diesel/LPG} = E\left(E_R^{NO_x} e_C^{NO_x} + E_R^{CO} e_C^{CO}\right) \quad (2.8)$$

$$e_C^{Electric} = E e_C^{electricity} \quad (2.9)$$

Summarizing, the target of the model, for a given forklifts engine type, consists of identifying a storage configuration allowing to minimize the average overall energy E (related to the decision variables n_x, n_y and n_z), required for the handling of items (of which quantity, sizes and weight are known) to be stored in a warehouse. The average overall energy (E) per item depends on the average energy required by forklift path and forks vertical movements (Eq. 2.2). The latter parameter depends on Eqs. 2.3 or 2.4 (for n_y even or odd number) and Eq. 2.5, calculated by means of Eqs. 2.6 and 2.7. After the minimization of E, it is possible to identify the minimum external cost by means of Eqs. 2.8 or 2.9.

3 Numerical Experiments

In order to test the model, it has been applied to a full case study. Three counterbalance forklifts have been considered in this case. They are equipped with a Diesel, a LPG and an electric engine, respectively (technical specifications are in Table 2).

Three items with different size have been considered (see Table 3), each of them characterized by four different weights.

Table 2 Technical specifications of forklifts

Specifications	Limits		
	Diesel	LPG	Electric
Load capacity (kg)	4000	4000	4000
Lift height (m)	7.18	7.18	7.18
Lift speed loaded/unloaded (m/s)	0.59/0.59	0.59/0.59	0.43/0.55
Travel speed unloaded (m/s)	5.83	5.83	5.83
Engine power (kW)	54	55	40

Table 3 Weight and sizes of items considered

ID item	Weights (kg)				Size (d_x, d_y, d_z) (m)
A500–A4000	500	1000	2000	4000	(2, 1.5, 0.5)
B500–B4000	500	1000	2000	4000	(6, 1.5, 0.5)
C500–C4000	500	1000	2000	400	(4, 1.5, 0.5)

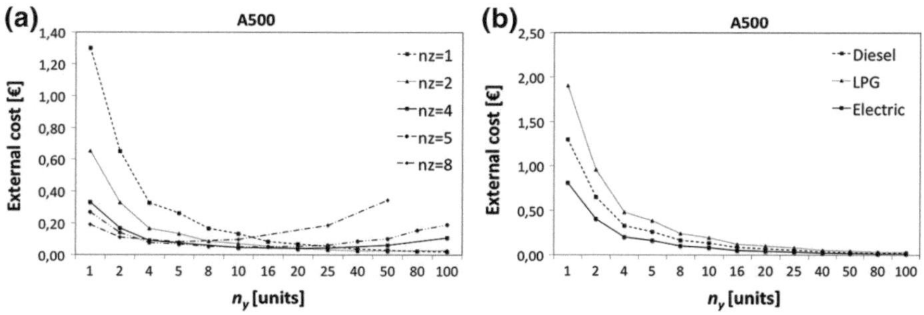

Fig. 2 External costs versus n_y values for different n_z and diesel engine (**a**) and for different engine types and $n_z = 1$ in (**b**) (weight = 500 kg)

The overall number of items adopted (Q) for the numerical experiments is equal to 10,000 units; the distance between the A_{LU} and the storage area considered is equal to 20 m (for both sides of the storage area). According to safety practices for material handling activities, two different values of v_v have been considered: high speed (4–5 m/s) for light load (500–700 kg) and low speed (1–2 m/s) for heavy load (3500–4000 kg); moreover, a maximum height of the stack of 4 m has been considered.

The average external costs for a storage configuration in case of item A500, when a diesel forklift is adopted, are in Fig. 2.

Results show how the average external costs are strongly related to the weight and the sizes of the items stored; moreover, for a given item, the storage configurations proved to affect strongly external costs. For example, in the case the B4000 item, external costs range in (5.00–45.00 €/unit) for layouts characterized by the same n_y value and different height of stack (n_z). In the case depicted in Fig. 2a referring to diesel engine and weight 500 kg, it is interesting to highlight how the minimum value of the external cost can be obtained for $n_z = 1$ and $n_y = 100$.

Among all parameters that characterize the storage configuration, a significant reduction in external cost value can be obtained by reducing the height of stacks (n_z); in particular lowest value of the external cost is generally identified for stockpiles characterized by low height and extended surface. The model identifies an optimal solution for $n_z = 1$ in case of items characterized by low weight and small sizes (A500 and A1000); for items characterized by higher sizes and weights the optimal n_z values slightly increases since a greater amount of energy is required for carrying the items, and more time is required for the material handling (higher distance for picking up and retrieving of items and slow speed travel of forklift).

As regards the type of forklift adopted for the material handling, minimal external costs values are not always obtained with the same storage configuration. Generally in case of electric forklifts the smallest value of the stack height ($n_z = 1$) allows minimizing external costs; in case of diesel and LPG engine powered forklifts, two values are identified as optimal. Solutions obtained in case of electrical forklift are characterized by environmental impacts lower than those found with diesel or LPG forklifts. However, they are obtained in the absence of any spatial constraint, and lead to different spatial requirements. A further comparison has been carried out between solutions obtained given n_z values. Results show how in case of $n_z > 1$ average external costs of electric forklift are higher compared to forklifts equipped with diesel or LPG engine (Fig. 2). This leads to conclude that the use of electric forklift does not ensure the minimal environment impact of material handling activities in a warehouse when configuration constraints (as stacking items on several layers in order to reduce the storage area) have to be met.

4 Conclusion

The external cost of internal logistic activities are affected by the type of the forklift engine adopted, by the number of movements required for the material handling, and by the forklift path. The model described in this paper is a tool for driving decision makers in identifying sustainable logistics solutions allowing minimizing the overall environmental impact due to all these factors.

The application of the model to a full-scale case study show its capabilities in identifying optimal logistic strategies ensuring a low environment impact of internal logistic activities. Moreover, in case of spatial constraints to be fulfilled, the model could help in finding the most suitable type of forklift to be adopted.

Acknowledgments The paper has been written within the framework of the project PON04a2_E "Smart Energy Master per il governo energetico del territorio - SINERGREEN - RES NOVAE". This project is supported by the Italian University and Research National Ministry research and competitiveness program that Italy is developing to promote "Smart Cities Communities and Social Innovation".

References

Accorsi R, Manzini R, Maranesi F (2014) A decision-support system for the design and management of warehousing systems. Comput Ind 65(1):175–186

Burinskiene A (2011) The travelling of forklifts in warehouses. Int J Simul Model 10:204–212

De Koster R, Le-Duc T, Roodbergen KJ (2007) Design and control of warehouse order picking: a literature review. Eur J Oper Res 182:481–501

Digiesi S, Mossa G, Mummolo G (2012) A loss factor based approach for sustainable logistics. Prod Plann Contr 23:160–170

ExternE Project-Series (2012) IER, University of Stuttgart. Available from http://www.externe.info/. Accessed 18 Feb 2015

Gaines LL, Elgowainy A, Wang MQ (2008) Full fuel-cycle comparison of forklift propulsion systems. Argonne National Laboratory—Energy Systems Division

Gnoni MG, Iavagnilio R, Mossa G, Mummolo G (2003) Modelling dynamics of a supply chain under uncertainty: a case from the automotive industry. Int J Automot Technol Manage 3(3–4):354–367

Queirolo F, Tonelli F, Schenone M, Nan P, Zunino I (2002) Warehouse layout design: minimizing travel time with a genetic and simulative approach—methodology and case study. In: 4th European simulation symposium

Sainathuni B, Parikh PJ, Zhang X, Kong N (2014) The warehouse-inventory-transportation problem for supply chains. Eur J Oper Res 237(2):690–700

Spadaro JV, Rabl A (1999) Estimates of real damage form air pollution: site dependence and simple impact indices for LCA. Int J Life Cycle Assess 4(4):229–243

Wu K-J, Liao C-J, Tseng M-L, Chiu ASF (2015) Exploring decisive factors in green supply chain practices under uncertain. Int J Prod Econ 159(1):147–157

A Reference Framework to Design Inventory Policies Using a Fill Rate Criterion in Lost Sales Contexts

Eugenia Babiloni, Ester Guijarro and Manuel Cardós

Abstract This paper suggests a reference framework to decide the best method to compute the base stock level in a lost sales and discrete demand context given a target fill rate. Under this context, only (Guijarro et al. in Eur J Oper Res 218 (2):442–447, 2012a) propose an exact fill rate expression, but it requires a huge computational effort. However, the literature shows several approximations which are simpler and easier to implement in practical environments. In this paper, we design a large experiment and analyze by means of data mining techniques under which conditions approximate expressions can provide an accurate enough approximation to the base stock level for the lost sales case. As a result, we propose a reference framework that allows the selection of the most suitable method to compute the base stock depending on the characteristics of the item.

Keywords Optimal policies · Fill rate · Lost sales · Discrete demand

1 Introduction

The three key questions that inventory managers need to answer for designing optimal policies are: (i) how often should the inventory status be reviewed; (ii) when should a replenishment order be placed and (iii) how large should the replenishment order be (Silver et al. 1998). When the system is reviewed periodically by the (R, S) inventory policy, answering the second question supposes to establish the review period, R, whereas the third question depends on the base stock

E. Babiloni · E. Guijarro (✉) · M. Cardós
Departamento de Organización de Empresas, Universitat Politècnica de València,
Camino de Vera, S/N, 46022 Valencia, Spain
e-mail: esguitar@doe.upv.es

E. Babiloni
e-mail: mabagri@doe.upv.es

M. Cardós
e-mail: mcardos@doe.upv.es

M. Amorim et al. (eds.), *Engineering Systems and Networks*, Lecture Notes
in Management and Industrial Engineering, DOI 10.1007/978-3-319-45748-2_12

level, S. In practical environments, the R is usually predefined. Then, the key question is to determine the optimal value of the base stock. In the literature, we find two possible approaches to determine the S: minimizing total inventory costs or guaranteeing the achievement of a target service level. As Bijvank and Vis (2011) point out the service approach is useful when a service level restriction is imposed by the replenishment process and it is easier to define a target service level than all the costs, especially the stock out cost. For this reason, practitioners tend to use the service level criterion to establish the base stock, S.

This paper focuses on the estimation of the optimal base stock level in a (R, S) periodic review stock policy when a target fill rate is given, demand is stochastic and discrete, and unfulfilled demand is lost. The fill rate is one of the service measures most used in practice since it considers not only the possibility that the system is out of stock, but also the size of the unfulfilled demand when it occurs (Tempelmeier 2007). This metric has been largely studied in the literature, but under periodic review (R, S) system and lost sales assumption there is only one exact method to compute it which requires an intensive and almost impractical computational effort to find the optimal base stock level, especially for large values of S. However, we find several approximations of the fill rate that present a good performance under some circumstances. We design a large experiment and analyze by means of Classification and Regression Trees the performance of approximate fill rates when they are used to establish the optimal S. The aim of this paper is to identify under which conditions approximate expressions can provide an accurate enough approximation to the exact S for the lost sales case and discrete demands. As a result, we propose a reference framework which can be used by managers to design optimal inventory policies.

The remainder of the paper is organized as follows. Section 2 describes the notation and assumptions. Section 3 is dedicated to presenting exact and approximate methods to compute the fill rate in a periodic system. The numerical experiment and the analysis carried out are explained in Sect. 4. Section 5 presents the proposed reference framework and its practical applications are summarized in Sect. 6.

2 Assumptions and Notation

This paper considers a periodic review, base stock (R, S) system that examines the status of an item every R fixed time periods and launches a replenishment order which raises the inventory position to the base stock S. In the rest we assume that: (i) time is discrete and organized in a numerable and infinite succession of equally spaced instants; (ii) the lead time is constant; (iii) the review interval is known; (iv) there is just one outstanding replenishment order at every time, i.e. $L < R$; (v) the replenishment order is added to the inventory at the end of the period in which it is received; (vi) demand during a period is fulfilled with the on-hand stock at the beginning of the period; and (vii) the demand process is stationary, i.i.d. and defined by any discrete function. Notation used in the rest of the paper is as follows:

S base stock (units),
L lead time for the replenishment order (time units),
R review period and replenishment cycle corresponding to the time between two consecutive deliveries (time units),
OH_t on hand stock in time t from the first reception (units),
NS_t net stock in time t from the first reception (units),
D_t accumulated demand during t consecutive periods (units),
$f_t(\cdot)$ probability mass function of D_t,
$F_t(\cdot)$ cumulative distribution function of D_t.

3 Methods to Compute the Fill Rate in a (*R*, *S*) Periodic System

The fill rate (*FR*) is a service metric that measures the fraction of demand that is directly satisfied with the on-hand stock and can be expressed as

$$FR = E\left(\frac{fulfilled\ demand\ per\ replenishment\ cycle}{total\ demand\ per\ replenishment\ cycle} | positive\ demand\ during\ cycle \right) \tag{1}$$

Guijarro et al. (2012a) propose an exact expression to compute (1) when demand follows any discrete demand distribution and the unfilled demand is lost:

$$FR_{Exact\_LS} = \sum_{i=0}^{S} P(OH_0 = i) \cdot \left\{ \frac{F_R(i) - F_R(0)}{1 - F_R(0)} + \sum_{j=i+1}^{\infty} \frac{i}{j} \cdot \frac{f_R(j)}{1 - F_R(0)} \right\} \tag{2}$$

The complexity of (2) is in how to compute the probability of every on-hand stock level at the beginning of the cycle, $\overline{P(OH_0)}$. Its estimation requires the convergence of the probability transition matrix, $\overline{\overline{M}}$, between the on-hand stock level at the beginning of the replenishment cycle to its ends (Cardos and Babiloni 2011). Then the number of iterations and the size of $\overline{\overline{M}}$ depend on S and for this reason the computation of the stock level turns very time-consuming.

3.1 Approximate Methods

Traditionally, the estimation of the *FR* has been simplified by terms of units short:

$$FR_{Trad} = 1 - \frac{E(unfulfilled\, demand\, per\, replenishment\, cycle)}{E(total\, demand\, per\, replenishment\, cycle)} \quad (3)$$

In the literature, we find quite a number works suggesting methods to compute (3) in different contexts, but most of them assume continuous demand distributions. Table 1 presents approximate methods considered in this paper. Note that expressions derived for continuous demand have been discretizated following (Guijarro et al. 2012b).

More concretely, we consider: (i) the traditional approach presented in classical inventory texts (FR_{Trad}), (ii) the FR expression based on the expected shortage per replenishment cycle suggested by Hadley and Whitin (1963) ($FR_{H\&W}$), (iii) the expression presented in Silver (1970) and reformulated by Johnson et al. (1995) (FR_{Silver}), (iv) the approximation derived by Teunter (2009) ($FR_{Teunter}$), (v) the approach suggested by Babiloni et al. (2012) ($FR_{Babiloni}$) and (vi) the approximation based on expression (3) derived by Guijarro et al. (2012a) ($FR_{Guijarro}$) for lost sales and discrete demand. Furthermore, we include as approximation the exact expression proposed by Babiloni et al. (2012) which computes the FR as expression (1) in the backordering context (FR_{Exact_Bk}).

4 Numerical Experiment

With the aim of analyzing how approximate FR expressions (Table 1) perform when they are used to establish the optimal S in a lost sales scenario, we design an experiment which combines an extensive range of values for R, L and target fill rate,

Table 1 Approximate methods to compute the fill rate for (R, S) systems and discrete demands

Author/s	Expression
Traditional	$FR_{Trad} = 1 - \frac{\sum_{i=S}^{\infty}(i-S)\cdot f_{R+L}(i)}{\sum_{k=1}^{\infty} k\cdot f_R(k)}$
Hadley and Whitin (1963)	$FR_{H\&W} = 1 - \frac{\sum_{i=S}^{\infty}(i-S)\cdot f_{R+L}(i) - \sum_{j=S}^{\infty}(j-S)\cdot f_L(j)}{\sum_{k=1}^{\infty} k\cdot f_R(k)}$
Silver (1970)	$FR_{Silver} = \frac{\sum_{i=-\infty}^{S} D_R\cdot f_{R+L}(i) + \sum_{i=S}^{S+D_R}(S+D_R-i)\cdot f_{R+L}(i)}{\sum_{k=1}^{\infty} k\cdot f_R(k)}$
Teunter (2009)	$\beta_{Teunter} = 1 - \frac{\sum_{j=0}^{S}(S-j)f_L(j) - \sum_{i=0}^{S}(S-i)f_{R+L}(i)}{\sum_{k=1}^{\infty} k\cdot f_R(k)}$
Babiloni et al. (2012)	$FR_{Babiloni} = \begin{cases} 1 - \frac{\sum_{i=1}^{S} f_L(S-i)\cdot \sum_{j=i+1}^{\infty}(j-i)f_R(j)}{\sum_{j=1}^{\infty} j\cdot f_R(j)} & \text{if } NS_0 > 0 \\ 0 & \text{if } NS_0 \leq 0 \end{cases}$
Guijarro et al. (2012a)	$FR_{Guijarro} = 1 - \frac{\sum_{i=0}^{S} P(OH_0=i)\cdot \sum_{j=i+1}^{\infty}(j-i)\cdot f_R(j)}{\sum_{j=1}^{\infty} j\cdot f_R(j)}$
Babiloni et al. (2012)	$FR_{Exact\_Bk} = \sum_{NS_0=1}^{S} f_L(S-NS_0)\cdot \left\{\frac{F_R(NS_0)-F_R(0)}{1-F_R(0)} + \sum_{D_R=NS_0+1}^{\infty} \frac{NS_0}{D_R}\cdot \frac{f_R(D_R)}{1-F_R(0)}\right\}$

Table 2 Experimental factors and values

Demand distribution	
Poisson (λ)	λ = 0.01, 0.05, 0.1, 0.2, 0.3, 0.4, 0.5, 0.75, 0.9, 1, 1.25, 1.5, 1.75, 2, 2.5, 3, 4, 5, 7, 10, 15, 20
Bin (n, θ)	n = 1, 2, 3, 4, 5, 6, 7, 8, 10, 12, 15, 20
	θ = 0.01, 0.05, 0.1, 0.15, 0.25, 0.5, 0.75, 0.9, 0.95, 0.99
NegBin (r, θ)	r = 0.05, 0.1, 0.2, 0.25, 0.3, 0.4, 0.5, 0.75, 0.9, 1, 1.25, 1.5, 1.75, 2, 2.5, 3, 3.5, 4, 5
	θ = 0.1, 0.15, 0.25, 0.3, 0.4, 0.5, 0.6, 0.7, 0.8, 0.9, 0.99
Inventory system	
Target fill rate	FR^* = 0.50, 0.55, 0.60, 0.65, 0.70, 0.75, 0.80, 0.85, 0.90, 0.95, 0.99
Review period	R = 2, 3, 4, 5, 7, 10, 15, 20
Lead time	L = 1, 3, 5, 7, 10, 15, 20

Total cases 89,760

FR^* to consider as many different realistic contexts as possible (Table 2). With regard to the demand distribution, this experiment considers binomial, Poisson and negative binomial distribution whose parameters have been selected in order to meet the four demand categories suggested by Syntetos et al. (2005).

The purpose of the experiment consists of calculating the minimum S that guarantees the achievement of a FR^* using methods presented in Sect. 2. Once the base stock is computed by each method, we apply Classification and Regression Trees (CART) as exploratory technique to identify under which conditions approximate methods are accurate enough to estimate the base stock. For each case we define a categorical dependent variable ("Best Method") which indicates the simplest and fastest method that leads to the exact S. We use the cyclomatic complexity to compute the simplicity. As independent variables we select: the random variable, the squared coefficient of variation of demand sizes (CV^2), the average inter-demand interval (p), the target fill rate (FR^*), the probability of zero demand ($P(0)$), and the average demand during R + L (μ_{R+L}).

5 Proposed Reference Framework

The application of CART reveals the underlying model behind the performance of the approximate FR methods and identifies zones showing an homogeneous performance when they are used to compute S. As a result, we identify four zones that are explained by two independent variables: μ_{R+L} and CV^2. For each zone, we analyze the percentage of cases where each method obtains the exact S and the percentage of cases in which approximate methods fall into errors. Deviations are measured by using the relative error (RE) expressed in terms of per 100 as follows

$$RE = \frac{S_{Exact} - S_{Approximate}}{S_{Exact}} \tag{4}$$

Computing errors as (4) enables not only to know the magnitude of the deviation but also the type of error by analyzing its sign. In this sense, when *RE* is negative (named *EC1* henceforth), approximate fill rate expression gives a base stock level greater than the exact one and therefore the systems reaches the target fill rate although increasing the average stock level and consequently the holding costs. If the *RE* is positive (*EC2*), the approximate base stock is lower than the exact one which means that the policy is not reaching the target fill rate and therefore the system is less protected against stockouts than managers expect.

The results of this analysis are summarized in Table 3 where the first two columns identify the zone where *FR* expressions present an homogeneous performance. The following columns indicate, for each approximation, the percentage of cases they compute the *S* accurately (% S_{Exact}) and the percentage of cases in which approximations fall into errors, indicating if it is *EC1* or *EC2* and its mean and standard deviation. This table becomes a reference framework for determining base stocks. In this sense, managers can select the most appropriate method while keeping in mind the magnitude and the type of error provided by the selected method. For example, if we are managing an item with μ_{R+L} = 1 (second node), the best method is the *Exact_Bk* which accurately computes the exact value of *S* in a 70.6 % of cases. However, there is a risk of falling into errors (29.4 % cases), being this error always *EC1* with μ = 64.4 and σ = 30.7.

6 Conclusions

This paper faces the traditional problem of determining the optimal base stock level, *S*, given a target *FR*. In a lost sales system, the exact method to compute the *FR* under discrete demand context requires an intensive computational effort that makes difficult its implementation in practice. For this reason, this paper studies under which conditions approximate *FR* expressions, which are simpler and more easily implementable, provide an accurate enough estimation of the *S*.

To accomplish the aim of this problem, we carry out a wide-ranging experiment and analyze the results by means of CART. As a result we propose a reference framework (Table 3) in which we identify zones where approximate methods present an homogeneous performance when are used to establish the *S*. In practical environments, the reference framework may help managers in the decision making process of selecting the best method to estimate the optimal *S*. This decision is a trade-off between the complexity of the method and the risks from using them. In this sense, risks associated to adopt any decision in terms of relative deviations

Table 3 Reference framework to establish the base stock level in a lost sales context

CV^2	μ_{R+L}	Method	% S_{Exact}	% EC1	μ	σ	% EC2	μ	σ
–	>2.99	*Exact_LS*	100	–	–	–	–	–	–
		Guijarro	23.2	76.8	−17.7	18.6	–	–	–
		Exact_Bk	10.1	89.9	−27.8	20.8	–	–	–
		Babiloni	1.2	98.8	−36	26.8	–	–	–
		Trad	1.2	98.8	−37.2	29.8	–	–	–
		H&W	1.2	98.8	−36	26.8	–	–	–
		Silver	2.6	97.3	−31.1	20.8	0.1	12.3	12.3
		Teunter	1.2	98.8	−36	26.8	–	–	–
–	0.85–2.99	*Exact_LS*	100	–	–	–	–	–	–
		Guijarro	38.5	61.5	−57.9	30.8	–	–	–
		Exact_Bk	70.6	29.4	−64.4	30.7	–	–	–
		Babiloni	22.4	77.6	−59	30.4	–	–	–
		Trad	18	82.0	−63.8	37.3	–	–	–
		H&W	22.4	77.6	−59	30.4	–	–	–
		Silver	64.1	30.1	−44.1	26	5.8	45.9	10.5
		Teunter	22.4	77.6	−59	30.4	–	–	–
≤0.13	≤0.85	*Exact_LS*	100	–	–	–	–	–	–
		Guijarro	87.4	12.6	−82.5	26.0	–	–	–
		Exact_Bk	96.9	3.1	−91.9	19.5	–	–	–
		Babiloni	84.6	15.4	−84.5	25	–	–	–
		Trad	78	20	−86.6	23.8	–	–	–
		H&W	84.6	15.4	−84.5	25	–	–	–
		Silver	86.5	0.1	−40.3	8.6	13.4	47.9	5.7
		Teunter	84.6	15.4	−84.5	25.0	–	–	–
>0.13	≤0.85	*Exact_LS*	100	–	–	–	–	–	–
		Guijarro	50.9	49.1	−70.5	30.4	–	–	–
		Exact_Bk	94.8	5.2	−82	27.2	–	–	–
		Babiloni	46.40	53.6	−71.5	30.2	–	–	–
		Trad	29.7	70.3	−78.3	33.7	–	–	–
		H&W	46.4	53.6	−71.5	30.2	–	–	–
		Silver	68.9	0.3	−23.9	5.2	30.8	44.6	10.4
		Teunter	46.4	53.6	−71.5	30.2	–	–	–

between exact and approximate base stock levels are also addressed in this paper. From a practical point of view, the proposed reference framework presents a twofold use. On one hand, it can be used as a decision tool for inventory managers who can choose the most efficient expression to establish the optimal base stock level. On the other hand, it can be used as a corrective tool since if a company is already using one of the approximations, the results of this work provide information about the risk derived from using it.

References

Babiloni E, Guijarro E, Cardos M, Estelles S (2012) Exact fill rates for the (R, S) inventory control with discrete distributed demands for the backordering case. Informatica Economica 16(3): 19–26

Bijvank M, Vis IFA (2011) Lost-sales inventory theory: a review. Eur J Oper Res 215(1):1–13

Cardos M, Babiloni E (2011) Exact and approximate calculation of the cycle service level in periodic review inventory policies. Int J Prod Econ 131(1):63–68

Guijarro E, Cardós M, Babiloni E (2012a) On the exact calculation of the fill rate in a periodic review inventory policy under discrete demand patterns. Eur J Oper Res 218(2):442–447

Guijarro E, Babiloni E, Cardós M (2012b) Determination of the order-up-to-level using the backordering fill rate expressions in a lost sales context and discrete demand. In: 4th production and operations management world conference (POM 2012). Serving the world, pp 1–10

Hadley G, Whitin T (1963) Analysis of inventory systems. Prentice-Hall, Englewood Cliffs

Johnson ME, Lee HL, Davis T, Hall R (1995) Expressions for item fill rates in periodic inventory systems. Naval Res Logistics 42(1):57–80

Silver EA (1970) A modified formula for calculating customer service under continuous inventory review. AIIE Trans 2(3):241–245

Silver EA, Pyke DF, Peterson R (1998) Inventory management and production planning and scheduling. Wiley, Hoboken

Syntetos AA, Boylan JE, Croston JD (2005) On the categorization of demand patterns. J Oper Res Soc 56(5):495–503

Tempelmeier H (2007) On the stochastic uncapacitated dynamic single-item lotsizing problem with service level constraints. Eur J Oper Res 181(1):184–194

Teunter RH (2009) Note on the fill rate of single-stage general periodic review inventory systems. Oper Res Lett 37(1):67–68

Contribution of Lean Principles in the Information Systems Development: An Experience Based on a Practical Case

Joana Pereira and Leonor Teixeira

Abstract Due to the increasing mechanization of industry, resultant from the technological changes combined with human best practices, it is possible to produce keeps getting better at reduced costs. The need to reduce costs and increase the service level to achieve continuous improvement in internal processes derives, on the one hand, from the strong competitiveness, and on the other, on consumers increasingly informed and demanding, requesting reduced delivery times, lower prices and quality levels of excellence. Efficiently manage resources and processes, reduce errors and maximize productivity should be the strategic foundation for organizations become more competitive and effective. Through improved information management mechanisms it is possible to make processes more efficient on which is included the development process of Information Systems (IS). In this context, this project aims to demonstrate the gains associated with good information management, combined with the *Lean Thinking* (LT) methodology in order to produce adequate and quality products (i.e. IS) by reducing the resources, time and information without added value associated.

Keywords Information system · Development process · Lean thinking · Continuous improvement

J. Pereira
Department of Economics, Management and Industrial Engineering, University of Aveiro, Campus Universitário de Santiago, 3810-193 Aveiro, Portugal
e-mail: rodrigues.pereira@ua.pt

L. Teixeira (✉)
IEETA/Department of Economics, Management and Industrial Engineering, University of Aveiro, Campus Universitário de Santiago, 3810-193 Aveiro, Portugal
e-mail: lteixeira@ua.pt

M. Amorim et al. (eds.), *Engineering Systems and Networks*, Lecture Notes in Management and Industrial Engineering, DOI 10.1007/978-3-319-45748-2_13

1 Introduction and Theoretical Framework

In an increasingly competitive environment, the information to the organizational scale is a critical resource that supports the daily activities and the organization decision-making. Information is a valuable resource for organizations so to be useful, information should be relevant, complete, accurate, and current (Oz 2009). Simultaneously it must be obtained economically, that is, cost effectively. According to O'Brien and Marakas (2010), Information Systems (IS) help businesses to improve the efficiency and effectiveness of its processes and help in decision-making and teamwork, which strengthens competitive position of the organization in a dynamic global environment. According to the authors, the greater role of the IS is to support the company's strategy to gain competitive advantage by reducing the associated resources, which translates into savings. Information has become a valuable asset, compared to production resources, material and financial therefore in a globalized society the information is a precious resource. Thus, have good quality information, reliable, in the right quantity and at the right time is a differentiating factor and value added, while the lack of information originates loss of competitive opportunities (Oz 2009). According to Hicks (2007) the way information is organized, visualised and represented can add value to flow to the end-user through the processes of exchange, sharing and collaboration, performed efficiently. The same author sustains that is difficult to align complex systems with the organization, so it can origin a significant detrimental effect on the organisation and its performance. Therefore, the capacity of an IS to store and manage data to produce useful information is the key to gain competitive advantage (Whitten and Bentley 2007).

Some authors suggest that the success of an IS depends in large part on the development approach, particularly in terms of involvement and commitment of stakeholders. One of the most traditional approach to develop IS is the Systems Development Life Cycle (SDLC), a stepwise approach for analysis, design, implementation and testing of the IS. According to Kendall and Kendall (2013) this cycle has seven phases: (i) identifying problems, opportunities, and objectives; (ii) determining user requirements; (iii) analysing system needs; (iv) designing the system; (v) developing and documenting software; (vi) testing and maintaining the system; and finally (vii) implementing and evaluating system. According to the same author this structured approach can be complemented with an agile approach based on principles of communication, simplicity and feedback, by making small adjustments during the development process.

Despite its advantages, sometimes IS are offset by the waste they generate, resulting in harmful consequences, including lost productivity, costly delays and errors, and unnecessary complexity, which is felt by everyone who interacts with the system (Bell and Orzen 2011). In this context, these authors identify three key waste: (i) excess information inventory, (ii) information overprocessing, and (iii) the poor data quality, which, besides not having value added, do not support an efficient workflow or an effective decision-making. Thus, it is possible to establish the analogy between these principles and the *Lean Thinking* (LT) philosophy.

Womack and Jones (1996) defined *Lean Thinking* as an "antidote to waste" in which waste refers to any human activity without value added. However, this concept should be more comprehensive and include other activities and resources misused that contribute to increase costs and time, with impact on the customers and other stakeholders satisfaction. Hines et al. (2008) sustains that to stay lean it is necessary to understand customers and what they value by defining the value streams or processes inside the organization and the supply chain. In order to satisfy customers, organizations need to eliminate or at least reduce the wasteful activities that costumers would not wish to pay for. Identifying and eliminating waste is fundamental to a lean organisation. There are seven types of waste reported by Womack and Jones (1996), first identified by Taiichi Ohno e Shigeo Shingo: (i) *overproduction*; (ii) *waiting*; (iii) *transport*; (iv) *extra processing*; (v) *inventory*; (vi) *motion*; and (vii) *defects*. Furthermore the authors identify an eighth waste related with *underutilisation of people*, in particular their ideas and creative input for improving the processes and practices. As a result of this global philosophy literature provides a set of methodologies, methods and tools categorized by: (i) approaches focused on continuous process improvement: PDCA—Plan, Do, Check and Act—Cycle (Basu 2009) and Visual Control (Feld 2001); (ii) approaches focused on tools: 5S's—Sort; Straighten; Shine; Standardize and Sustain —(Feld 2001); Value Stream Mapping (VSM) (Feld 2001); Kanban (Feld 2001); Ishikawa Diagram (Basu 2009); Standard Work (Feld 2001); 5W2H—Who, Where, What, When, Why; How; How much—(Tague 2005); (iii) approaches focused on methodologies: SMED—Single Minute Exchange of Die (Shingo 1985); Six Sigma (Basu 2009) that includes the DMAIC—Define, Measure, Analyze, Improve and Control—phases; Error Proofing: Poka Yoke (Basu 2009) to bring about such improvements (Hicks 2007).

Following this line of reasoning, this thought which aims to a systematic elimination of waste and value creation can be replicated in information management (Hicks 2007) allowing corporate earnings by increasing information quality. IS has a critical implication for business growth and customer engagement, so establish practices such as lean techniques which highlights the value of IS by reducing waste and increasing productivity (Roberts et al. 2010). According to Ibbitson and Smith (2011) lean information management is defined as an approach to improving organisational systems, to reduce waste and drive up value, reducing non-value added activities. This waste may include the effort necessary to overcome difficulties in retrieving or accessing information (or lack of), or the activities required to confirm and correct inaccurate information (Hicks 2007). Hicks (2007) and Ibbitson and Smith (2011) present an analogy between the seven deadly waste of manufacturing and information management: (i) *overproduction:* generation and maintenance of excess information with non-value added; (ii) *waiting:* time and resources used to identify information which needs to flow; information is not ready, which result in people waiting for the right information; (iii) *extra processing:* activities undertaken to overcome a lack of information that may include creation of new information or accusation of additional information; (iv) *defects:* waste is defined by the resources and activities used to correct and verify inaccurate information and inappropriate actions based on this; work containing errors, wrong information, and incomplete information; (v) *transport:* is not defined as a waste because it does not consume resources, however mass correspondence or e-mails can be defined as a waste; movement of information which does not add value to the user who receives it; (vi) *inventory:* is not defined as a waste because information is typically stored in a digital platform, which does not represent a significant financial cost, however legacy databases and file archives gives more information than the information consumer needs at a given moment in time; (v) *motion:* is not defined as a waste but gatekeepers/single seat licences leads to waiting and movement between computers; unnecessary steps taken due to improper organization and representation of information.

Many studies have been done to prove the benefits of application of LT in the information processes, including Goodman (2012) who applied the Lean and Six Sigma methodology in a pharmaceutical industry to access, analyse and present information (publically available); Tan et al. (1998) who applied the matrix House of Quality (HOQ) to IT-related system design and improvement; Chookittikul et al. (2008) used Six Sigma DMAIC and SIPOC to improve and control the quality of graduates in a university; Khodambashi (2014) who evaluated a IS in clinical process applying the VSM and the A3 method; and Soares and Teixeira (2014) who conducted a study based on lean methodologies (SIPOC and PDCA) in logistics field. In this way, the purpose of this article is exposing the gains measured by applying LT tools in the development process of a specific IS inserted in the context of a cork industry company.

2 Practical Case: Application of Lean Principles in the Development Process of an Information System

2.1 Problem Contextualization and Objectives

The purpose of this study is to presents the experience and procedures concerning the application of lean principles in the development process of an IS which aims to compile all information related with the life cycle of the capsule (an element of the capsulated cork stoppers, main product made in the company cork) and its production process. The necessity of a new IS was originated by the growth of the organization that culminated on the increase the complexity and variety of products and, consequently in the inefficiency of current information management mechanisms. The final objective of IS is to improve the organization communication, particularly by creating a "recipe" that allows the capsules to be produced within specifications in order to standardize the production process, avoiding variations in the final product. Other benefit that will be expected with implementation of this IS is the time saving on operators and machines that results in costs reducing. In order to ensure the success of the final solution (IS), the development process was performed based on lean principles, as will be presented in the next section.

2.2 Development of the Information System Using Lean Principles

This IS development was based on lean principles, methodologies and tools, in order to allow waste elimination, cost reductions and quality improvement. One of the main goals of this IS is to be consistent and have good quality information. According to Chookittikul et al. (2008), the DMAIC methodology sustains that the system should be *Defined* to ensure quality, all defects should be *Measured* and *Analysed* to be eliminated, which allows an *Improvement* on system quality, finally, it is necessary *Control* mechanisms to continuously ensure the quality of IS. The IS development process was incremental and iterative, and involved the key users to evaluate their needs and the critical points to act earlier. In order to follow a continuous improvement philosophy it was applied the PDCA cycle to address the problem and solve it effectively. Based on the integration of DMAIC methodology in the SDLC approach the following phases were developed, as hereinafter described (see Fig. 1).

Phase 1 aimed to work with users to identify their problems and expectations for the new IS and understand opportunities for improvement, as well as the critical points. This was a phase of observation and data collection. Initially, for analysing the current state and identify the main points to act, the *value stream mapping* (VSM) was applied in the process. Hence, it was used the *Ishikawa diagram* to identify the causes of problems related to access to information. Then the *seven wastes* in terms of information were identified. Finally, to summarize this phase, was applied the *5W2H* tool for planning work and to demonstrate the existing problem. Figure 2 presents the procedures and results of this phase, which contributed strongly for the understanding of the problem and know the opportunities for improvement.

Phase 2 and **phase 3** occurred simultaneously and were designed to determine the needs of the system, based on observation, interviews, complemented with the

SDLC Cycle	1 Identifying problems, opportunities, and objectives	2 Determining human information requirements	3 Analysing system needs	4 Designing the recommended system	5 Developing and documenting software	6 Testing and maintaining the system	7 Implementing and evaluating the system
DMAIC	Define, Measure, Analyse	Define, Measure, Analyse	Define, Measure, Analyse	Analyse, Improve	Improve	Improve	Control

Fig. 1 Integration of DMAIC methodology in the SDLC approach to develop a IS

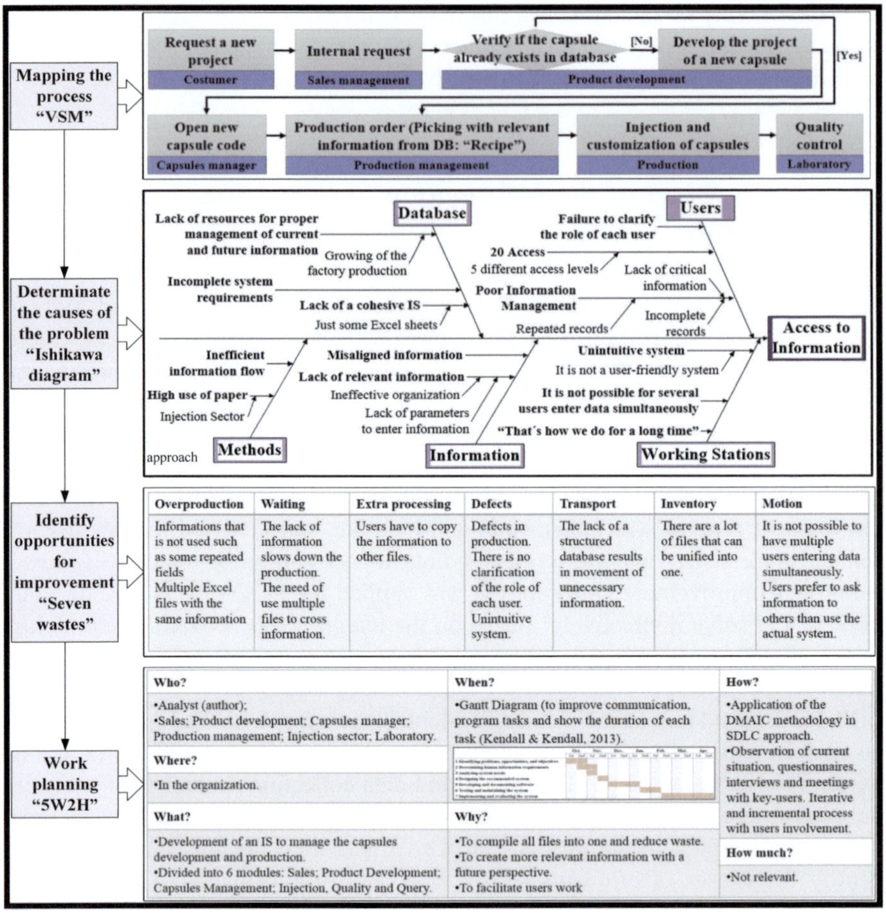

Fig. 2 Lean thinking applied on phase 1 of the SDLC: identifying problems, opportunities, and objectives

data collected on phase 1. In this phase the potential users were strongly involved, and the solution was conceptualized using the *Unified Modeling Language* (UML) notation, more specifically, the Use-Case Diagram and Class Diagram. In this stage was applied the 5S's tool to the existing information in order to separate the information with added value from the waste. This method was applied as follows: (i) *Sort:* separate the information, maintaining the necessary (with value) and discarding the unnecessary (waste); (ii) *Straighten:* structuring information with value and establish the different links between that (achieved with a UML class diagram); (iii) *Shine:* create mechanisms to ensure that there is no waste of information (achieved with controlled data entry mechanisms like ComboBox to prevent data replication); (iv) *Standardize:* develop standards to maintain and control the first 3S's (e.g. user manual); and (v) *Sustain:* bet on training to discipline users to follow the standards implemented. Is important to refer that some of these things will be applied in the following phases.

Phase 4 is the transition from abstract models (in this case in UML notation) to a tool closer to reality. Therefore, the IS design was done using a mock-up prototyping tool more specifically the ForeUI©. The operational model developed includes some features that the final system will have [Selected Features Prototype according to Kendall and Kendall (2013)]. This model incorporates a set of interfaces about the different modules (Injection, Capsules Management, Product Development, Quality, and Query) that will become part of the system, in order to facilitate the transition to the final solution—IS (see Fig. 2 left side). The interfaces were done in order to get closer to reality and had different iterative cycles (PDCA) to incrementally incorporate suggestions from users, to develop a user-friendly system.

The **phase 5** consists on the IS development by using Microsoft Access tool and the Visual Basic language, developed based on the prototype created and analysed in the previous phase (see Fig. 3). Firstly, it was decided to create a *Poka Yoke* system (i.e. Error Proofing). To this, the interface was mechanized, preventing replication of information, by incorporating ComboBoxes and CheckBoxes, with predefined values. Moreover automatic fields were also created. To force the user to complete the necessary data, were created required fields that do not leave the process go forward unless all fields are filled out correctly. We can establish an analogy with the *Jidoka* tool that stops production when an anomaly is detected. The visual control was particularly important in the construction of IS, by incorporating images and colours that appeal to different situations, making the process simple, logical and intuitive.

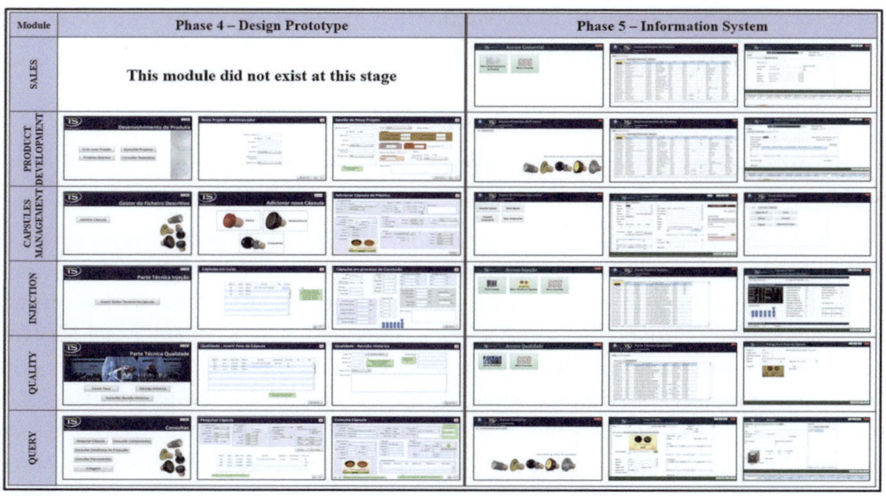

Fig. 3 Some interfaces of the prototype (*left*) and corresponding interfaces of real IS (*right*)

In order to reduce time regarding the future usage of the system were also created other tools such as the calculation of the cost of capsules automatically, the incorporation of the capsule photo in the file, the automation of projects indicators, direct links to folders, as well as an inventory of injection tools, among others. Finally, we can compare the different teams that contribute to the production of capsules with a series manufacturing production. In this case, notification menus (e.g. in the injection/laboratory capsules that require data update) act as a *kanban* that control the order of activities in a sequential manner. In this phase was done a user manual to work with the IS.

Phase 6 and **phase 7** occurred simultaneously because the system was implemented and tested with users at same time, and some changes were made. In a last step, the control of the IS was performed to ensure quality in a continuous improvement perspective. During these phases the users received training, either in group or individually, depending on their interaction with the system.

Since the implementation of the developed tool is still recent, there are no measurable results. However, the users appreciate the tool and recognize the benefits it brought by incorporating all the information related to the capsules on a single platform, which reduces errors and allows to increase the quality of information available. These also emphasize its simplicity of use and the advantages in terms of having information previously non-existent. On the other hand, due to the large amount of information and a network system that requires some improvements, the IS becomes slow when many people uses the network. Currently it is possible to have multiple users accessing the IS simultaneously, and has been clarified the role of each user, i.e., was created a standard work. It is expected (situation under study) that the creation of a most complete production "recipe" based on the system, will allow produce capsules within specifications in injection (initial goal). All things considered, the success or failure of this application is due to the cooperation of users, until now extremely positive, and their awareness that this IS is going to be an improvement in the organizational structure.

3 Conclusions and Future Work

Not always the most complex software is the best solution to the enterprise level, as this study demonstrates. The application of *Lean Thinking* (LT) tools in the context of information management and in the development process of the IS has shown many advantages, proven by some studies, although it is a recent practice. The continuous improvement of information processes and practices should be seen as part of the company culture and not something imposed. Involving key-users is a decisive factor in this process, first it is necessary to understand their needs and then integrate them in the development of IS to get positive results. In this study some lean techniques were applied in the development process of a IS, and results demonstrated the huge potential of LT, increasing activities with added value and reducing waste in the process. Finally, is essential to internalize that be lean is always a work in progress, a continuous improvement, requiring the involvement and commitment of all stakeholders.

References

Basu R (2009) Implementing six sigma and lean. Butterworth-Heinemann, Oxford

Bell S, Orzen M (2011) What is information waste? Lean Enterprise Institute, Inc

Chookittikul J, Busarathit S, Chookittikul W (2008) A six sigma support information system: process improvement at a Thai University. IEEE Computer Society

Feld W (2001) Lean manufacturing: tools, techniques, and how to use them. CRC Press, Boca Raton

Goodman E (2012) Information analysis: a lean and six sigma case study. Bus Inf Rev 29:105–110

Hicks B (2007) Lean information management: understanding and eliminating waste. Int J Inf Manage 27:233–249

Hines P, Found P, Griffiths G, Harrison R (2008) Staying lean: thriving not just surviving. Lean Enterprise Research Centre—Cardiff University, Cardiff

Ibbitson A, Smith R (2011) The lean information management toolkit. Ark Group

Kendall K, Kendall J (2013) Systems analysis and design, 9th edn. Pearson, New Jersey

Khodambashi S (2014) Lean analysis of an intra-operating management process—identifying opportunities for improvement in health information systems. Procedia Comput Sci 37: 309–316

O'Brien J, Marakas G (2010) Management information systems, 10th edn. McGraw-Hill Higher Education

Oz E (2009) Management information systems, 6th edn. Thomson Course Technology

Roberts R, Sarrazin H, Sikes J (2010) Reshaping IT management for turbulent times. McKinsey & Company New York. http://www.mckinsey.com/insights/business_technology/reshaping_it_management_for_turbulent_times. Cited 14 Feb 2015

Soares S, Teixeira L (2014) Lean information management in industrial context: an experience based on a practical case. Int J Ind Eng Manage 5:107–114

Shingo S (1985) A revolution in manufacturing: the SMED system. Productivity Press, Cambridge

Tague N (2005) The quality toolbox, 2nd edn. ASQ Quality Press

Tan K, Xie M, Chia E (1998) Quality function deployment and its use in designing information technology systems. Int J Qual Reliab Manage 15:634–645

Whitten J, Bentley L (2007) Systems analysis and design methods, 7th edn. McGraw-Hill/Irwin, New York

Womack J, Jones D (1996) Lean thinking: Banish waste and create wealth in your corporation. Simon & Schuster, Inc, New York

Data-Driven SKU Differentiation Framework for Supply Chain Management

Alexander Kharlamov, Luís Miguel D.F. Ferreira and Janet Godsell

Abstract Supply chains (SC) are a source of competitive advantage. However, managing SC's is hard due to its size, dynamism, complexity and mostly, context dependency. To overcome such difficulties and better match SC practices with the context characteristics, the adoption of analytical methods is required. Exploiting the availability of operational data, a four stage data-driven SKU differentiation framework is proposed relying on exploratory factor analysis for the criteria selection and cluster analysis for the stock keeping unit (SKU) classification. The application of the framework is illustrated using a B2B instrumental case. Further research is required to validate the framework on other cases, as well as exploring alternative types of SC entities, e.g. clients or suppliers.

Keywords Supply chain · Differentiation · Factor analysis · Cluster analysis

1 Introduction

In the competitive business world, a well-functioning supply chain (SC) is a source of competitive advantage (e.g. Fisher 1997; Payne and Peters 2004; Christopher 2011; Hofmann et al. 2013). However, supply chain management (SCM) depends on an heterogeneous context characterised by different markets, customers and products (e.g. Fisher 1997; Shewchuk 1998; Christopher 2011). The heterogeneity of the SC contexts has long made one-size SC approaches ineffective (Gattorna and

---

A. Kharlamov (✉) · J. Godsell
WMG, IIPSI, The University of Warwick, Coventry CV4 7AL, UK
e-mail: a.a.kharlamov@warwick.ac.uk

J. Godsell
e-mail: j.godsell@warwick.ac.uk

L.M.D.F. Ferreira
DEGEIT, Universidade de Aveiro, Campus Universitário de Santiago, 3810-193 Aveiro, Portugal
e-mail: lmferreira@ua.pt

M. Amorim et al. (eds.), *Engineering Systems and Networks*, Lecture Notes in Management and Industrial Engineering, DOI 10.1007/978-3-319-45748-2_14

Walters 1996; Shewchuk 1998). The mismatch between SC practices and the context characteristics increases the problem of supply and demand mismatch (Fisher 1997; Vitasek et al. 2003; Payne and Peters 2004; Christopher et al. 2009; Christopher 2011).

SCM is constituted by SC practices which must be tailored (adapted rather than adopted) to a particular and heterogeneous context (Lapide 2005). Matching SC practices with context characteristics requires understanding of the particularities of a specific SC, i.e. SC differentiation (Hofmann et al. 2013). Differentiation already made some of the adopting companies "*clearly more successful*" (Hofmann et al. 2013, p. 4). However, markets are going through an "*era of turbulence*" (Christopher and Holweg 2011, p. 63). SCs are getting increasingly complex and dynamic often dealing with a great variety of stock-keeping-units (SKU), clients and suppliers (Hofmann et al. 2013). Understanding the differences of large, dynamic and complex SCs is often beyond the human ability alone, which in practice often leads to what Fuller et al. (1993) calls *the averaging effect* in SCM, e.g. a one-size approach. In order to answer the demands of modern SCM, the large and complex reality must be simplified. SCM must adapt and evolve into greater analytical sophistication embracing the "*information technology (IT) revolution*" (Christopher et al. 2009, p. 460). Focusing specifically on SKU's, an effective integration of analytical methods into SKU differentiation for SCM must answer the three following questions:

RQ1—How can the relevant criteria for SC differentiation be selected?
RQ2—How can the SKU clusters be developed?
RQ3—How can the SKU clusters be operationalized?

2 Literature Review

The importance of finding SC practices matching product, market and customers' characteristics as a way to improve SCs have been accepted since Gattorna and Walters (1996) up to the current times (e.g. Hofmann et al. 2013). The seminal and pioneering contributions touched on most aspects now considered SCM. Considering procurement, one classical example is Kraljic's (1983) strategic matrix. Fuller et al. (1993) discussed the need to tailor logistics to fit the demand characteristics. Regarding SC end-to-end, Fisher (1997) raised significant awareness to the need to match the SC with the product characteristics by splitting innovative and functional products. Focusing on manufacturing strategy, the idea of differentiation was popularised in Shewchuk's (1998) claim that "one size does not fit all". Towards the development of market specific SC strategies, Christopher and Towill (2002) suggested the duration of life-cycle, lead-time, volume, variety and COV (DWV3) as the main criteria. Further, differentiation through SC segmentation based on demand profiles (Godsell et al. 2011, 2013).

Table 1 only summarizes possible criteria for SC differentiation when mentioned at least three times in the literature. Despite the variety of proposed criteria for differentiation, there is little consent about which ones should be selected when and the selection is essentially based on intuition or previous research. Additionally, the number of proposed groups or clusters is usually pragmatically low on average ranging 2–4 clusters. Regarding cluster formation, clustering is mentioned at its most basic form (e.g. Childerhouse et al. 2002; Payne and Peters 2004; Godsell et al. 2011). It is clear that there are many possible criteria for SC differentiation as well as many possibilities to cluster/group entities in a supply chain. However, a data-driven open-end SC differentiation framework that would be compatible with any possible criteria for differentiation is missing.

A four stage SC differentiation framework is proposed (Fig. 1), consisting of Collect, Select, Develop and Operationalise stages. The first *Collect* stage gathers possible criteria for differentiation (e.g. demand data, product characteristics) based on literature and managers' experience which are then extracted from companies' Information Systems (IS) records (given the availability). The second *Select* stage analyses the previously identified criteria and based on the specific data selects the significant ones. The third *Develop* stage identifies a range of clusters based on the data, which are profiled against the previously selected criteria. Finally, the *Operationalise* stage focuses on assigning appropriately tailored practices (Lapide 2005; Godsell et al. 2013) to the previously identified clusters based on its individual profiles, which translates into the individual cluster strategy. The output of the framework is a set of operationalised SKU clusters.

3 Research Approach

The framework is tested on a single case (Stake 1995) selected by "*planned opportunism*" (Pettigrew 1990, p. 174). A business-to-business (B2B) food industry specialised in processing fruit for dairy and operating on a make-to-order (MTO) further characterised in Table 2.

The detailed framework is illustrated in Fig. 2. Input data is collected through open interviews with managers and extraction of the past operational data from the ERP system. After the data collection (given the set is factorable) follows the factor analysis (FA), specifically principal component analysis (PCA) to select and reduces the original criteria into more meaningful dimensions. The new dimensions input a hierarchical cluster analysis (CA) routine. Once the appropriate range of clusters is identified relying on the dendrogram, each set is profiled against the selected criteria (the ones that constitute the dimensions) so appropriate tailored practices can be assigned. To ensure that each criterion is significant, a *t-test* is used. Each stage and iteration is validated with managers in order to achieve meaningful results.

Table 1 Differentiation criteria mentioned/used at least three times in literature (full table on provided on request)

	Frequency of appearance	Oliver and Webber (1982)	Gattorna and Walters (1996)	Fisher (1997)	Mason-Jones et al. (2000)	Naylor et al. (1999)	Lamming et al. (2000)	Li and O'Brien (2001)	Christopher and Towill (2002)	Lee (2002)	Childerhouse et al. (2002)	Vitasek et al. (2003)
Demand variability	17				1	1	1	1	1	1	1	1
Lead time/time window	15	1	1	1	1		1		1	1	1	
Product life cycle	14			1	1		1		1	1	1	
Volume	14				1		1		1	1	1	1
Product variety	9	1		1			1		1	1	1	
Profit margin	7			1	1			1	1	1		
Flexibility	6	1	1				1					
Point of product configuration	5						1	1				
Responsiveness	5	1			1		1					
Customer expectations	5				1		1	1				
Reliability of supply	5	1	1				1			1		
Demand Predictability	4			1						1		
Product complexity	4				1		1					

(continued)

Table 1 (continued)

	Frequency of appearance	Oliver and Webber (1982)	Gattorna and Walters (1996)	Fisher (1997)	Mason-Jones et al. (2000)	Naylor et al. (1999)	Lamming et al. (2000)	Li and O'Brien (2001)	Christopher and Towill (2002)	Lee (2002)	Childerhouse et al. (2002)	Vitasek et al. (2003)
De-coupling point	3				1							
Frequency of delivery/orders	3		1									

	Aitken et al. (2003)	Bruce et al. (2004)	Payne and Peters (2004)	Christopher et al. (2004)	Cigolini et al. (2004)	Lee (2004)	Holweg (2005)	Lovell et al. (2005)	Christopher et al. (2006)	Christopher et al. (2009)	Godsell et al. (2011)	Godsell et al. (2013)
Demand variability	1	1	1	1	1	1	1			1	1	1
Lead time/time window	1			1	1	1	1		1	1		
Product life cycle	1	1		1	1	1	1	1		1		
Volume	1		1		1	1	1			1	1	1
Product Variety	1						1			1		
Profit margin				1		1						
Flexibility				1	1	1						
Point of product configuration					1	1	1					
Responsiveness						1	1					
Customer expectations						1	1					
						1						

(continued)

Table 1 (continued)

	Aitken et al. (2003)	Bruce et al. (2004)	Payne and Peters (2004)	Christopher et al. (2004)	Cigolini et al. (2004)	Lee (2004)	Holweg (2005)	Lovell et al. (2005)	Christopher et al. (2006)	Christopher et al. (2009)	Godsell et al. (2011)	Godsell et al. (2013)
Reliability of supply												
Demand Predictability				1					1			
Product complexity					1	1						
De-coupling point						1	1					
Frequency of delivery/orders			1			1						

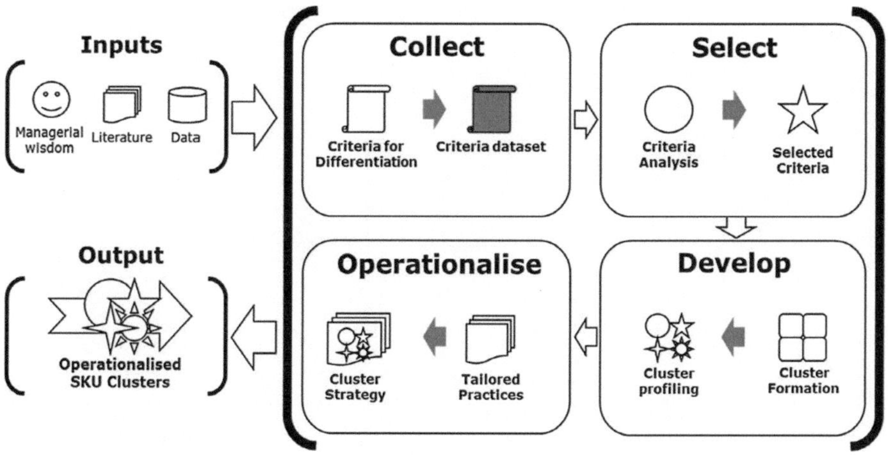

Fig. 1 Data-driven SKU differentiation framework

Table 2 Case characteristics (approximate values)

Characteristic	Description
Unit of analysis	Company (5 production units)
Time frame under analysis	2 years of sales (weekly)
Level of analysis	Stock-keeping-units (SKU)
Industry	Food processing industry
Business relationship	B2B
Current manufacturing strategy	MTO
Experience	20+ years
Key product	Fruit-based compound for dairy
Relationship with customers	Strategic supplier
Relationship with suppliers	Large customer (holds leverage)
Number of SKU	~1000
Number of suppliers	~250
Number of customers	~100
Average capacity utilisation	~75 %
Capacity utilisation during high season	~95 %
Seasonality	Peak between spring and summer
Demand volume trend	Positive growth
Geographic location	Europe and North of Africa

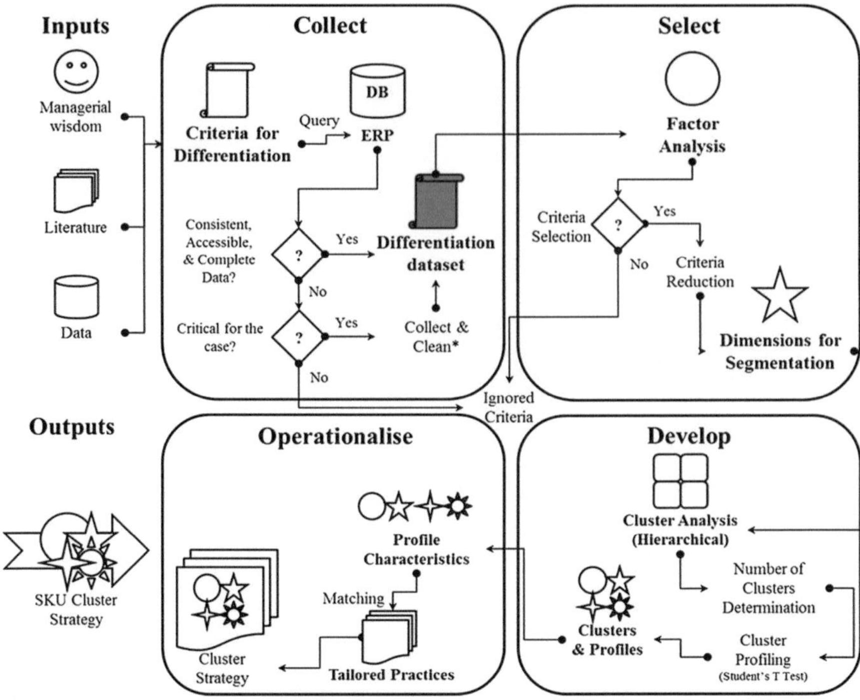

Fig. 2 Detailed SKU differentiation framework (*if reasonable/possible)

4 Findings

Starting with a pool of 30 potential criteria for differentiation (Table 3), 22 were considered due to data availability. After the PCA five criteria remained along with the product variety and revenue to characterise the clusters.

After CA, four distinct clusters are identified, detailed in Table 4. For example, the majority of the demand volume (74 %) is stable/predictable (Cluster 2), highlighting a problematic cluster (Cluster 3) of clients struggling with inventory planning for these specific SKU.

5 Contribution

The present framework is designed for practice, seeking to equip managers with robust statistical discovery methods in an intuitive framework. The use of discovery methods allows managerial input through the steps, e.g. PCA inputs and output interpretation, deciding on the appropriate number of clusters. Most parametric requirements can be bypassed since it deals with the 'population' rather than a

Table 3 Bases for differentiation

	Ignored	Considered	
		Dropped	Selected
Criteria for differentiation	Stock outs	Product life cycle	Demand variability
	Inventory costs	Profit margin	Time window for delivery
	Quality problems	Point of product configuration	Volume
	Obsolescence	Responsiveness	Order corrections ratio
	Number of supply sources	Customer expectations	Frequency of orders/deliveries
	Order line value	Reliability of supply	Product variety[a]
	Flexibility	Product complexity	Revenue[a]
	Product Value Density	De-coupling point	
		Demand Predictability	
		Demand Pareto analysis	
		Minimum run size	
		Change over	
		Substitutability of SKU	
		Range	
		Number of customers	

[a]Cluster characteristics

'sample'. Unlike previous contributions on SC differentiation, the framework is compatible with virtually any input (e.g. continuous, binary, ordinals) and open-ended as it relies on exploratory data analysis and classical clustering method. Such approach is only possible because previously unavailable operational data is now easily accessible. It is estimated that the approach will be applicable to most real-world SCs.

Academics are encouraged to 'borrow boldly' (Gartner et al. 1992; Burt 1992; Kellert 2008; Whetten et al. 2009) from other disciplines. The contribution to theory is successfully borrowing data-mining methods used in the field of marketing (e.g. Tsiptsis and Chorianopoulos 2009) into the field of SCM. The differentiation framework overcomes the known complexity of dealing with large and otherwise opaque data sets. This framework overcomes challenges or supports preconceived beliefs and biases managers might have.

RQ1—How can the relevant criteria for SC differentiation be selected?

To select criteria, it has been often suggested few equally weighted pre-selected criteria (e.g. Payne and Peters 2004; Vitasek et al. 2003; Godsell et al. 2011;

Table 4 Matching clusters with SC practices and principles

		Cluster 1	Cluster 2	Cluster 3	Cluster 4
Cluster profile	Variety	54 %	21 %	18 %	7 %
	Revenue	26 %	63 %	8 %	3 %
	Volume	10 %	74 %	8 %	8 %
	Demand variability	n.s.	Stable	n.s.	Unstable
	Order corrections	V. low	Average	High	Low
	Delivery time window	Short	V. short	Short	V. long
	Delivery frequency	Sporadic	Frequent	V. Rare	Average
Tailored SC practices	S&OP	✓			
	Minimise waste	✓			
	Reduce variety	✓			✓
	Postponement	✓			✓
	MTF (forecast statistically)		✓		
	MTO				✓
	Collaborative planning/VMI			✓	
	Improve visibility			✓	

Holweg 2005). In contrast, using FA (PCA), it is possible to analyse many criteria and effectively narrow down to few significant ones, easing the analysis as well as allowing additional insight through the process (e.g. what is the most distinctive criterion, correlations).

RQ2—How can the SC clusters be developed?

Clusters or groups have been mostly set intuitively delimited by linear boundary conditions and cut-off points ('boxes') where "*if too many levels are selected, then subsequent analysis becomes more complicated*" (Christopher et al. 2009, p. 465). However, CA overcomes the complexity and very large portfolios can be easily clustered as the machine performs the demanding computation.

RQ3—How can the SC clusters be operationalized?

The outputs of the proposed framework will depend on its inputs. Using criteria recommended from SCM literature allow an informed discussion about appropriate tailored practices accordingly to the individual cluster profiles. Through the stages, experienced managers recognise and discover different patterns. Such discovery process enables new insight and informed reasoning about how to best adapt to the particularities of each SC.

References

Aitken J, Childerhouse P, Towill D (2003) The impact of product life cycle on supply chain strategy. Int J Prod Econ 85(2):127–140

Bruce M, Daly L, Towers N (2004) Lean or agile: a solution for supply chain management in the textiles and clothing industry? Int J Oper Prod Manag 24(2):151–170

Burt RS (1992) Structural holes. Harvard University Press, Cambridge

Childerhouse P, Aitken J, Towill DR (2002) Analysis and design of focused demand chains. J Oper Manage 20(6):675–689

Christopher M (2011) Logistics and supply chain management, 4th edn. FT Press, Upper Saddle River

Christopher M, Holweg M (2011) Supply chain 2.0: managing supply chains in the era of turbulence. Int J Phys Distrib Logistics Manage 41(1):63–82

Christopher M, Lowson R, Peck H (2004) Creating agile supply chains in the fashion industry. Int J Retail Distrib Manag 32(8):367–376

Christopher M, Peck H, Towill H (2006) A taxonomy for selecting global supply chain strategies. Int J Logist Manag 17(2):277–287

Christopher M, Towill DR (2002) Developing market specific supply chain strategies. Int J Logistics Manage 13(1):1–14

Christopher M, Towill DR, Aitken J, Childerhouse P (2009) Value stream classification. J Manuf Technol Manage 20(4):460–474

Cigolini R, Cozzi M, Perona M (2004) A new framework for supply chain management. Int J Oper Prod Man 24(1):7–41

Fisher M (1997) What is the right supply chain for your product? Harvard Bus Rev 75(2):105–116

Fuller J, O'Conor J, Rawlinson R (1993) Tailored logistics: the next advantage. Harvard Bus Rev 71(3):87–93

Gartner WB, Bird BJ, Starr JA (1992) Acting as if: differentiating entrepreneurial from organizational behavior. Entrepreneurship Theory Pract 16(3):13–31

Gattorna JL, Walters DW (1996) Managing the supply chain—a strategic perspective. Macmillan, London

Godsell J, Diefenbach T, Clemmow C, Towill D, Christopher M (2011) Enabling supply chain segmentation through demand profiling. Int J Phys Distrib Logistics Manage 41(3):296–314

Godsell J, Kharlamov A, Vasishta S, Burdett J (2013) Challenging the volume-variability paradigm: an empirical investigation in a global beverage company. In: Proceedings of 2013 20th international annual EurOMA conference, Dublin

Hofmann E, Beck P, Füger E (2013) The supply chain differentiation guide. Springer, London

Holweg M (2005) The three dimensions of responsiveness. Int J Oper Prod Manage 25(7):603–622

Kellert S (2008) Borrowed knowledge: Chaos theory and the challenge of learning across disciplines. The University of Chicago Press, Chicago

Kraljic P (1983) Purchasing must become supply management. Harvard Bus Rev 61:109–117

Lamming R, Johnsen T, Zheng J, Harland C (2000) An initial classification of supply networks. Int J Oper Prod Man 20(6):675–691

Lapide L (2005) Benchmarking best practices. J Bus Forecast 24(4):29–32

Lee H (2002) Aligning supply chain strategies with product uncertainties. California Manag Rev 44(3):105–119

Lee H (2004) The triple-A supply chain. Harvard Bus Rev 82(10):102–112

Li D, O'Brien C (2001) A quantitative analysis of relationships between product types and supply chain strategies. Int J Prod Econ 73(1):29–39

Lovell A, Saw R, Stimson J (2005) Product value-density: managing diversity through supply chain segmentation. Int J Logist Manag 16(1):142–158

Mason-Jones R, Naylor B, Towill DR (2000) Lean, agile or leagile? matching your supply chain to the marketplace. Int J Prod Res 38(17):4061–4070

Naylor JB, Naim MM, Berry D (1999) Leagility: integrating the lean and agile manufacturing paradigms in the total supply chain. Int J Prod Econ 62(1–2):107–118

Oliver RK, Weber MD (1982) Supply-Chain management: logistics catches up with strategy. In: Christopher MG (ed) Logistics: the strategic issues. Chapman & Hall, London, pp 63–75

Payne T, Peters MJ (2004) What is the right supply chain for your products? Int J Logistics Manage 15(2):77–92

Pettigrew A (1990) Longitudinal field research on change: theory and practice. Organ Sci 1 (3):267–292

Shewchuk JP (1998) Agile manufacturing: one size does not fit all. In: Strategic management of the manufacturing value chain, 129, p 1998

Stake RE (1995) The art of case study research. Sage Publications, Thousand Oaks, CA

Tsiptsis K, Chorianopoulos A (2009) Data mining techniques in CRM: inside customer segmentation. Wiley, West Sussex

Vitasek BKL, Manrodt KB, Kelly M (2003) Solving the supply-demand mismatch. Supply Chain Manage Rev, 58–64

Whetten DA, Felin T, King BG (2009) The practice of theory borrowing in organizational studies: current issues and future directions. J Manage 35(3):537–563

Deploying "Packaging Logistics" in Paper Napkins

Jesús García-Arca, A. Trinidad González-Portela Garrido and José Carlos Prado-Prado

Abstract In this paper, the potential of "packaging logistics" for improving efficiency and sustainability in the supply chain is presented from an applied point of view. After synthesizing the main conceptual issues of "Packaging Logistics", the paper focuses on the analysis and improvement of a sample of 13 references of paper napkins.

Keywords Packaging · Logistics · Supply chain · Sustainability · Paper napkins

1 Introduction

Nowadays, supply chains should increase efforts to increase competitiveness and sustainability by deleting activities that do not generate value and developing innovations (Christopher 2005; Andersen and Skjoett-Larsen 2009). Simultaneously, different "stakeholders" show a growing interest in the adoption of sustainable policies in supply chain management. Seuring and Müller (2008) show the potential of adopting a sustainable supply chain in terms of environmental, economic and social development, as it saves resources, reduces waste and generates competitive advantages.

In this context, packaging should be considered as a key part of the supply chain that allows its efficiency and sustainability (Klevas 2005). Traditionally, packaging functions have been associated to the protection of the product. However, a wider vision of packaging functions should include the differentiation of the product and the search for efficiency in production and logistics, including the activities of reusing, recycling and recovering according to the new environmental regulations (for instance, the European Directive 94/62 and its update, 2004/12/EC). In this sense, authors such as Jönson (2000) or Garcia-Arca and Prado-Prado (2008)

J. García-Arca (✉) · A.T. González-Portela Garrido · J.C. Prado-Prado
Grupo de Ingeniería de Organización (GIO), Escuela de Ingeniería Industrial,
Universidad de Vigo, Campus Lagoas-Marcosende, c/Maxwell 9, 36310 Vigo, Spain
e-mail: jgarca@uvigo.es

M. Amorim et al. (eds.), *Engineering Systems and Networks*, Lecture Notes
in Management and Industrial Engineering, DOI 10.1007/978-3-319-45748-2_15

identified three main functions in packaging design: commercial, logistics and environmental.

On the other hand, the packaging system is distributed in three interconnected levels: the primary packaging (the consumer packaging), the secondary packaging (some primary packages grouped for easing handling or displaying) and the tertiary packaging (some primary or secondary packages stacked on a pallet or roll container for easing handling, warehousing and transport) (Jönson 2000).

Saghir (2002), Hellstrom and Saghir (2006), Garcia-Arca and Prado-Prado (2008), García-Arca et al. (2014) identify the different design requirements that packaging should accomplish. These requirements are linked to each level of the packaging system in different ways, so we need a new, integrated and coordinated perspective in relation to packaging system, logistics and product. This new perspective has promoted the appearance of the concept "packaging logistics".

Saghir (2002) proposed a complete definition of the term "Packaging Logistics" as "… the process of planning, implementing and controlling the coordinated Packaging system of preparing goods for safe, secure, efficient and effective handling, transport, distribution, storage, retailing, consumption and recovery, reuse or disposal and related information combined with maximizing consumer value, sales and hence profit".

Saghir (2002), Hellström and Saghir (2006), García-Arca and Prado-Prado (2008), Olander-Roase and Nilsson (2009), Vernuccio et al. (2010), Svanes et al. (2010), Hellström and Nilsson (2011), Azzi et al. (2012) have developed the conceptual basics of "Packaging Logistics".

Recently, García-Arca et al. (2014) proposed the three main pillars that could ease the adoption of "Packaging Logistics": the definition of the design requirements, the adoption of a coordinated structure for packaging design and new product development and, finally, the definition of an evaluation system, not only to value the impact on each packaging alternative, but also to compare different alternatives from an overall perspective (global costs, the sales, the customers' satisfaction, the environmental impact, …).

However, there are no so many techniques to compare quantitatively the efficiency and sustainability of the supply chain between different packaging alternatives (the third pillar). In fact, almost all efforts are focused on efficiency of some logistics and productive activities, without a sustainable supply chain perspective (Saghir and Jönson 2001; Saghir 2002; García-Arca et al. 2014). Finally, this global perspective should include the last step of the supply chain and that means the shops or the supermarkets.

To illustrate the potential of "Packaging Logistics" in the retail market, a random sample of 13 references of paper napkins has been selected and analyzed. These products are available in three of the most representative retail chains in Spain (see Fig. 1). This sample represents the heterogeneity of products available in the market, from the point of view of type of product (size, fold, layer, color, thickness, the number of units per package, …) and logistics. However, all of them have similar type of primary packaging: plastic bag. In parallel, we analyzed the logistical problems of these products in the supply chain, particularly, at the shops.

Fig. 1 The sample of 13 references of paper napkins

The first seven references are marketed in SRP ("Shelf Ready Packaging", type "box-pallet"). Four of them use EUR pallets (1200 * 800 mm), whereas the other three references use half pallets EUR (800 * 600 mm). The six remaining references require additional manipulation to place them on the shelves of points of sale. These last references require an additional handling of each package so, a priori, have a lower level of logistical efficiency. Nevertheless, this design decision could be justified from the point of view of rotation and adaptation to space available in shops.

2 Improving the Packaging

Regarding to the analysis of the primary packaging (plastic bag that wraps the paper napkins), the 13 references have a range of significant clearance in the three dimensions of the package: length (between 6 and 37 mm), width (between 5 and 20 mm) and height (between 3 and 19 mm). These clearances exceed the machinery manufacturers' recommendation (between 3 and 6 mm in each dimension) in order to avoid rejection and low productivity in the packing lines. In Fig. 1 some of these clearances can be sighted and in Table 1 the quantitative analysis about primary packaging is summarized.

These clearances generate both an extra plastic consumption (which affects the cost of materials) and an extra generation of waste (which means a higher environmental impact, including an increase in the cost of green dot). In this analysis, we found particularly big the clearances in references 2, 4, 9 (especially, in the length), 10 and 11. At the other extreme, the references 6, 7 and 12 have clearances according to the recommendations of machinery manufacturers.

If we compare the real clearances in the sample of products with the recommendations of machinery manufacturers, we conclude that 10 of the 13 packs (with the exceptions of references 6, 7 and 12) have significant plastic overruns. This extra plastic consumption ranges between 3 and 20 % (see Table 1).

Delving in this field, the analysis of the weight of the plastic packaging (including the weight of extra plastic) in connection with the amount of product containing (number of paper napkins) shows important differences. These differences range between 2.6 and 8.9 g per 100 napkins.

Obviously, these differences also are explained by the differences in thickness and quality of the plastic and the own volume of the napkins. In this sense, beyond the qualities and clearances applied in the plastic packaging, the other parameters that affect the size of the package are related to the product design decisions such as the number of layers of each napkin (1, 2 or 3) the thickness of this layer, the size and the number of napkins per pack (between 45 and 200 units).

This last situation hinders the analysis in order to extrapolate the differences among one reference and the others. However, in the case of similar quality napkin (layers, individual weight and size) it is confirmed that the amount of plastic per napkin decreases as the number of units per pack increases. Logically, we could also apply additional improvements to the product such as the compression of volume of napkins before packing. With this action, the global efficiency and sustainability improve even more.

However, the decision on the number of napkins per pack and their size, folding and quality, without neglecting its logistical and environmental impact, could have some commercial constraints which could hinder their change. Logically, it would be much easier, a priori, to act on reducing clearances such as a few references have changed according to the recommendations of machinery manufacturers.

Moving on to the analysis in tertiary packaging, we have studied the environmental and logistic efficiency of the first seven references marketed in "box-pallet" (EUR pallet and half EUR pallet). In this sense, it would be reasonable that these pallets contained the largest amount of products as possible. To carry out the analysis, however neither the size of the primary packaging, nor the napkin design, nor the type of pallet have been changed, since it could affect business needs, especially with respect to the image of the product, the rotation of stock and the availability of space at the shops.

As a reference of an "efficient" pallet, the RAL standard ("Recommendations AECOC for Logistics", chapter "Efficient Unit Load", 2012) has been used.

Table 1 Analysis and improvements of the primary packaging

Ref.	No units per pack	Weight plastic pack (g)	Real length (clearance, mm)	Real width (clearance, mm)	Real height (clearance, mm)	Ideal length (mm)	Ideal width (mm)	Ideal height (mm)	% extra plastic use (m^2) (%)
1	100	3.5	159 (12)	155 (10)	85 (8)	147	145	77	5.8
2	200	5.1	309 (19)	158 (18)	94 (16)	290	140	78	13.3
3	160	5.0	307 (17)	156 (9)	89 (5)	290	147	84	3.8
4	200	5.5	309 (16)	166 (18)	125 (17)	293	148	108	11.7
5	100	7.1	180 (17)	166 (10)	130 (9)	163	156	121	7.3
6	100	5.6	157 (6)	154 (5)	105 (3)	151	149	102	0.0
7	200	5.5	299 (6)	158 (7)	73 (6)	293	151	67	0.5
8	80	3.6	158 (9)	157 (19)	102 (11)	149	138	91	9.9
9	120	6.5	324 (37)	162 (16)	86 (10)	287	146	76	14.1
10	180	5.6	175 (32)	170 (20)	135 (19)	143	150	116	20.8
11	140	5.3	326 (17)	180 (16)	96 (13)	309	164	83	9.8
12	45	4.0	206 (7)	193 (5)	75 (4)	199	188	71	0.4
13	50	4.0	202 (9)	199 (6)	78 (11)	193	193	67	3.9

"% of extra plastic use" compares the real surface of plastic packaging with the ideal surface of plastic packaging according to recommendations of machinery manufacturers

Therein, it is stated that as a general rule the maximum height of a EUR pallet load should not exceed 2.6 m (maximum internal height in truck; 1.3 m if pallets are stacked). In the case of a half EUR pallet load, RAL proposes as the maximum height, 1.3 m. Likewise, the RAL standard covers the SRP (2007) proposing as the maximum height for a "box pallet" 1.3 m (including height of wood pallet); this is the limit we have used for improving the products. The main reason for this decision is that napkins are low density products, so it would be efficient to reach this limit for achieving a good rationalization of the transport.

The analysis of the seven references shows the logistical and environmental improvement in six of them (four with EUR pallet and two with half EUR pallet) as it is presented in Fig. 2. The potential of improvement of these alternatives ranges between 7 and 14 %, in the case of EUR pallet, and between 9 and 12 %, in the case of half EUR pallet.

All these improvements imply a significant reduction in the number of palletized loads; thanks to this, companies could achieve a significant reduction in their logistic costs (handling, storage and transport along the supply chain including shops). The source of these improvements varies by product, but includes actions such as the deployment of new patterns of palletization, harnessing the maximum height of pallet and, in some cases, even the adoption of mixed patterns of palletization (depending on the layer of pallet).

Additionally, these changes also supply benefits in terms of both material consumption (comparatively, less cardboard is needed in the "box-pallet") and environmental impact (comparatively, less waste of cardboard is generated). The proposed improvements in the use of cardboard and in the generation of waste fluctuate between 1.5 and 13 % (see Table 2).

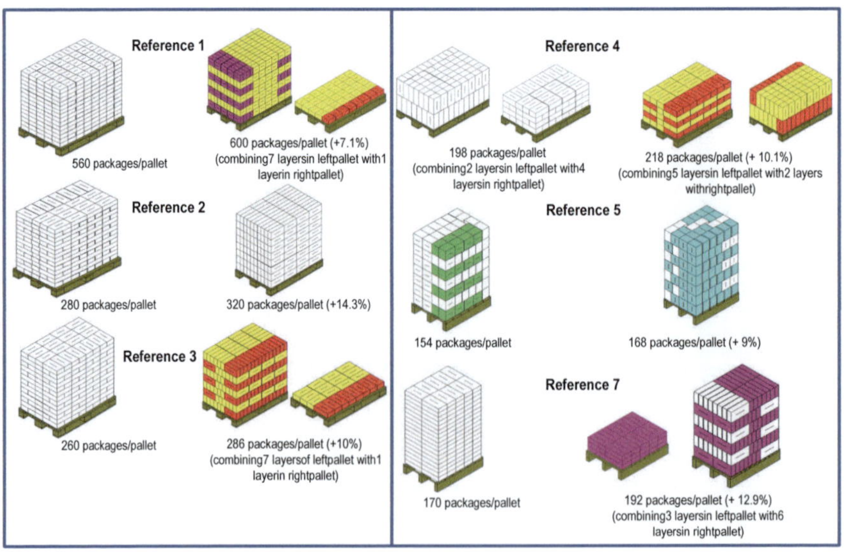

Fig. 2 The improvements in "box-pallets" of paper napkins

Table 2 Analysis and improvement of pallets of products

Ref.	Initial packs/pallet	Cardboard weight (g/m^2)	Initial pallet height (mm) (pallet included)	Improved packs/pallet	% of improvement (packs/pallets) (%)	Improved pallet height (mm) (pallet included)	% improvement in cardboard consumption in box pallet (m^2 cardboard per pack) (%)
1	560	800	1255	600	7.1	1252	6.9
2	280	1000	1255	320	14.3	1264	11.9
3	260	900	1237	286	10	1257	7.7
4	198	900	1163	218	10.1	1271	1.5
5	154	780	1280	168	9.1	1286	7.9
7	170	780	1290	192	12.9	1258	13.6

In this context of sustainability, improved palletization also contributes to re-duce the environmental impact of transport, thanks to the reduction in the number of pallets that are moved throughout the supply chain.

Likewise, some relevant differences between samples of cardboard used in "box-pallet", have been found; these differences range between 780 and 1000 g/m^2.

However, although a heavier cardboard implies a greater amount of waste (and a higher cost of green dot), this statement should not extended to the field of the cost of materials, since other aspects are involved, such as the external aesthetic design (for example, inks, colours, varnishes, text, …) or the combination of papers used in making corrugated cardboard.

Commercially, the negative impact of these improvements should not be relevant, although in some cases depending on the new pattern of palletization, it could be necessary to redesign the aesthetic design of primary packaging (plastic bag) to allow the visibility of the brand and other marketing features of the product.

Finally, the adoption in some improvements of mixed patterns of palletization could hinder the automation of packing the napkins in suppliers; anyway, this proposed solution is nowadays applied in some products, as it is presented in Fig. 1.

3 Conclusions

"Packaging Logistics" should be considered as an element that actively contributes to the efficiency and sustainability of supply chains. The potential of deploying the approach "packaging logistics" in all types of product and sectors, both retail and industrial, should be highlighted. In this context, this paper illustrates how it is possible to achieve competitive advantages thanks to the adoption of some changes in packaging of the usual products, such as the paper napkins.

References

AECOC (2012) RAL. Recomendaciones AECOC para la Logística. Unidades de Carga Eficientes. AECOC, Barcelona, Spain

Andersen M, Skjoett-Larsen T (2009) Corporate social responsibility in global supply chains. Supply Chain Manage: Int J 12(2):75–86

Azzi A, Battini D, Persona A, Sgarbossa F (2012) Packaging design: general framework and research agenda. Packag Technol Sci 25:435–456

Christopher M (2005) Logistics and supply chain management strategies for reducing cost and improving service. Financial Times Pitman Publishing, Londres

García-Arca J, Prado-Prado JC (2008) Packaging design model from a supply chain approach. Supply Chain Manage: Int J 13(5):375–380

García-Arca J, Prado-Prado JC, Gonzalez-Portela Garrido AT (2014) Packaging logistics: promoting sustainable efficiency in supply chains. Int J Phys Distrib Logistics Manage 44 (4):325–346

Hellström D, Nilsson F (2011) Logistics-driven packaging innovation: a case study at IKEA. Int J Retail Distrib Manage 39(9):638–657

Hellström D, Saghir M (2006) Packaging and logistics interactions in retail supply chain. Packag Technol Sci 20:197–216

Jönson G (2000) Packaging technology for the logistician, 2nd edn. Department of Design Sciences, Division of Packaging Logistics, Lund University, Lund

Klevas J (2005) Organization of packaging resources at a product-developing company. Int J Phys Distrib Logistics Manage 35(2):116–131

Olander-Roase M, Nilsson F (2009) Competitive advantages through packaging design—prepositions for supply chain effectiveness and efficiency. International conference on engineering design, ICED 2009. Stanford University, USA, pp 279–290

Saghir M (2002) Packaging logistics evaluation in the Swedish retail supply chain. University of Lund, Lund

Saghir M, Jönson G (2001) Packaging handling evaluation methods in the grocery retail industry. Packag Technol Sci 14:21–29

Seuring S, Müller M (2008) From a literature review to a conceptual framework for sustainable supply chain management. Int J Cleaner Prod 16(15):1699–1710

Svanes E, Vold M, Møller H, Pettersen MK, Larsen H, Hanssen OJ (2010) Sustainable packaging design: a holistic methodology for packaging design. Packag Technol Sci 23:161–175

Vernuccio M, Cozzolino A, Michelini L (2010) An exploratory study of marketing, logistics, and ethics in packaging innovation. Eur J Innov Manage 13(3):333–354

Differentiation of the Difficulty Level of Supply Chain Management Integration Actions

Fernanda Arantes, Maria Leite and Antonio Cezar Bornia

Abstract Integration is a key factor to ensure alignment of actions throughout the supply chain, increasing the competitiveness of companies. However, to integrate upstream and downstream actions is quite a complex task that involves different levels of difficulty. Several studies have been developed to assess the supply chain management integration (SCMI), with different actions corresponding to integration among chain members; nevertheless, little has been discussed about the level of integration required for each activity. Thus, this research proposes an initial assessment of the differentiation of the difficulty level between the actions of SCMI, in order to facilitate the planning of these actions gradually and consistently, using the Item Response Theory (IRT). The selected SCMI actions were ranked according to the difficulty level for their implementation, allowing differentiating the simplest tasks to be performed from those that require a higher level of integration between partners.

Keywords Supply chain management · Level of integration · Alignment · Item response theory

---

F. Arantes (✉)
Department of Production and Systems Engineering, Federal University of Santa Catarina, Rua Adalgisa Luna de Menezes, 681, apt 403E, João Pessoa-PB 58051-840, Paraíba, Brazil
e-mail: nandapaesarantes@gmail.com

M. Leite
Department of Production Engineering, Federal University of Paraiba, Centro de Tecnologia, Campus I, Bloco G, João Pessoa-PB, Brazil
e-mail: mariasileneleite@hotmail.com

A.C. Bornia
Department of Production and Systems Engineering, Federal University of Santa Catarina. Campus Universitário Trindade, Caixa Postal 476, Florianópolis-SC, Brazil
e-mail: cezar@deps.ufsc.br

M. Amorim et al. (eds.), *Engineering Systems and Networks*, Lecture Notes in Management and Industrial Engineering, DOI 10.1007/978-3-319-45748-2_16

1 Introduction

Integration is a key factor to ensure the alignment of actions along the supply chain. Only when there is integration among members it is possible to manage the upstream and downstream flows efficiently and thereby create value for the final customer and greater gains for all members. Thus, the success of SCM is directly related to the level of integration between its members (Lambert et al. 1998), which consists in the intensity at which they develop a joint work, combining their resources in a complementary manner to meet the needs of the final consumer.

The integration with suppliers and customers, eliminating waste and becoming more flexible, has become a necessity, since the optimization of internal structures is no longer enough for a good business strategy (Frohlich and Westbrook 2001).

The increase of the integration level improves sensitivity to customer needs and makes time response lower than in least integrated competitors, positively influencing the success of companies, as it consists in the proper use of SCM practices (Rangel 2012; Sezen 2008; Wong and Boon-Itt 2008). However, there is much to be done in order to identify how to raise the level of integration between the members of a supply chain.

In general, they are poorly integrated, becoming a problem for its management (Childerhouse et al. 2011), since there is less and less room for error, resulting in a higher need for planning regarding unexpected events (Knemeyer et al. 2009).

Some studies have been developed regarding the measurement of integration, among them are Frohlich and Westbrook (2001), Rosenzweig et al. (2003), Vickery et al. (2003), Aryee et al. (2008), Thun (2010), Chen et al. (2009), Kim (2009), Childerhouse and Towill (2011), Childerhouse et al. (2011), Zhao et al. (2011), Schoenherr and Swink (2012), Basnet (2013), Danese (2013), Marin-Garcia et al. (2013), He et al. (2014). These researches address the topic in different ways, with outstanding contributions to the measurement of the supply chain management integration (SCMI).

In these surveys, various integration indicators can be found, related to different types of integration: internal or external, with suppliers or customers. Issues as the influence of integration with suppliers in the integration with customers are addressed, as well as the impact of integration on the performance of the supply chain or in the development of new products, among other issues. However, one might not observe a list of SCMI indicators with the level of difficulty to its implementation in the SCM.

In view of this, this research aims to highlight some of the main SCMI indicators presented in the literature, establishing a relationship with the level of difficulty in implementing each of them.

2 Supply Chain Management Integration

The need for management of the relationship with suppliers and customers, in addition to the internal production process, is inherent to all companies (Stevens 1989). The management of supply chains consists in expanding this vision beyond the first-tier relationships, with the focused company also integrating with suppliers of their suppliers and their customers' customers. The purpose of this integration, according to Vickery et al. (2003), is to achieve a balance between the quality in the customer service and the cost efficiency, through the sync of customer requirements with the flow of materials and information along the chain.

This integration of activities along the supply chain enables the elimination of faults in the process, increasing flexibility and reducing costs, creating greater competitive advantage for all members (Rosenzweig et al. 2003), since it involves developed organizational routines between the companies, creating combinations of unique skills, broadening both the knowledge and the joint capabilities (Schoenherr and Swink 2012).

Researches have proven that, from an average level of internal integration, companies can already make significant gains, outperforming many of their competitors, since the integration reduces uncertainty and improves efficiency and efficacy of processes (Childerhouse et al. 2011).

However, there is still a barrier to achieve SCMI, because the settings are quite large, emphasizing the flows, but disregarding the strategic nature of the supply chain (Flynn et al. 2010). Furthermore, as difficult as defining SCMI is operationalizing it (Näslund and Hulthén 2012), as it must involve all members in search of a common goal.

It can be analyzed from two different perspectives: internal or external and upstream or downstream. Internal integration is one that involves the planning and control systems and external integration refers to the interaction with other members of the supply chain (Flynn et al. 2010; Näslund and Hulthén 2012; Topolsek et al. 2010). Some authors argue that SCMI should start internally and then extend to the rest of the chain (Flynn et al. 2010; Topolsek et al. 2010). However, there is another group of authors arguing that the characteristics of the external integration positively influence the internal integration, to the extent that the approach of the external integration induces companies to fit internally (He and Lai 2012; Primo 2010).

Regarding the direction, it can be upstream, more focused on the management of non-physical flows, mainly informational (Danese 2013; Frohlich and Westbrook 2001); it involves long-term relationship and joint planning, seeking to increase the flexibility and supply chain efficiency (Thun 2010; Wong and Boon-Itt 2008). On the other hand, the downstream integration is more related to the management of physical flows, of goods and services (Frohlich and Westbrook 2001; Wong and Boon-Itt 2008), and increases the sensitivity to variations demand (Flynn et al. 2010; Thun 2010; He et al. 2014).

Thus, both internal and external integration, as integration with suppliers and customers, complement each other. In both cases, as well as one needs the other to have better results, the existence of one positively influences the other in a cyclical process, raising the level of integration along the supply chain. Important factor to deal with the continuous increase in the market complexity (Frohlich and Westbrook 2001; van der Vaart and van Donk 2008).

3 Methodology

From the review of surveys that sought to assess the level of integration in SCM, a questionnaire was developed, applied to the managers of 41 Brazilian companies. This instrument was designed to evaluate how these companies relate to their suppliers. Therefore, each company answered an average of five questionnaires, one for each supplier. This generated a sample of 205 observations of company-supplier relationships.

These data were analyzed using the Item Response Theory (IRT), which allows the analysis of qualitative data, generating parameters of discrimination and difficulty for each item. IRT is a set of mathematical models with three major utilities: analyzing items and scales, creating and managing measures for qualitative factors and measuring individuals in this construct (Reise et al. 2005).

The questionnaire submitted to the managers had as response options: yes, no or not applicable. Thus, the data were analyzed using IRT two-parameter logistic model, represented by the following equation:

$$P_{ij} = P(\theta_j, a_i, b_i) = \frac{1}{1 + e^{-a_i(\theta_j - b_i)}} \tag{1}$$

Being a_i the discrimination parameter of item i, b_i the difficulty parameter of item i, and θ_j the latent trait of the respondent j. Thus, in addition to identifying the items that are most suitable to measure the SCMI, one can also identify the level of difficulty to perform each action.

With this, one can trace a plan to raise the level of integration between members of the supply chain, starting from the simplest action to the most complex.

4 Results and Discussion

Table 1 shows the items of the questionnaire used to assess the level of integration of surveyed companies with their respective suppliers, ordered by level of difficulty. For each item, parameters of discrimination (a) and difficulty (b) are presented, with their respective standard errors (SE). The b values are generated based on a scale with mean 0 and standard deviation 1 (0.1); therefore, negative values are

Table 1 Discrimination parameters and difficulty of items, ordered by difficulty level

No.	Item	a	SE (a)	b	SE (b)
1	Quality as expected	**0.872**	0.281	**−3.442**	0.969
2	Long-term relationship	**1.342**	0.315	**−2.854**	0.533
3	Exchange of products damaged in delivery	**0.969**	0.214	**−2.551**	0.549
4	Exchange of products under warranty	**1.132**	0.274	**−1.836**	0.414
5	Good personal relationships	**2.019**	0.580	**−1.772**	0.340
6	Access to information request	**0.922**	0.260	**−1.490**	0.375
7	Open communication	**2.152**	0.660	**−1.263**	0.231
8	Inform demand forecasting	**1.451**	0.385	**−0.682**	0.157
9	Solving problems together	**1.139**	0.330	**−0.592**	0.180
10	Product quality helps keep the relationship	**1.174**	0.225	**−0.446**	0.145
11	Tolerance to delay of eventual payment	**0.845**	0.189	**0.011**	0.180
12	Exchange information about the market	**0.950**	0.219	**0.042**	0.166
13	Exchange information to improve the process	**1.443**	0.319	**0.045**	0.130
14	Notice of future price increase	**1.045**	0.226	**0.126**	0.156
15	Mutual benefit	**1.130**	0.247	**0.168**	0.147
16	Notice of delayed request	**1.314**	0.218	**0.193**	0.166
17	Provide inventory information	**0.944**	0.204	**1.138**	0.251
18	Lowest price guarantee	**2.185**	0.366	**1.166**	0.164
19	Receive inventory information	**0.890**	0.221	**2.582**	0.566
20	Receive cost information	**1.262**	0.410	**2.797**	0.615
21	Using compatible information systems	**0.933**	0.334	**4.662**	1.483

Bold indicates discrimination and difficult parameters of items

generated. This does not imply absence of integration, but only the distance in relation to the sample mean. The important thing to note is the ordering of items based on the parameters of difficulty and the distance between them.

Based on the estimation of IRT, the delivery of products with quality as expected by the companies is the easiest aspect to be fulfilled by suppliers for the integration of the supply chain. This SCMI aspect was highlighted by Rosenzweig et al. (2003), Kim (2009), Jones et al. (2010) and Childerhouse et al. (2011).

The second item deals with the importance of developing long-term relationships, highlighted by Danese (2013) and Marin-Garcia et al. (2013), for one of the intrinsic characteristics to integration is the need for trust between partners, which can only be developed over time. Items 3 and 4 address the need for standard procedures to perform the agreed faster and more efficiently, highlighted by Fawcett et al. (2012).

Companies are made of people, making the existence of good interpersonal relationships also an initial factor for the existence of SCMI. Aspect discussed by Cai et al. (2010), Schoenherr and Swink (2012) and Basnet (2013). Subsequently, comes up factors such as access to information request (Lee and Whang 1998; Sahin and Robinson 2002; Danese 2013), open communication (Motwani et al. 1998; Chen and Wu 2010; Thun 2010), inform demand forecasting (He et al. 2014; Lee and Whang 1998; Sahin and Robinson 2002; Li et al. 2006; Ding et al. 2011), solving problems together (He et al. 2014; Thun 2010; Basnet 2013; Chen et al. 2009; Kim 2009; Lambert and Cooper 2000; Danese 2013; Frohlich and Westbrook 2001).

Delivering quality products (Rosenzweig et al. 2003; Childerhouse et al. 2011; Kim 2009), flexibility (Chen and Wu 2010; Rosenzweig et al. 2003; Kim 2009), sharing of information about the market conditions (Childerhouse et al. 2011) and the seek for process improvement (Danese 2013) are the next aspects for obtaining the SCMI, requiring a little more effort on the part of members of the supply chain.

Reporting anticipated a future price increase, seeking for mutual benefit and ensuring a lower price than competitors are win-win actions, emphasized by Sahin and Robinson (2002). Reporting delays in the request (Lee and Whang 1998), sharing inventory information (Frohlich and Westbrook 2001; Kim 2009; He et al. 2014), having suppliers cost information (Sahin and Robinson 2002; Li et al. 2006) and sharing information through compatible information systems (Frohlich and Westbrook 2001; Aryee et al. 2008; Childerhouse et al. 2011; Stevens 1989; Kim 2009) are the actions of integration more difficult to be performed by companies, among all shown.

Regarding the discrimination of items, 'Lowest price guarantee', 'open communication', 'good personal relationships', 'Inform demand forecasting', 'Long-term relationship', 'Notice of delayed request' are highlighted. These items allow better differentiation of the relationship position as to SCMI level.

5 Conclusion

Companies are dealing with increasingly competitive environments and this forces them to seek solutions to increase efficiency and, therefore, remain on the market. To manage the actions along the supply chain has become a necessity and this requires that companies can integrate their actions with their upstream and downstream members. This is a complex task because SCMI involves different aspects with different levels of difficulty in its implementation.

Thus, this research proposes an initial assessment of the difficulty level differentiation between SCMI actions, in order to facilitate the planning of these actions gradually and consistently. Future research can extend this list of SCMI actions and insert them in the same scale, providing more information for companies on how to integrate with their partners in the supply chain.

References

Aryee G, Naim MM, Lalwani C (2008) Supply chain integration using a maturity scale. J Manufact Technol Manage 19:559–575

Basnet C (2013) The measurement of internal supply chain integration. Manage Res Rev 36: 153–172

Cai S, Jun M, Yang Z (2010) Implementing supply chain information integration in China: the role of institutional forces and trust. J Oper Manage 28:257–268
Chen S-P, Wu W-Y (2010) A systematic procedure to evaluate an automobile manufacturer-distributor partnership. Eur J Oper Res 205:687–698
Chen H, Daugherty PJ, Roath AS (2009) Defining and operationalizing supply chain process integration. J Bus Logistics 30:63–84
Childerhouse P, Towill DR (2011) Arcs of supply chain integration. Int J Prod Res 49:7441–7468
Childerhouse P, Deakins E, Böhme T, Towill DR, Disney SM, Banomyong R (2011) Supply chain integration: an international comparison of maturity. Asia Pacific J Mark Logistics 23:531–552
Danese P (2013) Supplier integration and company performance: a configurational view. Omega-Int J Manage Sci 41:1029–1041
Ding HP, Guo BC, Liu ZS (2011) Information sharing and profit allotment based on supply chain cooperation. Int J Prod Econ 133:70–79
Fawcett SE, Jones SL, Fawcett AM (2012) Supply chain trust: the catalyst for collaborative innovation. Bus Horiz 55:163–178
Flynn BB, Huo B, Zhao X (2010) The impact of supply chain integration on performance: a contingency and configuration approach. J Oper Manage 28:58–71
Frohlich MT, Westbrook R (2001) Arcs of integration: an international study of supply chain strategies. J Oper Manage 19:185–200
He Y, Lai KK (2012) Supply chain integration and service oriented transformation: evidence from Chinese equipment manufacturers. Int J Prod Econ 135:791–799
He Y, Lai KK, Sun H, Chen Y (2014) The impact of supplier integration on customer integration and new product performance: the mediating role of manufacturing flexibility under trust theory. Int J Prod Econ 147:260–270
Jones SL, Fawcett SE, Fawcett AM, Wallin C (2010) Benchmarking trust signals in supply chain alliances: moving toward a robust measure of trust. Benchmarking Int J 17:705–727
Kim SW (2009) An investigation on the direct and indirect effect of supply chain integration on firm performance. Int J Prod Econ 119:328–346
Kim D, Cavusgil E (2009) The impact of supply chain integration on brand equity. J Bus Ind Mark 24:496–504
Knemeyer AM, Zinna W, Eroglu C (2009) Proactive planning for catastrophic events in supply chains. J Oper Manage 27:141–153
Lambert DM, Cooper MC (2000) Issues in supply chain management. Ind Mark Manage 29:65–83
Lambert DM, Cooper MC, Pagh JD (1998) Supply chain management: implementation issues and research opportunities. Int J Logistics Manage 1–19
Lee HL, Whang S (1998) Information sharing in a supply chain. Stanford University, Stanford, pp 1–22
Li J, Sikora R, Shaw MJ, Tan GW (2006) A strategic analysis of inter organizational information sharing. Decis Support Syst 42:251–266
Marin-Garcia JA, Alfalla-Luque R, Medina-Lopez C (2013) Supply chain integration scales validation and benchmark values. J Ind Eng Manage 6:423–440
Motwani J, Larson L, Ahuja S (1998) Managing a global supply chain partnership. Logistics Inf Manage 11:349–354
Näslund D, Hulthén H (2012) Supply chain management integration: a critical analysis. Benchmarking Int J 19:481–501
Primo MAM (2010) Supply chain integration mechanisms for alleviating supply problems in manufacturing firms. Oper Manage Res 3:43–59
Rangel DA (2012) Proposta de um procedimento para identificar, avaliar e priorizar riscos em cadeias de suprimentos. Universidade Federal da Paraíba, Mestrado
Reise SP, Ainsworth AT, Haviland MG (2005) Item response theory—fundamentals, applications, and promise in psychological research. Curr Dir Psychol Sci 14:95–101
Rosenzweig ED, Roth AV, Dean JW Jr (2003) The influence of an integration strategy on competitive capabilities and business performance: an exploratory study of consumer products manufacturers. J Oper Manage 21:437–456

Sahin F, Robinson EP (2002) Flow coordination and information sharing in supply chains: review, implications, and directions for future research. Decis Sci 33:505–536

Schoenherr T, Swink M (2012) Revisiting the arcs of integration: cross-validations and extensions. J Oper Manage 30:99–115

Sezen B (2008) Relative effects of design, integration and information sharing on supply chain performance. Supply Chain Manage Int J 13:233–240

Stevens GC (1989) Integrating the supply chain. Int J Phys Distrib Logistics Manage 19:3–8

Thun J-H (2010) Angles of integration: an empirical analysis of the alignment of internet-based information technology and global supply chain integration. J Supply Chain Manage 46:30–44

Topolsek D, Cizman A, Lipicnik M (2010) Collaborative behaviour as a facilitator of integration of logistic and marketing functions: the case of slovene retailers. Promet-Traffic & Transp 22:353–362

Van Der Vaart T, Van Donk DP (2008) A critical review of survey-based research in supply chain integration. Int J Prod Econ 111:42–55

Vickery SK, Jayaram J, Droge C, Calantone R (2003) The effects of an integrative supply chain strategy on customer service and financial performance: an analysis of direct versus indirect relationships. J Oper Manage 21:523–539

Wong CY, Boon-Itt S (2008) The influence of institutional norms and environmental uncertainty on supply chain integration in the Thai automotive industry. Int J Prod Econ 115:400–410

Zhao X, Huo B, Selen W, Yeung JHY (2011) The impact of internal integration and relationship commitment on external integration. J Oper Manage 29:17–32

Establishing a Link Between Lean Practices and Corporate Sustainability

Cristóvão Silva and Paulo Vaz

Abstract In this paper we describe how the concept mapping methodology, which combines a qualitative case study approach—based on interviews, focus groups and plant visits—and quantitative methods—using a software and data-driven mapping methods—was used to answer the following question: can lean manufacturing contribute to a company environmental and social sustainability? The conclusions of the research are empirically analysed. As expected the impact of lean on productivity and process efficiency was identified but the results also demonstrate that it has a positive effect on business sustainability.

Keywords Lean practices · Concept mapping · Corporate sustainability

1 Introduction

Different management approaches, such as lean and green have been adopted throughout the world. The lean approach is necessary to build an organizational culture of continuous improvement. Research in lean manufacturing has highlighted it ability to increase efficiency and customer response time, reduce costs and improve quality and profitability. The use of green approaches for manufacturing have sometimes created amazing reductions in energy consumption, waste generation and hazardous materials used while building the companies' images as socially responsible organizations (Bergmiller and McCright 2009a).

Several attempts to establish a relationship between lean manufacturing and green outcomes has appeared in the literature since the early 1990'

C. Silva (✉)
Department of Mechanical Engineering, CEMUC, University of Coimbra, Coimbra, Portugal
e-mail: cristovao.silva@dem.uc.pt

P. Vaz
Polytechnic Institute of Viseu, Viseu, Portugal
e-mail: pja.vaz@gmail.com

M. Amorim et al. (eds.), *Engineering Systems and Networks*, Lecture Notes in Management and Industrial Engineering, DOI 10.1007/978-3-319-45748-2_17

(Florida 1996; Sarkis 1995). King and Lenox (2001) establish a link between lean and green showing that adopters of ISO 9000 quality standards are more likely to also adopt ISO 14000 environmental standards. Azevedo et al. (2011) establish, based on a case study, a relationship between green and lean upstream supply chain practices and sustainable business development. Bergmiller and McCright (2009b) conclude that a significant synergy exists between lean and green programs and that firms pursuing leanness are more successful if they also pursue green objectives. A good literature review on the connection between lean and green practices is presented in Dües et al. (2013) where it is referred that most papers addressing this connection focus on the efficient use of energy and resources and the reduction of waste and pollution. In their paper, the authors refer that the goals set for achieving leanness will be a catalyst for successfully implement green practices and help in reaching green goals as well.

These research papers seem to prove the existence of synergies between lean and green practices or point to a positive impact of lean practices on green performance. Nevertheless, there is also some research pointing in the opposite direction, i.e., showing that not all lean practices are positively related to environmental performance. See, for example, the work of Rothenberg et al. (2001) which describe the implementation of lean practices in a company painting process that have resulted in better quality and cost-effective process but that proved to be more harmful to the environment. Zhu and Sarkis (2004) contributes to this discussion referring that, in the Chinese context, JIT practices had a negative impact on supply chain environmental outcomes.

In a recent paper, McDaniel and Vastag (2010) refer that the existing studies on the relationship between lean and green tend to focus on the USA and Asia region. They also refer some drawbacks existing in the previous lean green studies based on case studies or pure quantitative methods. To overcome these drawbacks they propose the use of a concepts mapping methodology, which combines a qualitative case study approach and quantitative methods, to establish a lean green relationship. Furthermore, they focus in the European region using the referred methodology in a Hungarian company.

In this paper we intend to contribute to the discussion about the positive/negative impact of lean practices on corporate sustainability, replicating the work of McDaniel and Vastag (2010) in a Portuguese company.

2 The Case Study Company

This project was conducted in a Portuguese manufacturer (company X) which integrates a world's leading company in the field of bicycle components production. The main product manufactured by company X is the bicycle roller chain.

Actually company X employs approximately 100 people. The combination of modern equipments and highly trained employs, and a large experience in roller chains production, places company X among the largest manufacturers worldwide.

Company X relies on the systematic use of quality, production and maintenance tools to achieve a system of continuous improvement. They have a Quality Management and Environmental System implemented and are certified by the ISO 9001:2008 and ISO 14001:2004 standards. They have an internal structure dedicated to the questions of occupational health, safety and hygiene.

Company X has been implementing lean manufacturing tools and techniques since 1995. They have training programs for all employees covering subjects like lean manufacturing, environmental best practices and occupational health, safety and hygiene.

3 Concept Mapping

Concept mapping is a structured process, focused on a topic or construct of interest, involving inputs from one or more participants, which produces an interpretable pictorial view (concept maps) of their ideas and concepts and how these are interrelated. There are several methods that all go by names like “concept mapping” or “mental mapping”. A clear description of the method used in this project can be found in Trochim (1989).

In concept mapping, ideas are represented in the form of pictures or maps. A concept map is a pictorial representation of a group thinking which display all of the ideas of the group relative to the topic at hand, shows how these ideas are related to each other and shows which ideas are more relevant (Trochim 1989).

To construct the maps, ideas have to be described and the interrelationships among them articulated. Multidimensional scaling and cluster analysis are applied to this information and the results are depicted in map form. The content, interpretation and utilization of the concept maps are determined entirely by the participating group.

The concept mapping methodology used in this project is composed by six steps. The first step is the preparation step and it consists in identifying who the participant will be and developing the focus for the project. The second step usually consists in a brainstorming session where the participants generate statements that describe the conceptual domain for the topic of interest. In the third step the participants are asked to do two things. First they provide information about how the statements are interrelated by sorting them into piles of similar ones. Then the participants are asked to rank the statements in a scale ranging from 1—the statement is relatively unimportant—to five—the statement is extremely important. The fourth step, called the representation step is the one where the analysis is done and the conceptual domain is represented graphically. A number of interrelated maps are produced using two major statistical analyses: multidimensional scaling and cluster analysis. In step five the participants interpret the generated maps. The last step involves the utilization of the maps. For this project, the concept mapping process was conducted using the software “Concept System Professional”—see www.conceptsystems.com for more information on the software.

4 Concept Mapping in the Case Study Company

This project was conducted as follow. A first visit to company X plant was made to explain the project to top managers, to plan the sessions required to achieve the desired results and to select the participants.

Eleven participants were selected, in agreement with the company top management, representative of several different company functions: product development, assembly, environmental management, health and safety management, quality and production planning and control.

A first session was held with the project participants to generate statements about the impact of lean manufacturing on environmental and social performance. In this brainstorming session the participants were asked to generate statements about the impact of Lean Manufacturing tools and techniques in helping the company to become more socially and environmentally sustainable. To do so they were requested to complete the following prompt: "One specific way in which Lean Manufacturing tools and techniques have helped our company to become more socially and environmentally sustainable is that they…". This sentence is the same as the one used by McDaniel and Vastag (2010) to allow a comparison of results among both studies. During the session 90 statements were generated. The authors analysed the generated statement, correcting wording and eliminating overlapping, reaching a list of 81 statements which describe the conceptual domain for the impact that Lean manufacturing tools and techniques have regarding company X ability to become more socially and environmentally sustainable. After randomizing the 81 statements and removing the personal identifiers they were send to each participant to rank them in a scale ranging from 1 to 5 as described in the previous section.

In a second project session the authors had individual meetings with each of the participants to collect information about how the statements were related to each other. To do so an unstructured card sorting procedure, as proposed in Rosenberg and Kim (1975), was used. The statements were printed in cards. In the session each participant received the set of 81 cards with the generated statements and was instructed to sort them into piles "in a way that makes sense to you".

The result of each participant sorting task was put in a similarity matrix which has as many rows and columns as there are statements. If the participant as placed statement i and statement j in the same pile, the matrix cell i, j takes the value 1, otherwise it takes the value 0. Then the individual sort matrix are added together to obtain a combined group similarity matrix. In this matrix, containing as many rows and columns as statements, the value for cell i, j indicates how many times the pair of statements i and j were put together in a pile. Thus in this combined group similarity matrix, the cell values can range from 0—statement i and j were never placed in the same pile—to the number of participants—all participants has placed both statements in the same pile.

The results of both described sessions are used to represent the conceptual domain. A two dimensional nonmetric multidimensional scaling of the combined group similarity matrix is performed to obtain a "point map". The point map generated with the 81 statements of this project is presented in Fig. 1.

The point map displays the location of all brainstormed statements as dots on a page. Ideas which are represented by dots which are close to each other are expected to be similar in meaning. For example statement 17 "they have contributed to improve work safety and hygiene conditions" and statement 44 "they have led to an improvement in the visual aspect (cleaning) of the installations" represented by dots close to each other, in the upper part of the map, should be very similar in meaning.

A second analysis, called hierarchical cluster analysis is performed to represent the conceptual domain. In this analysis the statements on the point map are arranged in clusters of statements which presumably reflect similar concepts. We generated five cluster solutions, considering 4 clusters to 9 clusters. A third session with the participants was held to present the six generated cluster solutions. They analysed the 4, 5, 6, 7, 8 and 9 cluster solutions, examining how the statement were grouped together in each solution. After some discussion a consensus was reached among the group considering that the 8 cluster solution, presented in Fig. 2, could be considered the "best one".

The cluster map shows the statement points enclosed by polygon shaped boundaries for the 8 generated clusters. The yellow bar near each dot represent the average importance rating, in a scale from 1 to 5 as previously referred, across participant for each generated statement. Furthermore, the importance rating was averaged across each statement and each cluster to obtain the average importance of each cluster represented by the number of cluster layers. After this step the participants were asked to agree in a name for each cluster, the result of this operation is also presented in Fig. 2.

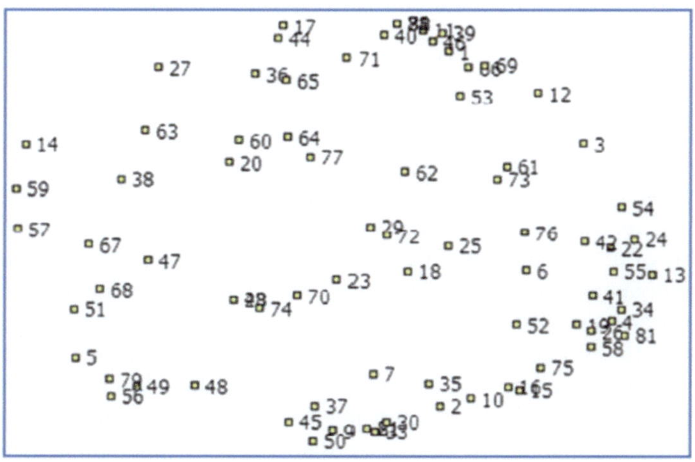

Fig. 1 Point map for the 81 generated statements

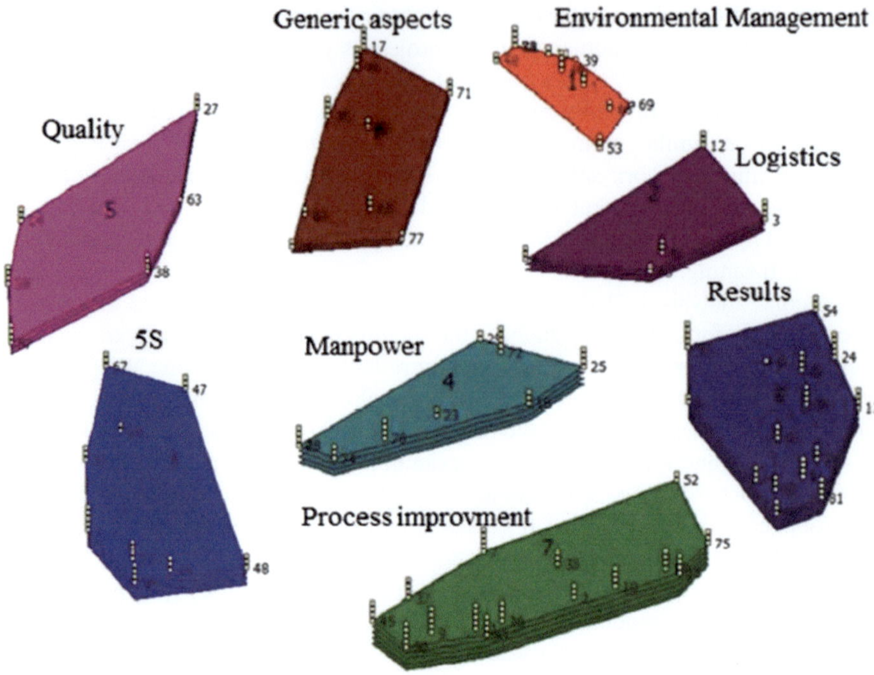

Fig. 2 Cluster map for the 81 generated statements

From Fig. 2 it is possible to find that the most important cluster was "process improvement", followed by "5S", "manpower" and "results". "Quality" was rated as 3 and the least important clusters were "generic aspects" and "environmental management", rated 2 and 1 respectively. The generated maps were analyzed. The results of this analyze is presented in the following section.

5 Discussion

In this section we discuss the results of the project, considering the cluster map presented in Fig. 2 and the sentences which integrate each cluster. This discussion is the result of a last brainstorming sessions held with the project participants to analyze the obtained cluster map. The participants started to concentrate themselves in the two highest ranked clusters: clusters 7 and 8. Both clusters integrate statements related to the company performance. Cluster 8 lead to the conclusion that, according to the participants, lean practices had contributed to an improvement of the economic performance of the company. Some sentences included in this cluster are related to the economic performance improvement of the company and to the improvement of customer service, essentially by reducing lead times. This cluster also integrates statements which refer an improvement in the company management practices. Thus, analyzing the cluster sentences, we can conclude that economic performance improvements were due to an increase of the efficiency of management methods. Sentences in cluster 7 indicate an improvement of the shop floor performance, in terms of: waste reduction, material flow improvement, equipment's availability and processes standardization. It is important to note that clusters 7 and 8 are close to each other, meaning that, according to the participants, the implementation of lean practices has led to new management practices, improving manufacturing processes which, in turn, has led to an improvement of the company economic performance and a better relationship with customers.

Near cluster 7, following the clockwise direction, we reach cluster 3 which includes a set of sentences essentially related to layout improvement. These two clusters are close to each other, meaning that the implementation of lean practices has contributed to a better layout which, in turn, allowed an improvement of manufacturing processes efficiency. Moreover, we verify that cluster 3 is near cluster 5 when following the clockwise direction. Cluster 5 is composed by statements related to the improvement of working conditions and to the reduction of quality problems. Thus, according to the participants, improvements in the shop-floor layout have contributed to a better working environment and a reduction of quality problems.

Following the same direction we reach cluster 6. In this cluster it is possible to find a set of sentences indicating improvements in working condition. Furthermore, this cluster already points to some aspects related to the social sustainability of the company, including statements that refer to influences that the implementation of lean practices had on the social recognition of the company and its positive influence on suppliers and customers. Finally, it is important to note that this cluster also starts to refer environmental results, resulting from the implementation of lean practices, mainly in what concerns the reduction of printed documents and of fuels consumption.

Thus, it is not strange, following the clockwise direction, to find cluster 1 which integrates statements associated to the company environmental performance. Cluster 1 is also near cluster 2, composed by a set of sentences related to the waste reduction achieved by the implementation of lean practices. Since cluster 2 is also near cluster 8, we can conclude that the improvement in management methodologies, resulting from lean practices implementation, has led to waste reduction which, in turn, contributed to a better company environmental performance.

Finally, it is important to note that cluster 4, composed by statements related to manpower, can be found in a central position of the cluster map, thus being close to all other clusters. This seems to indicate the importance of manpower both for the implementation of lean practices and to the achieved results.

Thus, interpretation of the generated cluster map allows to conclude that there is a direct relationship between the implementation of lean practices and the company social and environmental performance. The bottom left part of the map—clusters 5, 3, 7 and 8—enhance the performance improvements reached by the company with lean practices implementation, considering economics aspects, company/customer relationships and shop-floor processes. On the other hand, the upper part of the map—clusters 6, 1 and 2—indicates the improvement of company social and environmental performance achieved with lean practices implementation. All these clusters are near cluster 4—manpower—indicating it importance to an effective implementation of lean practices.

6 Conclusion

In concept mapping, ideas are represented in the form of pictures or maps. A concept map is a pictorial representation of a group thinking which displays all of the ideas of the group relative to the topic at hand, shows how these ideas are related to each other and shows which ideas are more relevant. In this paper we study the impact of lean practices implementation on business social and environmental sustainability, using a concept mapping approach. Eleven participants from a case study company were selected to create a concept map of their ideas about the relationship between lean practices and business sustainability. This study seems to point to a positive relationship between the adoption of lean practices and green practices already identified by other authors.

As referred previously this project replicates the work of McDaniel and Vastag (2010) which apply the same concept mapping methodology in a Hungarian company. The next step will consist in a detailed comparison between both studies to try to find some generalization of these project findings.

References

Azevedo S, Carvalho H, Cruz-Machado V (2011) The influence of green practices on supply chain performance: a case study approach. Transp Res Part E Logistics Transp Rev 47(6):850–871

Bergmiller GG, McCright PR (2009a) Parallel models for lean and green operations. In: Proceedings of the 2009 industrial research conference, Miami, FL, May 2009

Bergmiller GG, McCright PR (2009b) Are lean and green synergetic? In: Proceedings of the 2009 industrial research conference, Miami, FL, May 2009

Dües CM, Tan KH, Lim M (2013) Green as the new lean: how to use lean practices as a catalyst to greening your supply chain. J Clean Prod 40:93–100

Florida R (1996) Lean and green: the move to environmentally conscious manufacturing. Calif Manage Rev 39(1):80–105

King AA, Lenox MJ (2001) Lean and green? An empirical examination of the relationship between lean production and environmental performance. Prod Oper Manage 10(3):244–256

McDaniel TH Jr, Vastag G (2010) Is lean green? Measuring the impact of lean manufacturing on corporate sustainable development performance. In: 17th international Euroma conference—managing operations in service economies, Porto, Portugal, June 2010

Rosenberg S, Kim MP (1975) The method of sorting as a data-gathering procedure in multivariate research. Multivar Behav Res 10:489–502

Rothenberg S, Pil FK, Maxwell J (2001) Lean, green and the quest for superior environmental performance. Prod Oper Manage 10(3):228–243

Sarkis J (1995) Manufacturing strategy and environmental consciousness. Technovation 15 (2):79–97

Trochim WMK (1989) An introduction to concept mapping for planning and evaluation. Eval Program Plann 12:1–16

Zhu Q, Sarkis J (2004) Relationship between operational practices and performance among early adopters of green supply chain management in Chinese manufacturing enterprises. J Oper Manage 22(3):265–289

Explaining Alliance Success Factors in Spanish Food and Beverage Supply Chain: Case Analysis

Jesús Morcillo Bellido and Alfonso Durán Heras

Abstract Over the last decades, partnerships have captured great interest of both academics and practitioners, being a form of business management that could help companies to achieve their strategic goals beyond the level they can reach acting as isolated entities. But even though authors have been working for many years on the applicability of the different theories that could explain alliance success, still nowadays some questions about why the alliance success rate is much lower than failures rate arise. In this study, developed from an in-depth analysis of cases in successful partnerships experiences within Spanish Food and Beverage industry, authors have tried to identify relevant factors that could provide additional explanations to those causes that can generate a remarkable higher success rate in certain companies. The aim would be to explore if these factors could provide clues to understand what drives alliance success that have not been sufficiently studied yet.

Keywords Alliances · Food and beverage alliances · Spanish alliances

1 Introduction

In industries such as aeronautics, partnerships between supply chain members have been a standard practice for decades, so when two giants like Boeing and Lockheed have competed for US major military contracts many industry's analysts generally thought Boeing would win those contracts due to its high reputation in the Pentagon, its great financial strength and its ability to deliver complex projects, however in most cases Lockheed got the contract (Wittman et al. 2009).

J.M. Bellido (✉) · A.D. Heras
Escuela Politécnica Superior, Área de Ingeniería de Organización, Universidad Carlos III de Madrid, Avenida de la Universidad no 30, 28911 Leganés (Madrid), Spain
e-mail: morcillo@ing.uc3m.es

A.D. Heras
e-mail: duran@ing.uc3m.es

M. Amorim et al. (eds.), *Engineering Systems and Networks*, Lecture Notes in Management and Industrial Engineering, DOI 10.1007/978-3-319-45748-2_18

Why did it happen?, following some authors (Breen 2002), Lockheed top level skills in strategic alliance development and management, among others with Northrop Grumman and BAE systems, could be the main reason that explains it as they are able to continuously present solutions that exceeded their competitors. Following authors such as Das and Teng (2000) "business alliances" could be defined as any "collaborative efforts between two or more firms in which they pool their resources in a joint effort to achieve mutually consistent goals that they could not achieve easily acting alone". But despite the great importance given to partnerships, the percentage of failures is greater than 50 % as has been published in numerous studies (Day 1995; Morcillo and Duran 2014).

Being so important, it is advisable to study the theories that have been developed as "explanatory theories of successful partnerships" which can be listed as: (i) the resource-based view, focused on the role that partner's resources play in the alliance success and suggesting that the greater the complementarity that resources are, the higher the alliance success probability will be (Wernerfelt 1984), (ii) the competence—based view, which focuses attention on the availability of skills developed by partners that support and facilitate the alliance formation and success (Hamel and Prahalad 1994; Schoenmakers and Duysters 2006), (iii) the relational factors view, which concerns about internal factors of relationship between partners (Morgan and Hunt 1994) and (iv) the competitive advantage view that focuses on the competitiveness improvement through alliance formation, translating this into improved position of the partners in the market (Day 1995; Hunt and Morgan 1995, 1997).

Authors of this study consider that these four theories are not mutually exclusive but complementary like, among others, prominent authors such as Wittman et al. (2009) have pointed out. But also paper's authors consider that by combining the partial explanatory models, forming an "integrated model", some potential success factors catalysts may not have been sufficiently studied, such as the relevancy of "common strategic shared objectives" between alliance partners and the related factor of the existence of "long-term commitment" between partners, beyond the contracts that usually support partnerships relations.

Both are qualitative factors beyond those usually studied, such as issues related to trust, but perhaps they could become "strategic cultural ties", so that when they are present they could make specific partnerships more resilient to the difficulties inherent in any relationship.

2 Objective and Study Methodology

This study is part of a research line that aims to deepen on the understanding of the mechanism that lead to successful partnerships, and also to make proposals for improving management practices, potentially applicable to other companies. Initially, a prospective analysis (Morcillo and Duran 2014) was performed in order to identify if the mechanisms and good management practices that support alliance's management could be found within Spanish Food and Beverage (F&B) industry using a sample of nine multi-sized companies, now authors want to analyze in depth a sample of four big companies which based their strategy on alliances in order to make a deep analysis and understanding of their alliance practices. The sample covers alliance examples from Sovena, Logoplaste, Cargill and Mahou—San Miguel group.

Due to specific nature of this issue it was decided to carry out case studies based on in-depth interviews. This method, according to Eisenhardt (1989), Rialp (1998) and Voss et al. (2002), is very suitable for issues related to strategic management decisions. Yin (1994) suggests to use case study analysis when boundaries between the context and the phenomenon to observe are not obvious. Case studies have been gradually recognized, despite some criticism, as a correct way "to address contemporary organizational problems and established credibility" (Ketokivi and Choi 2014).

Information collection was done: (i) through senior company's executive in-depth interviews, using semi-structured questionnaires, guaranteeing that information would be used only for academic purposes and (ii) secondary published data; this approach was applied in all cases in a similar manner.

Given the nature of the topic and the difficulty of accessing the information source, often considered as highly confidential, companies selection was decided according to criteria such as: company relevancy within the sector and authors accessibility to organization's management.

3 Spanish Food and Beverage Case Analysis

Analysis has been focused on the following partnerships: Sovena with Mercadona, Logoplas with Sovena, Cargill with Hojiblanca and Grupo Mahou with several companies. Sovena is a fairly unknown company at consumer level and it could be a surprise when newspapers published that this company plans to grow one million new olive trees in Cordoba, Spanish province, in 2015–16. But it should not seem so strange considering that it is one of the oil industry leaders in Spain, being the inter-supplier for Mercadona cooking oil range and that this company seeks to be more self-sufficient in the supply of raw material to reduce its dependence on oil mills and cooperatives.

This oil group, based in Portugal, is the world's largest supplier of olive oil for distributor owned brands, with 1.2 billion euros total sales through this channel. In the Spanish market, Sovena sold 166.000 million litters of olive oil to Mercadona under Hacendado brand in 2013, out of 213.000 million litters sold by Sovena Spain.

Mercadona has developed a very special partnership model with around 100 first tier suppliers called "inter-suppliers" and in the case of Sovena the relationship started in 2009, when Mercadona took the decision of having only a solid and unique provider for oil. Sovena was contacted to discuss terms and conditions of this relationship, however Sovena was already a supplier from the moment in which its bought Agribética company few years earlier.

By analyzing capabilities that Sovena has developed in this inter-supplier relationship, some highly outstanding could be identified: (i) greater focus on quality excellence as the result of the requirements and standards from its client, (ii) strict process control, following a key point for Mercadona's strategy that established as not acceptable to maintain within inter-suppliers even the smallest area of waste, assuming this would most likely mean a higher cost, with subsequent impact on the final customer price. Suppliers are forced to systematically analyse potential improvement areas as part of its "Total Quality" programs, and (iii) active participation in the plan launched by Mercadona to integrate second and third tier suppliers, which goes up till the raw materials providers with the aim to integrate them into their improvement plans. This forces Sovena to develop upstream actions to set quality standards in the processes of its suppliers.

When searching into this partnership foundation, it became apparent that a very relevant reason for both companies, when they decided to enter into a model of collaborative relationship so tight and "for life", clearly was the fact that both companies had a shared strategic vision and very similar long-term objectives (they are seeking for growth and long-term economic sustainability, while avoiding decisions based on short-term results). This is largely influenced by the management style that both companies have implemented, being family businesses.

However, it should be pointed out that the implementation of this collaborative relationship has been complex due to Sovena's need of a gradual adaptation to Mercadona's highly demanding requirements which is continuously searching for maximum efficiency, and pressure on the whole organization to be able to support customer growth and high speed decisions. Sovena has not developed any formal structure for alliance's capacity development (such as formal procedures, partnership office, etc.) like some studies recommend (Kale et al. 2009), even alliance formal procedures do not exist and this could be explained by the fact that they do not have an alliance strategy to develop many partnerships that would require identification, assessment and control, but they have developed a specific "mono-alliance" applicable when there is a unique business opportunity.

This way of thinking made its "alliance strategy" so "lean" that no more infrastructure is required than a business involved CEO and his management team. CEO's involvement is absolute, from the first informal conversation with Mercadona until regular follow up meetings to discuss improvement plan implementation as inter-supplier. Contrary to what might be supposed, there is no perception of lost power or self-control due to the fact there is a single client in Spain, however it is understood that there was an opportunity for taking advantage through this relationship and this is reciprocal in the client side.

For both companies this experience was perceived as positive, both have grown and in the case of Sovena sales have increased sharply (675 million euros in Spain in 2013, 45 % over previous year), with narrow but stable margins. On top of that, they have developed alliance skills which are being applied to enter into other markets, such as the new relationship with other major distributors in USA.

Logoplaste is also a Portuguese family company whose business is to manufacture, in the client's plant, the plastic bottles needed to deliver product to final consumers. As an example of its way of doing business, in 2008 Logoplaste proposed bottle redesigns for Unicer's Vitalis at Caramulo (Portugal), so that these water brands could reduce the total amount of raw materials used for each unit.

This company has established its "production line" inside Sovena manufacturing's plants, so they produce from resins or preformed bottles in a completely synchronized manner finished products demanded by Sovena production lines in the plant, as a clear practice of "lean operations". The agreement is based on a type of contract called "open book", in which all information regarding the bottle's manufacturing costs is shared with the client and based on this, they jointly commit to: (i) improvement actions, usually in conjunction with the client and mainly focused on cost savings and (ii) a specific business net result, usually ranging between 3 and 5 % of the turnover.

This agreement includes implicit commitments to be assumed such as: (a) collaboration in the development of new products and improvement of the existing range, (b) to be able to compete on price and service, with any new supplier wishing to go into business at any moment, (c) participation in the internal client's processes, such as in the production planning, with the idea of supporting client improvements and (d) be willing to make new investments to grow along with the client's growth.

This company has developed, and continuously improved, this way of working that not only applies to Sovena's agreement but also to a few other customers including Sovena's competition. According to those interviewed, the model works based in certain pillars such as: (i) a shared business vision, (ii) a strong trust between the parties in daily business, not only at top level but also at the middle management level, and (iii) there is an excellent personal relationship between both organization's CEOs. These facts could be considered as the "glue" that solidifies the principles of the partnership, which certainly as it could be imagined is based on a long term agreement due to the huge specific investment amount and the implied risks involved.

Another case studied was a large company, Cargill, which "provides food, agriculture and services to the World", the company manages worldwide a broad business related to animal nutrition, cereals, starches and glucose, oils and various types of food ingredients, many of them performed under alliances. Cargill has developed a worldwide network of collaborative relationships based on seven platforms, according to their business groups, which acts as a catalyst, promoter and leader of many alliances.

Among tools to promote partnerships Cargill provides funding and technical advice to many small providers as part of its collaborative network. In this study, it has been studied particularly an important alliance held with Hojiblanca, one of the leaders in the Spanish olive oil industry. Through this alliance, Hojiblanca got access to an extensive distribution network and Cargill had complemented its wide product range with an excellent Spanish olive oil. So, it was a clear theoretical example that fits within the resource based—view and other explanatory theories of alliances success.

This alliance had serious development problems when factors that could weaken the relationship appeared, and one of them was the fact that oil was just something complementary for Cargill. According to authors who have worked on the relevancy of "resources complementarity" in the resource-based theory, the fact of needing the partner would be very important as a motivating factor, but in the opinion of the authors of this paper—after extensively discussing the issue with managers involved in such kind of situations—this complementary resources alliance must have a certain symmetry, so a similar level of partner priority is key as motivating factor.

Furthermore, both companies wanted to lead the alliance almost without the other partner, in the case of Hojiblanca because of their expertise in the sector and in the case of Cargill based on its business culture that could only accept the leadership role in alliances.

Another case studied was Grupo Mahou—San Miguel, leader in the Spanish beer market, they operate in a mature market in which the pursuit of growth through partnerships is a key strategic lever for this company, whose business is threatened by the growth of distributor brands ("white brands") which accounted for about a third of the Spanish market in 2014.

They have conducted numerous alliances in the last five years, such as: (i) Carlsberg, through which this brand tried to get Spanish market distribution and Mahou a margin for its role as distributor network, while Mahou could use Carlsberg's distribution network in the UK by paying a certain fee, (ii) Coors, managing manufacturing and distribution for this brand in Spain, (iii) Sara Lee, through an agreement for the capillary joint distribution in bars and restaurants, (iv) Intermalta, partnering with other competitors to improve the supply of raw materials, (v) Corona, joint marketing and distribution in Spain and Mexico, (vi) Warsteiner alliance for distribution in Spain, and (vii) joint venture in India to develop a comprehensive beer business in the country, which ended in 2014 with the acquisition by Mahou of the capital owned by Indian investors.

Its strategy is based on the development of a multitude of partnerships, for achieving specific strategic targets and currently they have plans to carry out alliances in geographical areas where their market share is still weak (like Asia and Latam) or to cover gaps in European innovative products.

4 Discussion and Conclusions

So far, theories related to resources, competences, competitive advantages and relational factors have been extensively studied, but we believe that probably the high relevance of the existence of "common and shared strategic objectives and long-term commitment" between partners that go beyond formal contracts has not been sufficiently studied. From the case studies it could be inferred that, apparently, a "different" collaborative relationship between some companies in the Spanish Food and Beverage industry is emerging and this could be considered a higher level of integration, and this goes beyond what could be captured through the partnership's explanatory models referenced in this paper's introduction.

Through analysis of the described case studies, it could be concluded that Cargill and Grupo Mahou are both companies with a broad partnership base, in which alliances are organized and work as a large and dynamic collaborative relationships network, they lunch and manage a partnership while there is a specific need for collaboration. Nevertheless, Cargill and Grupo Mahou's partners are far from participating in their strategic vision and long term objectives, as they belong to an extensive relationship network participating as merely necessary element but not critical at all in order to achieve their long term strategic objectives. Cargill possibly could replace many of them by companies with similar characteristics without jeopardizing the company's alliance results. From Sovena and Logoplaste's experiences, it follows another explanatory model of alliances including some refinements and with remarkable success, where partners are part of a network highly committed in the achievement of strategic objectives and integrate into its supply chain. They also share a long-term business vision, take joint risks, trust the others and expect long-term results. Their relationship could be defined as "for life" and it seems to be like this in reality.

It could be inferred that this relationship model, developed by companies like Sovena and Logoplaste (where they share a vision and long-term client strategic objectives) shows a much higher robustness than what could be found in Cargill and Mahou alliance case studies. This "strategic integration" is possible only from voluntary, indubitable and committed collaboration in achieving long term strategic objectives that always must be in line to the supply chain leader's vision—role is taken by the partner closest to the end customer—and the existence of a strong mutual commitment for their achievement.

Both factors could mean an additional added value in the explanatory models of successful partnerships, particularly within the relational model. This circumstance will be a main subject in a broader study to be carried out on these and other alliance cases, in order to prove that the explanatory power of this conclusion goes beyond the field study performed in these companies.

References

Breen B (2002) High stakes, big bets. Fast company, pp 66–78
Das TK, Teng B (2000) A resource-based theory of strategic alliances. J Manage 26(1):31–61
Day GS (1995) Advantageous alliances. J Acad Mark Sci 23(4):297–300
Eisenhardt KM (1989) Building theories from case research. Acad Manage Rev 14(4):532–550
Hamel G, Prahalad CK (1994) Competing for the future. Harvard Bus Rev 72(4):122–128
Hunt SD, Morgan RM (1995) The comparative advantage theory of competition. J Mark 59:1–15
Hunt SD, Morgan RM (1997) The resource-advantage theory of competition: a snake swallowing its tail or a general theory of competition? J Mark 61(3):74–82

Kale P, Singh H, Bell J (2009) Relating well: building capabilities for sustaining alliance network. In: Kleindorfer PR, Wind Y (eds) The network challenge: strategies for managing the new interlinked enterprise. Pearson Press, London

Ketokivi M, Choi T (2014) Renaissance of case research as scientific method. J Oper Manage 32:232–240

Morcillo J, Duran A (2014) Success factors in alliances: challenges in the Spanish Food and Beverage Industry. In: Proceedings 8th international conference on industrial engineering and industrial management, XX International conference on industrial engineering and operations management, International IIE conference proceedings 2014. Málaga, 23–25 July 2014

Morgan RM, Hunt SD (1994) The commitment-trust theory of relationship marketing. J Mark 58 (3):20–38

Rialp A (1998) El método del caso como técnica de investigación y su aplicación al estudio de la función directiva. IV Taller de Metodología. Sesión: Técnicas de la Investigación Cualitativas. Análisis de casos. La Rioja

Schoenmakers W, Duysters G (2006) Learning in strategic technology alliances. Technol Anal Strateg Manage 18(2):245–264

Voss C, Tsikriktsis N, Frohlich M (2002) Case research in operations management. Int J Oper Prod Manage 22:195–219

Wernerfelt B (1984) A resource—based view of the firm. Strateg Manage J 5(2):171–180

Wittmann CM, Hunt SD, Arnett DB (2009) Explaining alliances success: competences, resources, relational factors and resource-advantage theory. Ind Mark Manage 38:743–756

Yin RK (1994) Case study research: design and methods. Sage, Beverly Hill

How to Design an Efficient and Sustainable Box?

Jesús García-Arca, A.T. González-Portela Garrido, J.A. Comesaña-Benavides and J.C. Prado-Prado

Abstract Packaging is one of the transversal elements that supports an efficient and sustainable supply chain. However, there are few methods to objectively measure the impact of packaging design on this efficiency and sustainability. In this paper, a method for designing boxes is presented (ESB, "Efficient and Sustainable Box"). Going beyond proposing a theoretical method, the authors have testing it in a company following the "action research" approach.

Keywords Packaging · Logistics · Supply chain · Sustainability · Box

1 Introduction

Today, supply chains must propose initiatives to improve competitiveness and sustainability by reducing activities that do not provide value and deploying innovations in processes and products.

In this context, packaging seems to be a key element that supports an efficient and sustainable supply chain management (Klevas 2005). Going beyond the important protection of the product, packaging should be designed, not only to support differentiation, but also to increase efficiency in logistics and for helping the activities of reusing, recycling and recovering according to legislation (for instance, the European Directive 94/62 and its update, 2004/12/EC).

Jönson (2000) and Garcia-Arca and Prado-Prado (2008) proposed in relation to packaging design three main functions: commercial, logistics and environmental. Saghir (2002), Hellstrom and Nilsson (2011), García-Arca and Prado-Prado (2008), Azzi et al. (2012) and García-Arca et al. (2014) identify the different requirements that the companies and the supply chains require the packaging. However, these

J. García-Arca (✉) · A.T. González-Portela Garrido ·
J.A. Comesaña-Benavides · J.C. Prado-Prado
Grupo de Ingeniería de Organización (GIO), Escuela de Ingeniería Industrial,
Universidad de Vigo, Campus Lagoas-Marcosende, c/Maxwell 9, 36310 Vigo, Spain
e-mail: jgarca@uvigo.es

M. Amorim et al. (eds.), *Engineering Systems and Networks*, Lecture Notes
in Management and Industrial Engineering, DOI 10.1007/978-3-319-45748-2_19

requirements are not distributed equally in the different levels of packaging, which require an integrated view of packaging, logistics and product. This new approach has promoted the development of the term "packaging logistics" (Saghir 2002).

García-Arca et al. (2014) propose the three main pillars that should ease the implementation of "Packaging Logistics":

- The definition of all design requirements;
- The adoption of a coordinated organizational structure for packaging and new product design;
- The definition of a system, both to measure the impact on one packaging design and to compare different alternatives from a global view.

Unfortunately, there are few techniques to assess the impact of packaging design on the efficiency and sustainability of the supply chains.

Thus, this paper proposes an analysis method for designing efficient and sustainable boxes. Going beyond, the authors have testing it in a company following the "action research" approach; that means, that the authors participate in the "packaging logistics" implementation in the company, leading the project. Thanks to this involvement, the researchers have the opportunity to witness the process, not only as mere observers, but also as real "agents of change" in intervention and knowhow compiling processes (Coughlan and Coghlan 2002).

2 A Method for Designing an Efficient and Sustainable Box

When we talk about packaging, we are really talking about a complex system deployed around three levels: primary packaging (packaging in contact with the product), secondary packaging (a level for containing some primary packages; for example, a box) and tertiary packaging (some primary or secondary packages stacked on a pallet or roll container) (Jönson 2000).

Likewise, a good packaging should contribute to decrease global supply chain costs. In many situations, costs are a way to value a choice among different alternatives. These costs can be both direct (for instance, purchase of packaging and waste management) and indirect (packing, losses due to bad packaging, handling, warehousing, and transport throughout the supply chain, including the shops and supermarkets) (García-Arca et al. 2014). These last indirect costs prevent some companies from understanding the importance of selecting a good packaging; going beyond, many companies even do not know the money involved in all these costs (García-Arca and Prado-Prado 2008).

However, it is not easy to measure in economic terms some type of decisions associated to packaging design from other points of view; for instance, the analysis of the losses due to a bad packaging, the customers' satisfaction or the environmental impact. Regarding the environmental impact, we can find some specific techniques such as Life Cycle Assessment (LCA; ISO 14040 2006) and, even, we can value partially some environmental costs thanks to Green Dot or the returnable packaging systems.

In literature, we can find different assessment proposals in order to face the difficulties associated to an objective comparison among packaging alternatives from an overall point of view. These proposals often include quantitative and qualitative scales (Grönman et al. 2013). The most well-known of these proposals is the "Packaging Scorecard" (Olsmats and Dominic 2003), adopted by companies such as IKEA or Wal-Mart.

However, Saghir (2002), Azzi et al. (2012) highlight the lack of techniques for assessing global efficiency and sustainability, in order to select the best packaging alternative. Thus, a good packaging alternative should simultaneously include a sustainable design with an economic cost. That means, in practice, the implementation of different assessment techniques (Grönman et al. 2013; Pålsson et al. 2013).

The authors consider global costs a good way to value packaging options, although not all type of design requirements could be analyzed under the costs perspective; for example, when assessment considers environmental, ergonomic or commercial requirements. However, the authors propose an assessment technique based on cost measurement. We think that this approach eases the selection of alternatives, supplying good solutions, although not necessarily the best one.

Our method proposes how to design a box that supports the development of a sustainable supply chain, combining both logistical efficiency (mainly through cost reduction) and environmental efficiency. The proposed method is based on these two approaches ensuring, as a previous hypothesis, that all the considered alternatives meet the protective requirements of the product and the commercial needs of the market. The method is named ESB ("Efficient and Sustainable Box").

Firstly, the logistical efficiency of a box (secondary packaging) would be related to the optimization of the packaging system, in terms of an efficient use of volume between the primary packaging, the secondary packaging and the tertiary packaging. The indicator to measure this efficiency could be synthesized in the amount of product per pallet, secondary packaging (box) or/and primary packaging. This approach, called "Efficient box", would reduce the costs of packaging, handling, storage, picking and transport throughout the chain, including shops.

On the other hand, the environmental efficiency of a box would be related to the rationalization of cardboard consumption required to market the products. This second approach ("Sustainable box") would not only reduce the consumption of raw materials and waste, but also the rationalization of energy consumption throughout the chain, including fuel and transport emissions. The indicator to measure this efficiency could be the number of cardboard surface per box volume.

In relation to the first approach ("Efficient Box"), ISO (International Standardization Organization) has proposed some standard pallets; among them, the pallets 800 $*$ 1200 mm (or pallet EUR) and 1000 $*$ 1200 mm should be highlighted. In this context, the selection of packaging dimensions (in boxes) based on module 600 $*$ 400 mm (including multiples) allows the adjustment to the above two pallets

Table 1 Modular dimensions in the base of a box according to ISO 3394 (2012)

600 * 400(4)	600 * 200(8)	600 * 133(12)	600 * 100(16)
300 * 400(8)	300 * 200(16)	300 * 200(24)	300 * 100(32)
200 * 400(12)	200 * 200(24)	200 * 133(36)	200 * 100(48)
150 * 400(16)	150 * 400(16)	150 * 133(48)	150 * 100(64)
120 * 400(20)	120 * 400(20)	120 * 133(60)	120 * 100(80)

The number of boxes per layer in an EUR pallet is presented between brackets

(ISO 3394 2012; see dimensional alternatives in Table 1). Furthermore, going beyond the EUR pallet implementation in the European retail market, the 600 * 400 module also extends to the dimensions of other elements such as the trolley (or "roll"), the shelves or the returnable packaging.

Obviously, besides the combinations of the modular system 600 * 400, there are other efficient solutions for a specific type of pallet; for example, the dimensions of packaging 266 * 400 mm or 266 * 600 mm (and its submultiples) are not included in the modular system 600 * 400 but, however, they place efficiently in an EUR pallet (although not in the 1200 * 1000 pallet). In the proposed model these other alternatives are not included. Besides the length and width in the box, we also should pay attention to the height of the box. In this sense, we have chosen as a reference the standard RAL ("AECOC Recommendations for Logistics") in the chapter "Efficient Load Units" (2012). In this reference, AECOC (Spanish Association for Commercial Coding) proposes, as a general rule, that the maximum height for a EUR load should not exceed 2.6 m (maximum height for a pallet in a truck; 1.3 m if the EUR loads are stacked in the truck). In practice nowadays, efficient height for a EUR load (without stacking) would be in a range between 2 and 2.1 m depending on the sector and the type of the supply chain.

With these dimensional constraints in mind, it would be possible to find the "efficient box", combining both the modular base, for fitting on the pallet, and the useful height of the pallet. This comment implies that, after discounting the wood height of an empty pallet (145 mm in EUR pallet), it is possible to fix the height of the box depending on the number of layers of boxes stacked on the pallet.

On the other hand, the second approach of the proposed model is the concept "Sustainable Box". In this second approach, the outer dimensions of each of these

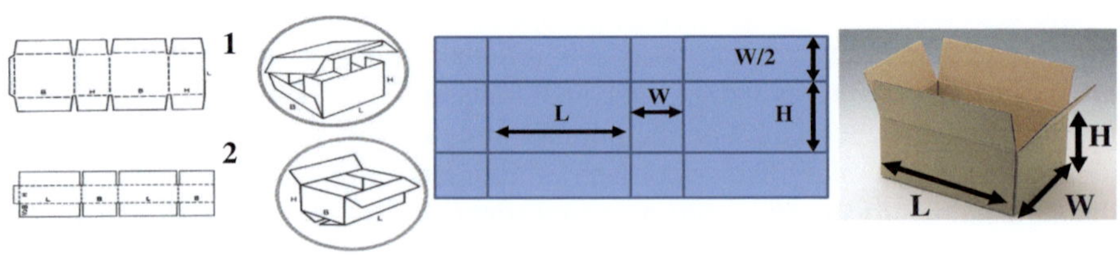

Fig. 1 The most common types of boxes: "wrap around" *box* (*1*) and "manual" *box* (2). *Source* AFCO. The cardboard surface and dimensions are presented on the *right*

boxes (length (L); width (W), height (H)) should rationalize the needed cardboard surface for building the box of a specific volume (V). This volume is used for containing the products and the primary packaging. In Fig. 1, the most common formats and shapes of boxes are presented: the manual one and the automatic one (built in "wrap around" machines). The theoretical expression for this cardboard surface would be: $2 * (W * H + L * H + W^2 + W * L)$; obviously, in this last expression, the final volume (V) is $L * W * H$. Likewise, for a specific volume (V) it would be possible to identify mathematically the optimal relationship among length (L), width (W) and height (H) that minimizes the consumption of cardboard surface in the box. In the most common cases (manual or automatic boxes), these optimal dimensions are: $L = (2 * V)^{1/3}$; $W = (V/4)^{1/3}$; $H = (2 * V)^{1/3}$.

With the above approach, it is possible to develop a graph presenting the relationship between the optimum cardboard surface and the external volume in a box (see Fig. 2). Below this optimal line, there is a no feasible area where it is not possible to find better solutions than the optimal alternative; therefore all feasible alternatives will be above this line that marks the optimal alternatives, including some alternatives of "efficient boxes" in terms of pallet efficiency.

Furthermore, the analysis of the ratio m^2/m^3 according to each volume allows to identify that this value decreases as the volume increases; in this sense, ensuring as a prior step the fulfillment of the commercial and ergonomic constraints, the implementation of bigger secondary boxes helps to reduce not only the consumption of cardboard and the generation of waste, but also the costs of handling and picking throughout the supply chain, including the shops.

For calculation purposes, in the practical development of the ESB method only the theoretical outer dimensions of the boxes are considered; in fact, each of these theoretical dimensions should be decreased in a range between 2 and 7 mm depending on the level of bulging and stiffness of the box (cardboard type) to avoid protruding from both the pallet base and the top height of the load. Obviously, in the proposed method could prevail the rationalization of the material (purchases of packaging, costs of waste management and picking,...) over the efficient use of the loading unit (costs of handling, storage,

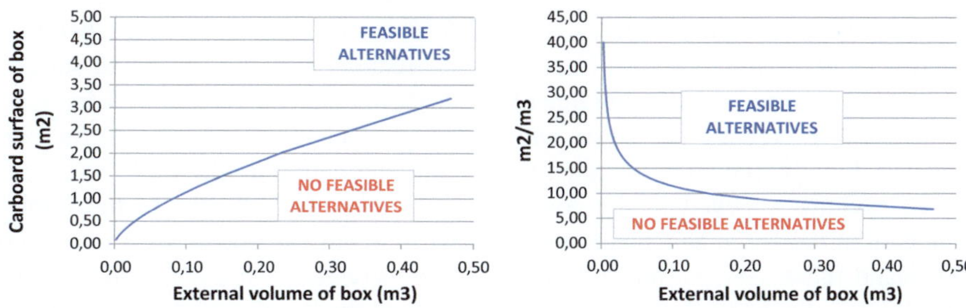

Fig. 2 Relationship between volume and cardboard surface of a box (*left*); Relationship between the ratio m^2/m^3 and the volume of a box (*right*)

transportation,…). In the application of ESB method, a qualitative consensus that improves the initial box will be sought.

3 Testing the Method

The Spanish company is specialized in the processing of frozen products. In 2014, the company employed 300, and its revenue reached 80 million euros, thereby positioning itself in the top 15 Spanish frozen food companies.

To illustrate the potential of the proposed method, we selected 2 boxes of 2 relevant products for its main client. The 2 products follow the same structure of packaging; as primary packaging, the bag; as secondary packaging, the cardboard box; and as tertiary packaging, the EUR pallet. We have chosen products with bags because this format allows more flexibility for analyzing alternatives in comparison with other alternatives of primary packaging such as the paper case, the can or the plastic tray. With the proposed method are simulated alternatives between 1 and 10 layers of boxes per pallet and a pallet maximum height of 2.1 m (including the height of pallet). This height is set by the client (see Table 2).

In the first box ("Box 1"), the starting solution was efficient in theory; a combination of modular dimensions is used and an efficient height has been chosen; however, with the consent of the client, other alternatives contemplating the increase of the number of bags in each box have been explored, following the ESB method. In this regard (see Fig. 3), a better box has been selected and implemented. This solution allows an increase of volume per box of 14 %; likewise, the number of bags per box is increased by 20 % and the number of bags per pallet by 5 %. To achieve these results, it has only been necessary to increase a 7.9 % the amount of cardboard on the new box, improving the ratio m^2/m^3 by a 5.7 %.

In the second box ("Box 2"), the starting situation was not as efficient as the "Box 1", due to the "no modular" dimensions of the box. With the ESB method two different types of improvements were discussed with the company and the client (see Fig. 3). On the one hand, keeping the number of bags per box (improvement 1); this solution implies a 19.2 % reduction in the consumption of cardboard surface. On the other hand, increasing a 62.5 % in the number of bags per box (improvement 2); this proposal implies a 3 % reduction of the cardboard surface with a 42.7 % increase of the volume of the box; simultaneously, a 21.9 % increase in the number of bags per pallet and a 32 % reduction in the ratio m^2/m^3 are achieved. Finally, the client selected the second alternative in "Box 2".

The savings achieved in the company and in its client thanks to the implementation of these two new boxes are estimated at about 146,000 euros per year. To understand these savings, the following data are presented: the company handles annually 315,000 boxes less than before throughout the supply chain (including the outlets); the company stores and transports 1330 pallets less than before; the company needs 77 cardboard tonnes less than before.

Table 2 Comparison between initial boxes and improved ones (the percentage of improvement in relation to initial box is presented between brackets)

Box (mm) (L*W*H)	Boxes/pallet	Bags/box	Bags/pallet	Box volume (m^3)	Surface of box (m^2)	m^2/m^3
Initial Box 1 600 * 200 * 244	64	15	960	0.0293	0.710	24.26
Improved Box 1 600 * 200 * 279	56	18 (+20 %)	1008 (+5 %)	0.0335 (+14.3 %)	0.766 (+7.9 %)	22.89 (−5.7 %)
Initial Box 2 375 * 300 * 139	112	8	896	0.0156	0.593	37.90
Improved Box 2 (1) 300 * 200 * 279	112	8 (0 %)	896 (+0 %)	0.0168 (+7.1 %)	0.479 (−19.2 %)	28.61 (−24.5 %)
Improved Box 2 (2) 400 * 200 * 279	84	13 (+62.5 %)	1092 (+21.9 %)	0.0223 (42.7 %)	0.575 (−3 %)	25.75 (−32 %)

	Efficient Box (Modular System according to ISO 3394; maximun pallet height: 2.1 m.)									Sustainable box				
	1.- Length	2.- Width	3.- Layers/ pallet	4.- Height ((2100-145 mm.)/[3])	5.- Boxes / layer	6.- Boxes/ pallet ([3]*[5])	of efficient Box (m3) ([1]*[2]*[4])	Cardboad surface in efficient box (m2)	9.- m2/m3 ([8]/[7])	10.- Optimal lenght	11.- Optimal width	12.- Optimal Height	13.- Volume of Box (m3) ([10]*[11]*[12])	14.- Cardboad surface in sustainable box (m2)
	...				...	...				...			...	
Box 2	120	133	2	977,5	60	120	0,0156	0,56	36,02	249,87	157,41	249,87	0,0098	0,33
	300	133	5	391,0	24	120	0,0156	0,45	29,09	249,87	157,41	249,87	0,0098	0,33
	600	133	10	195,5	12	120	0,0156	0,48	30,87	249,87	157,41	249,87	0,0098	0,33
	200	200	5	391,0	24	120	0,0156	0,47	30,23	250,08	157,54	250,08	0,0099	0,33
	120	400	6	325,8	20	120	0,0156	0,75	48,27	250,08	157,54	250,08	0,0099	0,33
	120	400	6	325,8	20	120	0,0156	0,75	48,27	250,08	157,54	250,08	0,0099	0,33
	150	400	7	279,3	16	112	0,0168	0,75	44,59	255,90	161,21	255,90	0,0106	0,35
	300	200	7	279,3	16	112	0,0168	0,48	28,60	255,90	161,21	255,90	0,0106	0,35
	150	400	7	279,3	16	112	0,0168	0,75	44,59	255,90	161,21	255,90	0,0106	0,35
	600	100	7	279,3	16	112	0,0168	0,53	31,69	255,90	161,21	255,90	0,0106	0,35
	...				...	...				...			...	
	600	133	7	279,3	12	84	0,0223	0,60	27,12	281,42	177,28	281,42	0,0140	0,42
	200	400	7	279,3	12	84	0,0223	0,82	36,48	281,65	177,43	281,65	0,0141	0,42
	...				...	...				...			...	
	150	100	1	1950,0	64	64	0,0293	1,03	35,04	308,11	194,10	308,11	0,0184	0,50
	300	100	2	977,5	32	64	0,0293	0,86	29,39	308,38	194,26	308,38	0,0185	0,51
	150	400	4	488,8	16	64	0,0293	0,98	33,34	308,38	194,26	308,38	0,0185	0,51
	300	200	4	488,8	16	64	0,0293	0,69	23,49	308,38	194,26	308,38	0,0185	0,51
	150	400	4	488,8	16	64	0,0293	0,98	33,34	308,38	194,26	308,38	0,0185	0,51
	600	100	4	488,8	16	64	0,0293	0,82	28,11	308,38	194,26	308,38	0,0185	0,51
Box 1	300	400	8	244,4	8	64	0,0293	0,90	30,76	308,38	194,26	308,38	0,0185	0,51
	600	200	8	244,4	8	64	0,0293	0,71	24,25	308,38	194,26	308,38	0,0185	0,51
	120	133	1	1950,0	60	60	0,0311	1,05	33,87	314,55	198,15	314,55	0,0196	0,53
	600	133	5	391,0	12	60	0,0312	0,77	24,62	314,82	198,32	314,82	0,0197	0,53
	120	400	3	651,7	20	60	0,0313	1,09	34,97	315,08	198,49	315,08	0,0197	0,53
	120	400	3	651,7	20	60	0,0313	1,09	34,97	315,08	198,49	315,08	0,0197	0,53
	200	400	5	391,0	12	60	0,0313	0,95	30,35	315,08	198,49	315,08	0,0197	0,53
	300	400	7	279,3	8	56	0,0335	0,95	28,38	322,41	203,11	322,41	0,0211	0,55
	600	200	7	279,3	8	56	0,0335	0,77	22,88	322,41	203,11	322,41	0,0211	0,55
	...				...	...				...			...	

Fig. 3 Alternatives of *two boxes* proposed by the ESB method (in the table, only the range of volume of *two boxes* is included)

4 Conclusions

The packaging design for its "multi-functionality" and "mainstreaming" should be considered as an element that actively contributes to the efficiency and sustainability of supply chains. The deployment of the approach "packaging logistics" demands simple measurement systems that provide the "objective" selection of efficient and sustainable boxes. This is the objective of the method presented in this paper, which proposes a box design from two perspectives: firstly, from the product until the pallet; on the other hand, conversely, from the pallet until the product.

References

AECOC (2012) RAL. Recomendaciones AECOC para la Logística. Unidades de Carga Eficientes. Edited by AECOC (Barcelona), Spain

Azzi A, Battini D, Persona A, Sgarbossa F (2012) Packaging design: general framework and research agenda. Packag Technol Sci 25:435–456

Coughlan P, Coghlan D (2002) Action research for operations management. Int J Oper Prod Manage 22(2):220–240

García-Arca J, Prado-Prado JC (2008) Packaging design model from a supply chain approach. Supply Chain Manage Int J 13(5):375–380

García-Arca J, Prado-Prado JC, Gonzalez-Portela Garrido AT (2014) "Packaging logistics": promoting sustainable efficiency in supply chains. Int J Phys Distrib Logistics Manage 44(4):325–346

Grönman K, Soukka R, Järvi-Käriäinen T, Katajajuuri J-M, Kuisma M, Koivupuro H-K, Ollila M, Pitkänen M, Miettinen O, Silvenius F, Thun R, Wessman H, Linnanen L (2013) Framework for sustainable food packaging design. Packag Technol Sci 26(4):187–200

Hellström D, Nilsson F (2011) Logistics-driven packaging innovation: a case study at IKEA. Int J Retail Distrib Manage 39(9):638–657

ISO (2006) ISO 14040:2006. Environmental management—Life cycle assessment—Principles and framework. Edited by ISO, Geneva

ISO (2012) ISO 3394:2012. Packaging. Complete, filled transport packages and unit loads. Dimensions of rigid rectangular packages. Edited by ISO, Geneva

Jönson G (2000) Packaging technology for the logistician, 2nd edn. Department of Design Sciences, Division of Packaging Logistics, Lund University, Lund

Klevas J (2005) Organization of packaging resources at a product-developing company. Int J Phys Distrib Logistics Manage 35(2):116–131

Olsmats C, Dominic C (2003) Packaging scorecard—a packaging performance evaluation method. Packag Technol Sci 16:9–14

Pålsson H, Finnsgård C, Wänström C (2013) Selection of packaging systems in supply chains from a sustainability perspective: the case of volvo. Packag Technol Sci 26:289–310

Saghir M (2002) Packaging logistics evaluation in the Swedish retail supply chain. Edited by University of Lund, Lund

Interoperability Frameworks in Public Administration Domain: Focus on Enterprise Assessment

Alexandre da Costa Castro, José Marcelo Almeida Cestari, Eduardo Rocha Loures, Edson Pinheiro de Lima and Eduardo Alves Portela Santos

Abstract For over a decade, the advent of e-Government increased the need to address its typical scenario's issues with a significant number of interoperability conceptual frameworks, considering aspects related to social, political and regional factors. However, such frameworks rely mostly on services and technological concerns, despite the existence of works in the literature that identify different perspectives in the Public Administration (PA) and governmental aspects of interoperability. In both approaches, the literature does not have a particular contribution relating PA domains with interoperability frameworks concerning the coherent definition of models, structured interoperability assessment (IA), concerns and barriers. This paper presents a correlation analysis of the existing frameworks, considering the elements related to PA and aspects associated to the observation and assessment of an organizational interoperability. This analysis can be used by decision makers as a tool for knowledge management concerning the prioritization and selection of attributes that best support the interoperability assessment in PA domain.

Keywords Interoperability frameworks · e-Government · Public administration

A. da Costa Castro (✉) · J.M.A. Cestari (✉) · E.R. Loures (✉) ·
E.P. de Lima · E.A.P. Santos
Industrial and Systems Engineering Postgraduate Program,
Polytechnic School, Pontifical Catholic University of Parana,
Rua Imaculada Conceição, 1155, Prado Velho, Curitiba-PR, Brazil
e-mail: costa.castro@pucpr.br

J.M.A. Cestari
e-mail: jose.cestari@pucpr.br

E.R. Loures
e-mail: eduardo.loures@pucpr.br

E.P. de Lima
e-mail: e.pinheiro@pucpr.br

E.A.P. Santos
e-mail: eduardo.portela@pucpr.br

M. Amorim et al. (eds.), *Engineering Systems and Networks*, Lecture Notes in Management and Industrial Engineering, DOI 10.1007/978-3-319-45748-2_20

1 Introduction

With the political reform launched in the late 1990s, the government effectiveness issue has been in vogue, and thus, actions related to administrative probity, citizen-government relationship and process intelligence have been brought to discussion. Projects deriving from the availability of Information and Communication Technology resources (ICT) have empowered public administrators. In the same period, already in the year 2002, the term e-Government/e-Gov (Electronic Government) had one of its definitions emerged as "the use of internet and its subjacent resources for the distribution and delivery of governmental services to citizens." (United Nations 2002). The possibilities thus evinced through the advent of e-Gov led public administration to prospect an increase in process intelligence, taxation, national security and defense. Indeed, better distributed systems as well as effective, transparent, and integer administration are aspects that impelled its adoption (e-Gov) (Bertot et al. 2010). Another benefit of e-Gov consists of the transformation and improvement of processes and public administration (PA) structures, aiming to better address the citizens' demands. Interoperability relevance in PA is very evident, given the various governmental programs, action plans, and policies that have taken this function into account in recent years (Soares and Amaral 2011). All the investments in e-Gov typically consider the technological factors associated to its feasibility, and some work has been conducted in order to establish standards and adjust semantic aspects. Considering the fact that many governments work with resources restricted by local budgets, and that 40 % of the IT budget in enterprises is admittedly dedicated to solving non-interoperability related issues (Ducq et al. 2012), the appropriate management of interoperability aspects is crucial. The identification of its barriers can increase the rational use of public budgets and allow the assessment of its adherence to some specific model or maturity level (Cestari et al. 2014). This paper aims to present the typical issues behind Public Administration regarding the interoperability demand, its specific contextual problematic and, on the other hand, to present an adherence and correspondence correlation with the existing frameworks that are meant to deal with interoperability in the e-Gov context.

2 Theoretical Background

The theoretical background of this paper was based on both *interoperability* and *public administration* related references. About 280 papers, whose content met relevant criteria of the areas under research, were selected from the main research database available (ISI-Wok, IEEEXplore, Science Direct). A detailed analysis of their abstracts was conducted, and just after that, a deeper contextual analysis was performed, resulting in the selection of 114 papers. The complete list of the selected papers, together with the corresponding reference presented in this paper, can be accessed on the permanent link http://goo.gl/7Jifsx, or on the university repository at http://www.produtronica.pucpr.br/ppgeps.

2.1 Concerns of Public Administration

The classical public administration concerns covers several aspects such as finance, geopolitics, and technology. The use of adequate interoperability models results in tangible benefits, promoting the alignment of internal processes, knowledge and information sharing, creation of common value, and alignment of public strategies. Investments in interoperability in the PA context allow a better perception over governmental actions by entities and citizens. However, from a traditional perspective, performance is hard to measure, given that the involved parts see it through different value perspectives (Sather 2011). Tangible and intangible results are perceptible and highly valuable to the public administrator, and the effort for such transformation has hold the attention of politicians, who started focusing on the formulation and implementation of interoperable strategies. These strategies may lead to public entities transformation (Soares and Amaral 2011) in order to overcome interoperability barriers and facilitate the adoption and implementation of the e-Gov concept at the organizational level. This transformation scenario mainly happens in the relation with local spheres, in which the central administration can define the rules of the adoption and implementation process (Nurdin et al. 2011). Some models and frameworks has covered a number of issues required for PA approaches, including the categorization of specific challenges of each context, such as interoperability of processes, semantic interoperability aspects in governments and inter-sectorial issues.

2.2 Interoperability Frameworks

An interoperability framework consists of a specific approach to enterprises willing to work conjointly, and thus provide integrated services. Under the application perspective, considering the e-Government inherent issues, it establishes a set of common elements as terms, concepts, principles, policies, guidelines, recommendations, rules, specifications, and practices (Naudet et al. 2010). According to (C4ISR Architecture Working Group 1998), although public services usually involve data exchange among ICT systems, interoperability is a wide concept and embodies the ability to operate conjointly, targeting at mutual benefits and common objectives (Athena 2007). An evident example is the principles presented in the second version of the EIF, in which aspects related to social inclusion and interoperability services in public health are covered. Such aspects does not ensure, separately, that this framework fulfills the typical needs in public health. This dimension, however, together with other interoperability aspects of PA, can help to justify its adaptation to the context. Other frameworks, which emerged as an answer to PA demands, may likewise justify their adoption by dealing with specific aspects, and thus supporting managers with a priori and posteriori actions. This paper depicts some of the main models recommended by the academy and by the industry in the PA context, besides considering their most extensive application. The *LISI* (*Levels of Information System Interoperability*) Framework, which aims providing the US Department of Defense (DoD) with a maturity model. Its approach, although pursuing generic concepts and models, focuses on the development of interoperability in the defense sector. Nevertheless, it is also employed in the design of other maturity models (Daclin et al. 2008) resulting in a detailed interoperability model, and thus the mapping between modeling and technological implementation. The *EIF* (*European Interoperability Framework*) which arose from an EC initiative (European Commission) with the responsibility to develop researches oriented to the new economic issues, focusing on costs, flexibility, and safety (Ducq et al. 2012). Its main purposes are: (i) promote and support availability of public services in the European community, encouraging trans-border actions and multi-sectional interoperability; (ii) serve as a guide to public administration so as to provide public services to enterprises and citizens; (iii) complement and unify the various frameworks of the member states (National Interoperability Frameworks—NIFs). The *MMEI framework* (Guédria et al. 2011), which allows enterprises to assess its interoperability maturity and provides a set of best practices, e.g. tasks, activities, whose execution enables the achievement of the required interoperability maturity levels. In this way, MMEI contribute to the inter-organizational research through insights on how the organization can be able to interact, collaborate, and share information with its business partners efficiently while prevents potential conflicts. The *LCIM* (*Levels of Conceptual Interoperability Model*) addresses levels of conceptual interoperability, exceeding models focused on technology, e.g. the LISI. It covers processes and information semantic dimensions in the interoperation process. Finally, the *OIMM* (*Organizational Interoperability Maturity Model*) which deals with the possibility of organizations to interoperate from an extension of the model preconized by the LISI, and thus establishing five maturity levels. It takes into account, in organizational assessment process, the systemic quality of the components in the interaction.

3 Literature Review Approach

Aiming to evince the correlation between Interoperability Frameworks and the main e-Gov issues, a summarized review of specific literature is presented. Some authors produced significant related works, listing the main frameworks applicable to the public administration reality (2009; Rezaei et al. 2014; Guijarro 2007; Gottschalk 2009; Lisboa and Soares 2014; Soares and Amaral 2011). In this way, exploratory works in this context were used as well as specific literature in Interoperability.

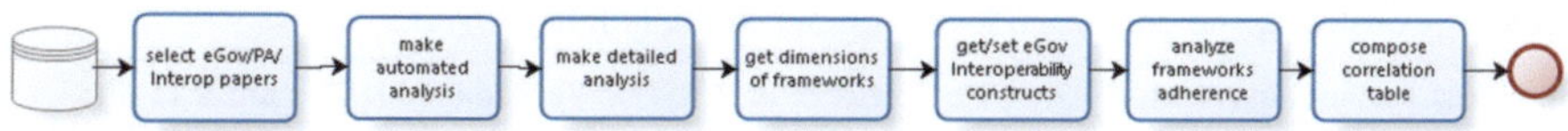

Fig. 1 Research workflow

Theoretical approaches to interoperability were considered, independently of its direct adaptation to the e-Gov context. The main objectives of this action consisted in obtaining works that considered, even indirectly, the required approaches to the typical issues of the e-Gov context, defining constructs that express the demands and challenges typical of public administration. At the same time, establishing correlations when the framework-oriented action is adherent to the construct. Figure 1 presents the workflow of this study and its results perspectives.

4 Interoperability Attributes in PA Domain—A Relational Approach

The classical attributes covered by interoperability frameworks represent both typical demands and issues faced by organizations. Differently from non-governmental enterprises, governmental entities comprehend required specificities and, according to literature, may be valid only in the e-Gov context, given responsibilities regarding politics, budget, and administrative probity issues, mainly. Although being a demand of legal order for the characterization of the attributes of a specific governmental reality, studies suggest that such characterization is much more related to an abstraction of reality than to legal and ordinary factors (Tambouris et al. 2011). The authors in (Cestari et al. 2014), suggested a pressing need for the identification of elements and their qualification in the domain. Each of the studied models and/or frameworks elaborates on these aspects in a particular way.

The aim of this section lies in semantically correlating similar attributes /demands, or those referred to aspects representing the Public Administration problematics, in order to be assessed under the light of a suggested model or framework. An analysis of the attributes identified may characterize a similarity with other attributes, and the combination of these attributes can evidence a reinterpretation through a construct (e.g. semantic cluster, interoperability concern) whose definition meets the attributes that compose it. Table 1 illustrates the constructs generated from the categorized attributes and with the references of their content. Each work shows relevant information about the definition of each of the constructs. As it is possible to observe in the course of the present paper, some frameworks seek, in their approach and structure, the fulfillment of existing expectations in the e-Gov context expressed in the constructs.

As illustrated in Table 2, some negative correlations (−, −−) can be observed and were appropriately presented, even though most models studied evinced

Table 1 Constructs and adherence level of frameworks in related works

Semantic service, invocation and sharing of service	Peristeras et al. (2009), Vernadat (2010), Furdík et al. (2011), Naudet et al. (2010), Rezaei et al. (2014), Zutshi et al. (2012)
Agility and flexibility	Gong and Janssen (2013), Pankowska (2008), Janssen et al. (2011), Janssen (2012)
Architecture-based and model-driven approaches	Peristeras et al. (2009), Gong and Janssen (2013), Wang et al. (2009), Chen et al. (2008)
Process management and life-cycle approaches	Ray et al. (2011), Soares and Amaral (2011), Zutshi et al. (2012)
Technological artifacts	Klischewski (2011), Janssen (2012), Soares and Amaral (2011)
Domain ontology for PA	Gong and Janssen (2013)
Semantic compatibility	Panetto and Cecil (2013), Vernadat (2010)

A full Table 1 (and its references) and Table 2 are available at http://goo.gl/7Jifsx or http://www.produtronica.pucpr.br/ppgeps

neutrality () or positive correlation (+, ++) with the Public Administration constructs and the covered dimensions. In general, it was detected a good service level as well as the ability to deal with and propose resolutions to interoperability issues in public administration. It was also possible to attest that the second version of the Europe Interoperability Framework presents a significant maturity and fulfillment level with issues beyond the interoperability dimensions such as society fulfillment expectations, inclusion, multi-languages, and interdisciplinarity. Through this analysis, supported by Table 2, it was possible to extend the evaluation of each construct related to specific dimensions of each selected framework, resulting in a more specific and detailed correlation, reinforcing the understanding potential of each model and dimension.

5 Conclusion and Future Works

The present study suggests that the correlation between interoperability frameworks in the PA context as well as the related problems can serve as start point for deeper analysis of interoperability in the e-Gov context. In the literature it is possible to identify the use of interoperability frameworks in several countries, although some key aspects were not identified. Documenting which frameworks were adopted and, consequently, the reasons that support the decision to adopt them would be very beneficial to a common perspective in the e-Gov context. This paper presented a direct, non-exhaustive correlation between attributes and existing problems in the e-Government scenario as well as specific relations of the main organizational interoperability frameworks, aspects commonly appraised by managers and decision makers on interoperability assessment dimensions (e.g. potential and

Table 2 Constructs and detailed adherence level in frameworks approaches

Construct/Principle/Attribute	EIF2									
	Subsidiarity and	User-centricity	Inclusion and accessibility	Security and privacy	Multilingualism	Administrative simplification	Transparency	Preservation of information	Openness	Reusability
Trans formation and re-engineering approaches	++	++	++	++	++	++	++	++	+	+
Process and organizational interoperability	++	+	+	+	+	++	++	++	+	++
Semantic service composition, integration, invocation and sharing of service	++	++	++	++	++	++	++	++	++	++
Enterprise application integration	++	++	++	++	++	++	++	++	++	++
Agility and flexibility	+	++	+	++	+	+	++	++	+	+
Next generation infrastructure	++	++	++	++	++	++	++	++	++	++
Cloud infrastructures	++	+	+	+	+	+	+	+	+	+
Architecture-based and model-driven approaches	+	+	+	+	++	+	+	++	+	++
Process management and life-cycle approaches	++	++	++	++	++	++	++	++	++	++

Construct/Principle/Attribute	EIF2		LISI				OIMM			
	Technological neutrality and	Effectiveness and efficiency	Data	Architecture	Technology	Infrastructure	Procedures	Architecture	Data	Infrastructure
Trans formation and re-engineering approaches	+	++	+	++	++	++	++	++	+	+
Process and organizational interoperability	++	+	++	++	++	+	++	+	+	+
	++	++	+	++	++	++	+	+	+	+

(continued)

Table 2 (continued)

Construct/Principle/Attribute	EIF2		LISI				OIMM			
	Technological neutrality and	Effectiveness and efficiency	Data	Architecture	Technology	Infrastructure	Procedures	Architecture	Data	Infrastructure
Semantic service composition, integration, invocation and sharing of service										
Enterprise application integration	++	++	+	++	+	++	+	+	+	+
Agility and flexibility	+	+	+	+	+	+	+	+	++	+
Next generation infrastructure	++	+	+	+	++	++	+	+	+	+
Cloud infrastructures	+	++	+	+	+	++	+	+	+	+
Architecture-based and model-driven approaches	++	+	+	++	+	++	+	+	+	++
Process management and life-cycle approaches	+	++	+	+	++	+	+	+	+	+

A full Table 1 (and its references) and Table 2 are available at http://goo.gl/7Jifsx or http://www.produtronica.pucpr.br/ppgeps

compatibility). The decision process deriving from this correlation, however, must take specific questions into account as questions regarding technology, processes, services, and information in the interoperation context. Other aspects related to inter/intra-organizational interoperability must be taken as crucial in the decision making process, justifying the adaptation or partial application of features of a framework. Local independent initiatives solve semantic and technological issues, but the choice of the best approach remains as the administrators' responsibility, given the aspects of the public services process. The academy and the industry has come across the decision making issue, reinforcing the fact that Frameworks themselves do not express the direct adaptation of their application to several contexts. In order to establish local/contextual questions, correlating them to the specific issues of each framework and pondering their adoption, represent an open issue and thus a valid and suggested object to future studies. Additionally, it is possible to make use of mathematical models for supporting decision and interoperability assessment so as to consider the interoperability dimensions and attributes in the public administration context.[1]

References

Athena (2007) The ATHENA Framework—Deliverable number : D.A4.2 Specification of interoperability framework and profiles, guidelines and best practices. Integrating and strengthening the european research, Mar 2007, pp 1–215. Available at: http://interop-vlab.eu/ei_public_deliverables/athena-deliverablesATHENA

Bertot JC, Jaeger PT, Grimes JM (2010) Using ICTs to create a culture of transparency: e-government and social media as openness and anti-corruption tools for societies. Gov Inf Q 27(3):264–271. Available at: http://linkinghub.elsevier.com/retrieve/pii/S0740624X10000201

C4ISR Architecture Working Group (1998) Levels of information systems interoperability (LISI), 30 Mar 1998

Cestari JMAP, Lezoche M et al (2014) A method for e-Government concepts interoperability assessment

Chen D, Doumeingts G, Vernadat F (2008) Architectures for enterprise integration and interoperability: past, present and future. Comput Ind 59(7):647–659. Available at: http://linkinghub.elsevier.com/retrieve/pii/S0166361508000365

Daclin N et al (2008) Methodology for enterprise interoperability. Available at: http://www.nt.ntnu.no/users/skoge/prost/proceedings/ifac2008/data/papers/2896.pdf

Ducq Y, Chen D, Doumeingts G (2012) A contribution of system theory to sustainable enterprise interoperability science base. Comput Ind 63(8):844–857. Available at: http://linkinghub.elsevier.com/retrieve/pii/S0166361512001248

Furdík K, Tomášek M, Hreňo J (2011) A WSMO-based framework enabling semantic interoperability in e-Government solutions. Acta Polytech Hung 8(2):61–76. Available at: http://epa.niif.hu/02400/02461/00028/pdf/EPA02461_acta_polytechnica_hungarica_2011_02_061-079.pdf. Accessed 4 Oct 2014

[1]A full list of references can be accessed in http://goo.gl/7Jifsx or http://www.produtronica.pucpr.br/ppgeps.

Gong Y, Janssen M (2013) An interoperable architecture and principles for implementing strategy and policy in operational processes. Comput Ind 64(8):912–924. http://doi.org/10.1016/j.compind.2013.06.018

Gottschalk P (2009) Maturity levels for interoperability in digital government. Gov Inf Q 26 (1):75–81. Available at: http://linkinghub.elsevier.com/retrieve/pii/S0740624X08000683

Guédria W et al (2011) On the use of an Interoperability framework in coopetition context. Available at: http://infoscience.epfl.ch/record/169640/files/Ontheuseofaninteroperabilityframeworkincoopetitioncontext.pdf

Guijarro L (2007) Interoperability frameworks and enterprise architectures in e-government initiatives in Europe and the United States. Gov Inf Q 24(1):89–101. Available at: http://linkinghub.elsevier.com/retrieve/pii/S0740624X06000864

Janssen M, Charalabibis Y, Kuk G, Cresswell T (2011) Guest editors' introduction: e-government interoperability, infrastructure and architecture: state-of-the-art and challenges. J Theor and Appl Electron Commer Res 6(1):I–VIII. http://doi.org/10.4067/S0718-18762011000100001

Janssen M (2012) Sociopolitical aspects of interoperability and enterprise architecture in e-government. Soc Sci Comput Rev 30(1):24–36. Available at: http://ssc.sagepub.com/cgi/doi/10.1177/0894439310392187. Accessed 4 Oct 2014

Klischewski R (2011) Architectures for tinkering? Contextual strategies towards interoperability in e-government. J Theor Appl Electron Commer Res 6(1):26–42. Available at: http://www.scielo.cl/scielo.php?script=sci_arttext&pid=S0718-18762011000100004&lng=en&nrm=iso&tlng=en. Accessed 4 Oct 2014

Lisboa A, Soares D (2014) E-government interoperability frameworks: a worldwide inventory. In: 6th Conference on Enterprise information systems—aligning technology, organizations and people, CENTERIS 2014, vol 16(0), pp 638–648. Available at: http://www.sciencedirect.com/science/article/pii/S2212017314002394

Naudet Y et al (2010) Towards a systemic formalisation of interoperability. Comput Ind 61 (2):176–185. Available at: http://linkinghub.elsevier.com/retrieve/pii/S0166361509002073. Accessed 2 Oct 2014

Nurdin N, Stockdale R, Scheepers H (2011) Understanding organizational barriers influencing local electronic government adoption and implementation: the electronic government implementation framework. J Theor Appl Electron Commer Res 6(3):5–6. Available at: http://www.scielo.cl/scielo.php?script=sci_arttext&pid=S0718-18762011000300003&lng=en&nrm=iso&tlng=en

Panetto H, Cecil J (2013) Information systems for enterprise integration, interoperability and networking: theory and applications. Enterp Inf Syst 7(1):1–6. http://doi.org/10.1080/17517575.2012.684802

Pankowska M (2008) National frameworks' survey on standardization of e-Government documents and processes for interoperability. J Theor Appl Electron Commer Res 3(3):64–82. Available at: http://www.scielo.cl/scielo.php?pid=S0718-18762008000200006&script=sci_arttext&tlng=en. Accessed 4 Oct 2014

Peristeras V, Tarabanis K, Goudos SK (2009) Model-driven e-Government interoperability: a review of the state of the art. Comput Stand Interfaces 31(4):613–628. Available at: http://linkinghub.elsevier.com/retrieve/pii/S0920548908001372

Ray D et al (2011) A critical survey of selected government interoperability frameworks. Transform Gov People Process Policy 5(2):114–142. Available at: http://www.emeraldinsight.com/10.1108/17506161111131168. Accessed 2 Oct 2014

Rezaei R, Chiew T, Lee S (2013) A review of interoperability assessment models. J Zhejiang Univ Sci C 14(9):663–681. Available at: http://link.springer.com/10.1631/jzus.C1300013. Accessed 4 Oct 2014

Rezaei R, Chiew TK, Lee SP et al (2014) Interoperability evaluation models: a systematic review. Comput Ind 65(1):1–23. Available at: http://linkinghub.elsevier.com/retrieve/pii/S0166361513001887

Sather HS (2011) A framework for analysing interoperability in electronic government. Int J Electron Finan 5(1):32. Available at: http://www.inderscience.com/link.php?id=38221

Soares D, Amaral L (2011) Information systems interoperability in public administration: identifying the major acting forces through a delphi study. J Theor Appl Electron Commer Res 6(1):61–94. Available at: http://www.scielo.cl/scielo.php?script=sci_arttext&pid=S0718-18762011000100006&lng=en&nrm=iso&tlng=en. Accessed 4 Oct 2014

Tambouris E et al (2011) Study on interoperability at local and regional level. Available at: http://ec.europa.eu/idabc/servlets/Doc53fa.pdf?id=28787

United Nations (2002) Benchmarking e-government: a global perspective. (ASPA, ed). United nations division for public economics and public administration

Vernadat FB (2010) Technical, semantic and organizational issues of enterprise interoperability and networking. Ann Rev Control 34(1):139–144. Available at: http://linkinghub.elsevier.com/retrieve/pii/S1367578810000155

Wang W, Tolk A, Wang W (2009) The levels of conceptual interoperability model: applying systems engineering principles to M&S. Soc Comput Simul Int 168. Available at: http://dl.acm.org/citation.cfm?id=1655398

Zutshi A, Grilo A, Jardim-Goncalves R (2012) The business interoperability quotient measurement model. Comput Ind 63(5):389–404. Available at: http://www.sciencedirect.com/science/article/pii/S0166361512000036. Accessed 2 Oct 2014

Main Factors Affecting the Development of Interorganizational Partnerships in Biodiesel Supply Chain in Brazil

Eliene Cristina Barros Ribeiro, António Carrizo Moreira, Luís Miguel Domingues Fernandes Ferreira, Leonardo Leocádio Coelho de Souza and Aldara da Silva César

Abstract Partnership in supply chain stands out for the need to coordinate the productive activity between different economic agents who often have conflicting goals. Thus, the aim of this paper is to analyse how the partnerships producer/supplier are developed in biodiesel supply chain, through the motivating and facilitating factors that affect this relationship. The methodological aspect involves a multiple case study conducted in three supply chains located in the South Region of Brazil, which have used semi-structured interviews. We involved in this investigation the three largest biodiesel producer plants in the South Region and the two most representative cooperatives in providing raw material for each one of the selected plants, totalling six cooperatives. We realized through the found motivating and facilitating factors that, despite the existing conflicts, both plants and cooperatives have shown the desire to renew the partnership over time.

Keywords Supply chain · Partnership · Biodiesel

E.C.B. Ribeiro (✉) · A.C. Moreira · L.M.D.F. Ferreira
Industry Engineering, Management and Economics Department, Universidade de Aveiro, University Campus of Santiago, 3810-193 Aveiro, Portugal
e-mail: elienecbr@hotmail.com

A.C. Moreira
e-mail: amoreira@ua.pt

L.M.D.F. Ferreira
e-mail: lmferreira@ua.pt

L.L.C. de Souza
Universidade Federal do Maranhão, Campus of Imperatriz, Rua Urbano Santos s/n, Centro, Imperatriz-MA, Brazil
e-mail: leoleocadio@gmail.com

A. da Silva César
Agribusiness Department, Universidade Federal Fluminense, Avenida dos Trabalhadores, 420, Vila Santa Cecília, Volta, Redonda-RJ, Brazil
e-mail: aldaracesar@id.uff.br

M. Amorim et al. (eds.), *Engineering Systems and Networks*, Lecture Notes in Management and Industrial Engineering, DOI 10.1007/978-3-319-45748-2_21

1 Introduction

The way organizations build their capacity to effectively manage partnerships in the supply chain has attracted researchers' attention over the past few years. Companies growing interest in investing in partnerships of producer/supplier type has been motivated by issues such as fierce competition, technological processes acceleration, as well as the growth of the social phenomena inside organizations (Ellram and Krause 2014).

Due to the issue complexity and the lack of investigations in various sectors of the economy, this study sought to characterize the producer/supplier partnership in the biodiesel supply chain in Brazil, focusing on the motivating and facilitating factors. According to Lambert et al. (1996), these aspects are justified because, to understand how the producer/supplier partnership influences the results, first, one must know the motivating factors, i.e. aspects that induced its formation in order to understand which the partners' expectations were. Enabling factors are used to identify features that stimulate or inhibit partnership, making possible a better understanding of the success or failure of the relationship.

The agribusiness sector, like other sectors of the economy, is subjected to several market rules, within a systemic and not isolated approach, adding more and more value to the generated products. Through this development, the sector is rethought, presenting itself as a segment of great economic and social importance in the global context (Watanabe et al. 2012). Family farming professionalization, management of cooperatives as important raw material suppliers, supply partnerships between large agribusinesses plants, cooperatives, and family farmers are aspects which are present and mark the management in agribusiness supply chains.

Partnership in the supply chain stands out as one of the possible ways to address the problems created by the need to coordinate the productive activity between different economic agents who often have conflicting objectives. Thus, new types of agribusiness relationships may have consequences for the stabilization of supply chains, managerial and technological qualification and territorial displacement of production companies as well as for generating a better response to the demands of the chain.

This paper is justified by the virtually absence of this kind of study in Brazil and in the world involving the biofuel production sector. Works done by Moharana et al. (2012) point out that, despite the interest of many agribusiness companies to get a closer relationship with their suppliers, factors such as the lack of commitment between the parties and the lack of confidence and difficulties in foreseeing gains have hindered such an approach, often leading to promising projects disruption.

In this scenario, this paper objective is to analyse how the producer/supplier partnerships are developed in biodiesel supply chain, through the motivating and facilitating factors that affect this relationship. For this, we have selected the partnership attributes that better suit this sector analysis and from them we identified the points which motivate, stimulate and inhibit partnerships development in order to suggest areas for improvement. Ellram and Krause (2014) report that the relationship in the producer/supplier partnership has been more prominent in recent years, and this has led to the need of obtaining corroborative analysis of its application in various sectors of the economy.

2 Methodological Aspects

This study was based on the Brazilian biodiesel program called "National Program for Production and Use of Biodiesel". We analysed the producer/supplier partnership between biodiesel producer plants that are certified with the "Social Fuel Seal" (SFS) and cooperatives of family farmers that are raw material suppliers. SFS is a governmental mechanism that aims to put family farming in the supply chain of biodiesel producer plants. Thus, plants receive tax benefits and guarantees of selling their end product (B100 biodiesel) in auctions of the National Petroleum, Natural Gas and Biofuels Agency and they promote financially and technologically the strengthening of family farming and cooperatives.

As contributions of this paper, first this research provides a practical-theoretical approach to the selection of the partnership attributes that suit the evaluation of the biodiesel supply chain in Brazil. Partnership attributes were selected using the following steps: (a) survey of the partnership attributes most commonly used in agro-industrial chains; (b) in loco observation in the biodiesel producer plants, family farmers cooperatives, and eighty family farms and public institutions of the sector; (c) after the in loco observation, 25 attributes were selected among the most frequently used in the literature which suit the bioenergy sector; (d) the selected attributes were sent to biodiesel industry experts, who indicate the ones which are more related to the sector in accordance to this paper objectives.

These experts are two researchers from the biodiesel industry in Brazil; two agronomy engineers who provide technical assistance to the SCS program and two supply managers of two biodiesel plants located in the South and Northeast Regions; (e) after the experts analysis, we selected nine attributes which were used in this research.

The second contribution of this study is the application of the selected partnership attributes using semi-structured interviews, in order to evaluate the producer/supplier partnership development process in the biodiesel supply chain.

2.1 Units of Analysis

Units of analysis consist of three supply-chains located in the South Region of Brazil, involving: (a) the three biodiesel producer plants located in this region, which coincidentally are among the four largest biodiesel plants in Brazil (Granol, BsBios Petrobras Marialva, BsBios/Petrobras Passo Fundo)—these companies together account for an average of 40 % of the biodiesel produced in the country;

(b) the two most significant cooperatives supplying raw oilseed, which stand out for feedstock volume, delivery regularity and length of partnership, totalling six cooperatives. We justify the choice for the South Region because it presents the most developed family farming and cooperatives in the country, and it is also the region where 95 % of the biodiesel plants are certified with SCS, thus demanding raw material from the family farmers.

Interview respondent agents from the biodiesel plants were the supply managers and the agricultural technicians responsible for the technical assistance, in SCS parameters. Regarding the interviewed agents from cooperatives, we detach the cooperative managers and the agricultural technicians responsible for the technical assistance in accordance to SCS rules. The instrument used for collecting data was the semi-structured interview which addressed the "key informants". According to Malhotra and Grover, this approach means collecting data from people who work directly in the investigated area, since the research tool content requires specific knowledge of the sector.

Information analysis procedure is characterized by two analyses types: data triangulation and content analysis. Data triangulation involved the interview, the document analysis and the literature review. Based on the content analysis or data treatment, we tried to make the obtained data significant and valid. So inferences and interpretations were provided in order to achieve this investigation objective.

3 Analysis of the Partnership Attributes in Biodiesel Supply Chain

Partnership attributes were incorporated by Mohr and Spekman (1994), Walton (1996) when they investigated current and expected satisfaction in partnerships as perceived by businessmen and their suppliers and customers. In our study, we have incorporated coordination mechanisms referred by these authors as partnerships attributes, which are widely accepted in the literature, in order to analyse current and expected satisfaction of interorganizational suppliers and customers. From there, some investigations aiming to analyse the producer/supplier partnership has emerged, using attributes that were adapted to the analysed sector. This paper's objectives are focused on motivating and facilitating aspects adapted from Lambert model (2008), which will be analysed based on the partnership selected attributes for the biodiesel supply chain in Brazil, with its idiosyncrasies.

3.1 Trust

Based on authors like Morgan and Hunt (1994), Dyer and Chu (2000), for this analysis, trust is based on the disposition to count on a partner believing that he will perform actions that will have positive results for both parties. In the analysed supply chains, trust was considered by cooperatives as the most important attribute for the development of a successful partnership. The action/result highlighted by cooperatives is directed to the fulfilment of the signed contract by plants. Regarding plants, we detected that trust is still incipient, because cooperatives have presented behaviours which are considered opportunistic.

3.2 Commitment

In this investigation, this attribute is seen as a continuing desire of maintaining a valuable relationship, with the appropriate behaviours as well as obligations and norms fulfilment (Morgan and Hunt 1994). For plants, the main lack of commitment action happens when cooperatives do not deliver the soybeans in the agreed volume, in favour of better export prices. So, due to this situation, plants have reported they do not commit to give up signing contracts with individual farmers until cooperatives deliver the amount of soybeans negotiated in the contract. This behaviour is seen by cooperatives as a competition of plants with cooperatives for family farmers and a decline in commitment.

3.3 Adaptation

For this investigation, adaptation attribute is analysed according to the concepts of Crotts and Turner (1999), Helfert et al. (2002) as the adequacy of suppliers to specific needs of clients and customers as well as the adjustments of clients to suppliers' capabilities. The main action regarding adaptation was the adjustment of the way they fill in the report on technical assistance that has to be sent to MDA every year. For cooperatives, reports are extensive and unviable, because each cooperative has between three and six thousands family farmers producing via SCS. Plants reported that the adaptation to the report template required by MDA is still an obstacle to the partnership relationship. The process of making five visits to each family farmer per harvest has not been done by most cooperatives. Plants understand that there is not a need for this type of procedure in the South Region, where soybeans farmers already have tradition and productivity, but they claim this is a requirement of MDA that needs to be fulfilled.

3.4 Information Sharing

In this research case, information sharing attribute, according to Mohr and Spekman (1994) vision, is the formal and informal sharing of meaningful and opportune information between partners. A point highlighted by plants referred to the information given in advance about the soybean volume to be effectively delivered by cooperatives every harvest, which allows an effective internal planning. Some cooperatives have recognized this imperfection in the development of actions regarding information sharing and we perceived a conflict between plants and cooperatives on the necessity to give information, in advance, about the volume of soybeans to be delivered.

Another point regarding information sharing which was considered important by plants is the service reports to be sent to MDA. These are the main information-sharing tools that report all the phases of the technical assistance provided by the cooperatives to family farmers, as well as the amount of raw material produced and productivity. Plants do not consider this report important just because it is sent to MDA for SCS renewal, but also because it is a source of productivity and production information and because the production and marketing actions for the coming harvests are scheduled based on these data.

3.5 Cooperation and Conflict

To the cooperation and conflict dimension, we addressed Dwyer et al. (1987), Morgan and Hunt (1994) principles in which cooperation can be defined as a combination of efforts that reflect the joint expectations of partner companies in achieving mutual and individual goals over time. In this partnership, the authors consider the conflict pertinent in order to measure the total level of disagreement between partners, and it is considered beneficial in a cooperative relationship. In the analysed supply chains, we can verify the existence of cooperation, but always with the presence of conflicts, that are recognized by plants and cooperatives as inherent to the activity, in a region where the production of biodiesel and soybeans is very competitive.

3.6 Power and Dependence

The analysis of power and dependence attributes was based on the concepts of Mohr and Spekman (1994) in which power is the ability of a company to impose itself over the partner. The lower the dependence between the partners, the greater is the probability of a fair power application. Thus, we noticed that plants are more dependent on cooperatives. The large number of plants certified under SCS generates great competition for soybeans produced by family farming in the region. In accordance to the interviews, both plants and cooperatives understand that the partner has a certain degree of importance in the partnership, which gives each party different powers and a certain degree of dependence, and this, in the perception of respondents, reduces the possibility of predatory relationships.

3.7 Satisfaction in Partnership

Spekman et al. (1998) understand satisfaction as a key variable to determine whether a producer/supplier partnership will continue. Based on these authors, satisfaction was considered as a judgment of all the previous experiences obtained with a partner, reflecting the relationship results. We found that plants and cooperatives realize that the partnership has been developed over time, so that both parties report they are satisfied and understand that conflicts are manageable. Plants and cooperatives are interested in contract renewal and in increase the volume of raw material negotiated, assuming that adjustments and conflicts are inherent to the partnership.

4 Conclusion

We found out that the major inhibiting factor that is preponderant for biodiesel plants is the fact that cooperatives do not fulfil the delivery of the volume of soybeans agreed in the contract. In this sector, the economic gain directs more options that are aimed at cooperatives, because there is a great competition for soybeans produced by family farming in the region. Thus, it was noticed that the transactions are partly based in opportunism and not in trust and commitment building. As opportunism, in this investigation, we cited the partnership relationship in which one chooses to breach the signed contract to transact with another market player that offers a value higher than the one agreed.

It was also found out that plants are highly dependent on cooperatives, because there is a great competition for soybeans in the South Region. However, it was realized that there is a certain degree of dependence of cooperatives on plants, because there is an interest in a customer portfolio diversification and there is also the interest of family farmers and cooperatives in the financial bonuses received per sack of soybeans. Financial bonuses and contract fulfilment by plants were considered by cooperatives as the main motivating and stimulating factors for the partnership. Plants highlighted the quality of the soybean produced by cooperatives and the favourable distribution logistics as stimulating factors for the partnership, which reduces transportation costs for both sides.

Partnership assumes that conflict will occur, but this does not mean the absence of errors, since dealing with *commodities* that come from *agribusiness* is dealing with bad weather issues and consequently commercial ones. However, both sides have reported that it is important to give clear information. Partnership was considered satisfactory for plants and cooperatives, which have interest in contracts renewal, despite the existing conflicts, which both parties have considered as inherent to the sector and manageable. Partners exchange was considered a commercial setback both for plants and cooperatives.

References

Crotts J, Turne G (1999) Determinants of intra-firm trust in buyer-seller relationships in the international travel trade. Int J Contemp Hospitality Manage 2:116–123
Dwyer FR, Schurr PH, Oh S (1987) Developing buyer–seller relationships. J Mark 51:11–27
Dyer JH, Chu W (2000) The determinants of trust in supplier-automaker relationships in the U.S., Japan and Korea. J Int Bus Stud 31(2):259–285
Ellram LM, Krause D (2014) Robust supplier relationships: key lessons from the economic downturn. Bus Horiz 57(2):203–213
Helfert G, Ritter T, Walter A (2002) Redefining market orientation from a relationship perspective: theoretical considerations and empirical results. Eur J Mark 36(9):1119–1139
Lambert DM, Emmelhainz MA, Gardner JT (1996) Developing and implementing supply chain partnerships. Int J Logistics Manage 7(2):1–18
Moharana HS, Murty JS, Senapati SK, Khuntia K (2012) Coordination, collaboration and integration for supply chain management. Int J Intersci Manage Rev (IMR) 2(2):46–50
Mohr J, Spekman R (1994) Characteristics of partnership success: partnership attributes, communication behavior and conflict resolution techniques. Strateg Manage J 15:135–152
Morgan RM, Hunt SD (1994) The commitment-trust theory of relationship marketing. J Mark 58:23
Spekman RE, Kamauff JW Jr, Myhr N (1998) An empirical investigation into supply chain management: a perspective on partnerships. Supply Chain Manage Int J 3(2):53–67
Watanabe K, Bijman J, Slingerland M (2012) Institutional arrangements in the emerging biodiesel industry: case studies from Minas Gerais—Brazil. Energy Policy 40:381–389. doi:10.1016/j.enpol.2011.10.023

Productivity Improvement, Considering Legal Conditions and Just in Time Principles in the Mixed-Model Sequencing Problem

Joaquín Bautista-Valhondo, Rocío Alfaro-Pozo and Cristina Batalla-García

Abstract In this work a new variant of the Mixed-Model Sequencing Problem with Work overload Minimization is formulated and evaluated. The new variant regularizes production, in addition to increase the amount of completed work through the workers' activation at certain moments of their workday and to fulfill the conditions from collective agreements in terms of operating level of workers. Thus, we seek to reduce the inventory cost by the balance of the component consumptions and maximizing productivity by the reduction of non-completed work, within the legal field. We check is possible to improve both production regularity and productivity at once and without exceeding the operation level of workers, by means of a case study.

Keywords Mixed-model · Sequencing · Just-in-time · Work pace · Saturation

1 Introduction

The market performance, specifically the automotive sector, has moved from unskilled manufacturers, whose vehicles were manufactured based on customer requirements at a high cost, to a mass production with a broad range of options, through assembly lines.

This work was funded by the Ministerio de Economía y Competitividad (Spanish Government) through the FHI-SELM2 (TIN2014-57497-P) project.

J. Bautista-Valhondo (✉) · R. Alfaro-Pozo · C. Batalla-García
BarcelonaTech (UPC), Research Group OPE-PROTHIUS, Universitat Politècnica de Catalunya, Barcelona, Spain
e-mail: joaquin.bautista@upc.edu

R. Alfaro-Pozo
e-mail: rocio.alfaro@upc.edu

C. Batalla-García
e-mail: cristina.batalla@upc.edu

M. Amorim et al. (eds.), *Engineering Systems and Networks*, Lecture Notes in Management and Industrial Engineering, DOI 10.1007/978-3-319-45748-2_22

Since then, flexibility, cost reduction, either by reducing labor costs or costs of handling and inventory, and productivity and quality increases have been the main objectives in pull manufacturing systems whose paradigms are the Just in Time (JIT) manufacturing philosophy and the Mixed-Model Assembly Lines (MMALs).

The *MMALs* features lead to two main problems: (1) the assignment of the operations necessary for the assembly of products to the workstations that make up the line; and (2) the determination of the manufacturing sequence of the different product types on the line in order to minimize the effect of the product diversity in terms of resource utilizations and component consumptions.

In this work, the last problem known in the literature as Mixed-Model Sequencing Problem (*MMSP*) is studied, considering the general optimization criteria proposed by Boysen et al. (2009): (1) to minimize the work overload or uncompleted work; and (2) to evenly distribute the material requirements on planning horizon. This work also includes working conditions for human resources working on the stations in terms of operating and activity levels.

Specifically, we propose and evaluate a new mathematical model for the Mixed-Model Sequencing Problem with Work Overload Minimisation (*MMSP-W*), from the $M4 \cup 3\_\dot{\alpha}I\_\eta$ reference model proposed by Bautista et al. (2015a). The new model, besides considering the activation of workers to improve productivity and saturation conditions for workers, considers the regularity of production benefiting the second objective defined by Boysen et al. (2009) that is in line with the Just In Time philosophy (*JIT*).

2 The Sequencing of Mixed-Model in JIT Systems

Since the success of Japanese companies after the crisis of the seventies, production systems known as Just In Time (*JIT*), whose origin is in Toyota, have boomed spectacularly. Indeed, the *JIT* philosophy, also called Douki Seisan (*DS*) by Nissan, governs most of the current flexible production systems with *MMALs*.

This production philosophy translates into a system that produces just what is needed, when is needed, with excellent quality and without wasting the system resources. Therefore, one of its general principles is to keep as regular as possible the consumption rates of components in all stations. This principle also promotes the balance of workloads required to the stations (Pinedo 2009).

Hence, to satisfy the industrial need of keeping as constant as possible the component consumptions, we use the deriving properties of maintaining the production mix over the manufacturing of products (Bautista et al. 2012). This requires the definition of $X_{i,t}^{*}$ that is the number of units of product type $i \in I$, out a total of t units that should ideally be manufactured to maintain the production mix:

$$X_{i,t}^{*} = \frac{d_i}{T} \cdot t \quad (i = 1, \ldots, |I|;\ t = 1, \ldots, T) \tag{1}$$

Cumulative production must be integer, thus it must be limited between closest values to the ideal cumulative production that maintains the production mix.

$$\left\lfloor \frac{d_i}{T} \cdot t \right\rfloor \leq X_{i,t} \leq \left\lceil \frac{d_i}{T} \cdot t \right\rceil \quad (i = 1, \ldots, |I|;\ t = 1, \ldots, T) \tag{2}$$

Specifically, when t units have been sequenced, the cumulative production of product types is limited to the proportion of each type in regard with the production plan. Thereby, the production mix will be kept throughout the sequence as far as possible, and both the component consumption and the requiered work will be regular. Thus, the *MMSP-W* is in line with another classic problem of literature that is named Product Rate Variation (*PRV*) and it was propose by Miltenburg (1989).

3 The *MMSP-W* with Labor Conditions and Production Mix Restrictions

From the $M4 \cup 3\_\dot{\alpha}I\_\eta$ (Bautista et al. 2015a) and the $M4 \cup 3\_pmr$ (Alfaro 2015) models, a new model to improve both productivity and working conditions is proposed. This model incorporates limitations established by collective agreements on the operating level or saturation of operators; the chance to complete more required work by increasing the activity factor of operators; and the cumulative production limitation to keep production mix along the sequence. The parameters and variables of $M4 \cup 3\_pmr\_\dot{\alpha}I\_\eta$ are the following:

Parameters	
K	Set of orkstations, arranged in series, that makes up the line $(k = 1, \ldots, \lvert K \rvert)$
b_k	Number of homogeneous processors at workstation $k (k = 1, \ldots, \lvert K \rvert)$
I	Set of product types that must be manufactured in the line $(i = 1, \ldots, \lvert I \rvert)$
d_i	Programmed demand of the product type $i (i = 1, \ldots, \lvert I \rvert)$
$p_{i,k}$	Processing time (at normal activity) required by one unit of a product type $i (i = 1, \ldots, \lvert I \rvert)$ at workstation $k (k = 1, \ldots, \lvert K \rvert)$ for each homogeneous processor
T	Total demand. Obviously: $\sum_{i=1}^{\lvert T \rvert} d_i = T$
t	Position index in the sequence $(t = 1, \ldots, T)$
c	Cycle time. Normal time assigned to each homogeneous processor in the workstations $(k = 1, \ldots, \lvert K \rvert)$ to process any product unit
l_k	Temporal window. Maximum time that each homogeneous processor of workstation k $(k = 1, \ldots, \lvert K \rvert)$ is allowed to work on any unit of product; once the cycle has been completed, the maximum time that a unit of product can be retained in station k is $l_k - c > 0$

(continued)

(continued)

$\dot{\alpha}_{k,t}$	Dynamic factor of the work pace or activity associated with the tth operation of the product sequence ($t = 1,\ldots,T$) at the workstation $k(k = 1,\ldots,\lvert K\rvert)$
$\dot{\alpha}_t$	Dynamic factor of the work pace or activity associated with period t ($t = 1,\ldots,T+\lvert K\rvert-1$) of the extended workday. This extended workday includes T manufacturing cycles (total demand) and $\lvert K\rvert - 1$ additional cycles, which are required to complete the required work by the production units in all the workstations. Note that if we associate the same dynamic factor with each moment of the workday in all of the workstations, we will have: $\dot{\alpha}_{k,t} = \dot{\alpha}_{t+k-1}$ ($k = 1,\ldots,\lvert K\rvert$; $t = 1,\ldots,T$)
η^{∞}_{med}	Allowable average saturation by the processors of each workstation ($k = 1,\ldots,\lvert K\rvert$)
$\eta^{\infty}_{\max}$	Allowable maximum saturation by the processors of each workstation ($k = 1,\ldots,\lvert K\rvert$)
Variables	
$x_{i,t}$	Binary variable equal to 1 if the product unit $i(i = 1,\ldots,\lvert I\rvert)$ is assigned to the position t ($t = 1,\ldots,T$) of the sequence and 0 otherwise
$\hat{s}_{k,t}$	Relative start instant. Positive difference between the start instant and the earliest start instant of the tth operation at workstation $k(k = 1,\ldots,\lvert K\rvert)$. It is fulfilled $\hat{s}_{k,t} = \max\{0;\ s_{k,t} - (t+k-2)\cdot c\}$
$v_{k,t}$	Processing time applied by each homogeneous processor (at normal activity: $\dot{\alpha}_{k,t} = 1\forall k\forall t$) to the tth product unit sequenced in the workstation $k(k = 1,\ldots,\lvert K\rvert)$
$\hat{v}_{k,t}$	Processing time reduced by the dynamic factor of activity $\dot{\alpha}_{k,t}$. It is established that $v_{k,t} = \dot{\alpha}_{k,t}\cdot\hat{v}_{k,t}$ ($k = 1,\ldots,\lvert K\rvert$; $t = 1,\ldots,T$)
$w_{k,t}$	Work overload measured in units of time (at normal activity) generated by the tth product unit sequenced in each homogeneous processor of the workstation $k(k = 1,\ldots,\lvert K\rvert)$

And the mathematical model, $M4\cup 3\_pmr\_\dot{\alpha}I\_\eta$, is the following:

$$\min W = \sum_{k=1}^{|K|}\left(b_k\sum_{t=1}^{T} w_{k,t}\right) \Leftrightarrow \max V = \sum_{k=1}^{|K|}\left(b_k\sum_{t=1}^{T} v_{k,t}\right) \quad (3)$$

$$\sum_{t=1}^{T} x_{i,t} = d_i \quad (i = 1,\ldots,|I|) \quad (4)$$

$$\sum_{i=1}^{|I|} x_{i,t} = 1 \quad (t = 1,\ldots,T) \quad (5)$$

$$v_{k,t} + w_{k,t} = \sum_{i=1}^{|I|} p_{i,k}\cdot x_{i,t} \quad (k = 1,\ldots,|K|;\ t = 1,\ldots,T) \quad (6)$$

$$\dot{\alpha}_{t+k-1}\cdot\hat{v}_{k,t} - v_{k,t} = 0 \quad (k = 1,\ldots,|K|;\ t = 1,\ldots,T) \quad (7)$$

$$\sum_{t=1}^{T} \hat{v}_{k,t} \leq \eta_{med}^{\infty} \cdot c \cdot T \quad (k = 1, \ldots, |K|) \tag{8}$$

$$\hat{v}_{k,t} \leq \eta_{\max}^{\infty} \cdot c \quad (k = 1, \ldots, |K|;\ t = 1, \ldots, T) \tag{9}$$

$$\sum_{\tau=1}^{t} x_{i,\tau} \geq \left\lfloor \frac{d_i}{T} \cdot t \right\rfloor \quad (i = 1, \ldots, |I|;\ t = 1, \ldots, T) \tag{10}$$

$$\sum_{\tau=1}^{t} x_{i,\tau} \leq \left\lceil \frac{d_i}{T} \cdot t \right\rceil \quad (i = 1, \ldots, |I|;\ t = 1, \ldots, T) \tag{11}$$

$$\hat{s}_{k,t} \geq \hat{s}_{k,t-1} + \hat{v}_{k,t-1} - c \quad (k = 1, \ldots, |K|;\ t = 2, \ldots, T) \tag{12}$$

$$\hat{s}_{k,t} \geq \hat{s}_{k-1,t} + \hat{v}_{k-1,t} - c \quad (k = 2, \ldots, |K|;\ t = 1, \ldots, T) \tag{13}$$

$$\hat{s}_{k,t} + \hat{v}_{k,t} \leq l_k \quad (k = 1, \ldots, |K|;\ t = 1, \ldots, T) \tag{14}$$

$$\hat{s}_{k,t}, v_{k,t}, \hat{v}_{k,t}, w_{k,t} \geq 0 \quad (k = 1, \ldots, |K|;\ t = 1, \ldots, T) \tag{15}$$

$$x_{i,t} \in \{0, 1\} \quad (i = 1, \ldots, |I|;\ t = 1, \ldots, T) \tag{16}$$

$$\hat{s}_{1,1} = 0 \tag{17}$$

The objective function (3) denotes the equivalence between the minimization of the total dynamic work overload (*W*) and the maximization of the total completed work (*V*). The sets (4) and (5) make sure the satisfaction of the demand plan and the assignment of each product to one position of the sequence. The set (6) establishes the relationship between the required processing time, $p_{i,k}$, the applied processing time at normal work pace, $v_{k,t}$, and the work overload generated at workstations, $w_{k,t}$, considering the same work pace factor for all workstations. The set (7) reduces the processing times, $\hat{v}_{k,t}$, by the work pace factor, $\dot{\alpha}_t$. Constraints (8) and (9) limit the average and maximum saturation values of processors. The sets (10) and (11) regulate the cumulative productions by product types to preserve the production mix. The sets (12)–(14) determine the relative start instants of operations at workstations. Constraints (15) and (16) establish the non-negativity and binarity of variables and (17) fixes the start instant of the operations.

4 Computational Experience

To assess the impact of production mix restrictions on the *MMSP-W* when the saturation limitations and workers' activation are considered, a case study linked to the engine assembly line of Nissan in Barcelona is used. Specifically, we use a production plan with a total daily demand of 270 engines, which are grouped into 9 types depending on the type of vehicle: 4 × 4, vans and trucks. The line has $|K| = 21$ workstations, with a cycle time[1] of $c = 175$ s and a time window[2] of $l_k = 195$ s by station and processor.[3]

Taking into account the collective agreement of Nissan, we set the allowable values for the saturation levels; these values are: $\eta^{\infty}_{\max} = 1.20$ and $\eta^{\infty}_{med} = 0.95$. We can similarly set the values for the minimum ($\alpha^{\circ}_{i,k} = 0.90$), normal ($\alpha^{N}_{i,k} = 1.00$) and maximum ($\alpha^{*}_{i,k} = 1.20$) activity factor allowed by the company.

The above together with the relationship between the operators' performance and their level of activation have allowed us to define three functions for the dynamic work pace factor (stepped, triangular and trapezoidal) (Bautista et al. 2015b). Figure 1 shows the stepped function for the study case in question.

We used the Gurobi v4.6.1 solver, on a Apple Macintosh iMac computer with an Intel Core i72.93 GHz processor and 8 GB of RAM using MAC OS X 10.6.7, limiting the CPU time to 2 h, to run the $M4 \cup 3\_pmr\_\dot{\alpha}I\_\eta$ model.

Once the results have been obtained, these were compared with the results given by: the $M4 \cup 3$ model that does not consider activation nor saturation, nor regular production; the $M4 \cup 3\_pmr$ model that only considers the regular production; the $M4 \cup 3\_\dot{\alpha}I$ model that considers the variable work pace factor; the $M4 \cup 3\_\eta$ model with saturation constraints; and the $M4 \cup 3\_\dot{\alpha}I\_\eta$ model that is a combination between the above two. Thereby we evaluate the influence of the labor conditions and the regularity constraints within the *MMSP-W* through the following indicators:

$$RPD_1 = \frac{\psi(S^*_{M4\cup 3}) - \psi(S^*_{M4\cup 3\_pmr})}{\psi(S^*_{M4\cup 3})} \cdot 100 \quad (\psi \in \Psi) \tag{18}$$

$$RPD_2 = \frac{\psi(S^*_{M4\cup 3}) - \psi(S^*_{M4\cup 3\_\dot{\alpha}I(f)})}{\psi(S^*_{M4\cup 3})} \cdot 100 \quad (\psi \in \Psi;\ f \in \Im) \tag{19}$$

$$RPD_3 = \frac{\psi(S^*_{M4\cup 3}) - \psi(S^*_{M4\cup 3\_\eta})}{\psi(S^*_{M4\cup 3})} \cdot 100 \quad (\psi \in \Psi) \tag{20}$$

$$RPD_4 = \frac{\psi(S^*_{M4\cup 3}) - \psi(S^*_{M4\cup 3\_\dot{\alpha}I(f)\_\eta})}{\psi(S^*_{M4\cup 3})} \cdot 100 \quad (\psi \in \Psi;\ f \in \Im) \tag{21}$$

$$RPD_5 = \frac{\psi(S^*_{M4\cup 3}) - \psi(S^*_{M4\cup 3\_pmr\_\dot{\alpha}I(f)\_\eta})}{\psi(S^*_{M4\cup 3})} \cdot 100 \quad (\psi \in \Psi;\ f \in \Im) \tag{22}$$

[1]Time assigned to workstations to perform its workload.

[2]Maximum time that a unit may be retained for a workstation, it is fulfilled $l_k - c \geq 0$.

[3]Team of two workers with identical skills and tools.

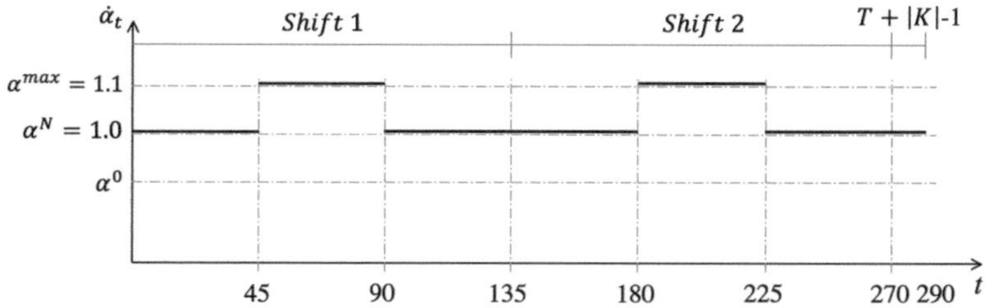

Fig. 1 Stepped function for the work pace factor, $\dot{\alpha}^S$, considering the case study from Nissan

$$RPD_6 = \frac{\psi(S^*_{M4\cup3\_\dot{\alpha}I(f)\_\eta}) - \psi(S^*_{M4\cup3\_pmr\_\dot{\alpha}I(f)\_\eta})}{\psi(S^*_{M4\cup3\_\dot{\alpha}I(f)\_\eta})} \cdot 100 \quad (\psi \in \Psi;\ f \in \Im) \quad (23)$$

where $\Psi = \{W, \Delta_Q(X), \Delta_Q(P)\}$ is the set of functions evaluated for each solution obtained by one specific model, S^*_M; and $\Im = \{\dot{\alpha}^S, \dot{\alpha}^T, \dot{\alpha}^Z\}$ is the set of functions for the dynamic work pace factor.

Tables 1 and 2 show the results. First, Table 1, where the values of overall work overloads are collected, shows the negative effect of saturation limitations on the amount completed work. Indeed, the incorporation of the saturation constraints into the model increases considerably the work overload (187 vs. 12315 s). However, when the activity factor increases by 3.33 % (stepped function), 5 % (triangular function) and 6.66 % (trapezoidal function), on average, the work overload is reduced. The incorporation of regularity does not produce an important increase of work overload.

Table 2 shows the relative percentage deviations in terms of work overload, cumulative production discrepancy (24) and required work discrepancy (25).

$$\Delta_Q(X) = \sum_{t=1}^{T}\sum_{i=1}^{|I|}\left(X_{i,t} - X^*_{i,t}\right)^2 = \sum_{t=1}^{T}\sum_{i=1}^{|I|}\left[\left(\sum_{\tau=1}^{t} x_{i,\tau}\right) - \left(t \cdot \frac{d_i}{T}\right)\right]^2 \quad (24)$$

$$\begin{aligned}\Delta_Q(P) &= \sum_{t=1}^{T}\sum_{k=1}^{|K|}\left(P_{k,t} - P^*_{k,t}\right)^2 \\ &= \sum_{t=1}^{T}\sum_{k=1}^{|K|}\left[\left(b_k \sum_{i=1}^{|I|} p_{i,k}\left(\sum_{\tau=1}^{t} x_{i,\tau}\right)\right) - \left(t \cdot \frac{b_k}{T}\sum_{i=1}^{|I|} p_{i,k} \cdot d_i\right)\right]^2 \end{aligned} \quad (25)$$

Table 1 Comparison between the work overload values, $W(s)$, obtained with the proposed model, $M4\cup3\_pmr\_\dot{\alpha}I\_\eta$, and those obtained with the reference models from literature

$M4\cup3$	$M4\cup3\_pmr$	$M4\cup3\_\dot{\alpha}I$	$M4\cup3\_\eta$	$M4\cup3\_\dot{\alpha}I\_\eta$			$M4\cup3\_pmr\_\dot{\alpha}I\_\eta$		
		$\dot{\alpha}^S, \dot{\alpha}^T, \dot{\alpha}^Z$		$\dot{\alpha}^S$	$\dot{\alpha}^T$	$\dot{\alpha}^Z$	$\dot{\alpha}^S$	$\dot{\alpha}^T$	$\dot{\alpha}^Z$
187	186	0	12,315	4602	796	0	4669	839	0

Table 2 Relative percentage deviations (%) of work overload (W) and non-regularity of both cumulative production ($\Delta_Q(X)$) and required work ($\Delta_Q(P)$)

$\psi \in \Psi$	RPD_1	RPD_2			RPD_3	RPD_4			RPD_5			RPD_6		
		$\dot{\alpha}^S$	$\dot{\alpha}^T$	$\dot{\alpha}^Z$		$\dot{\alpha}^S$	$\dot{\alpha}^T$	$\dot{\alpha}^Z$	$\dot{\alpha}^S$	$\dot{\alpha}^T$	$\dot{\alpha}^Z$	$\dot{\alpha}^S$	$\dot{\alpha}^T$	$\dot{\alpha}^Z$
W	0.5	100	100	100	−6485.6	−2360.8	−325.7	100	−2396.7	−348.7	100	−1.5	−5.4	0.0
$\Delta_Q(X)$	97.7	14.5	31.2	43.4	27.4	−83.5	−213.8	−36.7	97.7	97.7	97.7	98.8	99.3	98.3
W	96.4	−61.8	−75.4	−56.8	−19.5	−193.3	−467.5	−224.9	95.4	95.7	96.3	98.4	99.2	98.9

The production mix restrictions improved by more than 95 % regularity of the sequences in terms of cumulative production and work required to the line. This regularity gain offsets the overload increase given by the $M4 \cup 3\_pmr\_\dot{\alpha}I\_\eta$ model in regard with $M4 \cup 3\_\dot{\alpha}I\_\eta$ model, with values of 1.5 and 5.4 %, when the stepped or triangular function are considered.

5 Conclusions

Simultaneous integration, of *JIT* concepts and aspects of human resources, has led to the $M4 \cup 3\_pmr\_\dot{\alpha}I\_\eta$ model. This model, in addition to complying with legal provisions on the operating level of the workers, contained in the collective agreements, can eliminate the work overload by increasing the activity factor of workers at certain moments of their working day. This makes possible to respect working conditions without compromising the line productivity. Moreover, the properties of maintaining constant the production mix favour the decrease of non-regularity of cumulative production and required work and, therefore, the regular consumption of components is also favoured.

Specifically, with an average increase by 6.66 % of normal activity, the work overload is completely eliminated while the saturation conditions are satisfied. In addition, the regularity of both cumulative production and required work improve the solution given by the reference model $M4 \cup 3\_\dot{\alpha}I\_\eta$, by 98.3 and 98.9 %, respectively.

In future research the gains will be economically evaluated in order to verify if regularity is compensated with the loss in completed work when the stepped, $\dot{\alpha}^S$, or triangular, $\dot{\alpha}^T$, functions are considered.

References

Alfaro R (2015) Modelado y resolución de variantes del problema de secuenciación de modelos mixtos con minimización de la sobrecarga (MMSP-W) con factores de actividad y regularidad en la producción. PhD thesis, Universitat Politècnica de Catalunya. doi:10.13140/2.1.1691.9362

Bautista J, Cano A, Alfaro R (2012) Modeling and solving a variant of the mixed-model sequencing problem with work overload minimisation and regularity constraints. An application in Nissan's Barcelona Plant. Expert Syst Appl 39(12):11001–11010

Bautista J, Alfaro-Pozo R, Batalla-García C (2015a) Consideration of human resources in the mixed-model sequencing problem with work overload minimization: legal provisions and productivity improvement. Expert Syst Appl 42(22):8896–8910. doi:10.1016/j.eswa.2015.07.044

Bautista J, Alfaro R, Batalla C (2015b) Modeling and solving the mixed-model sequencing problem to improve productivity. Int J Prod Econ 161:83–95. doi:10.1016/j.ijpe.2014.11.018

Boysen N, Fliedner M, Scholl A (2009) Sequencing mixed-model assembly lines: survey, classification and model critique. Eur J Prod Econ 192(2):349–373

Miltenburg J (1989) Scheduling mixed-model assembly lines for just-in-time production systems. Manage Sci 35(2):192–207

Pinedo ML (2009) Planning and scheduling in manufacturing and services. In: Springer series in operations research and financial engineering, 2nd edn, vol XVIII, 536. ISBN: 978-1-4419-0909-1

Proposal of a Framework for Assessing Environmental Performance of Supply Chains

C. Silva, L.M.D.F. Ferreira and S.G. Azevedo

Abstract Companies need to excel in many areas to achieve a competitive advantage. Supply Chain Management is critical for company's overall performance, while its operations can lead to a significant impact on the environment. It is therefore crucial that organizations measure the environmental performance of their supply chains in order to define strategies that contribute to minimize the negative impact of their operations. This paper aims to suggest a framework for the assessment of environmental performance of an upstream supply chain integrating Analytic Hierarchy Process with a modified Balanced Scorecard.

Keywords Environmental performance · Supply chain management · Composite index

1 Introduction

Organizations are increasingly aware and concerned with the environmental and social impact of their business activities. The focus on supply chains is a forward step into a broader adoption and development of sustainability. Supply chain managers must address a complex assortment of factors that include the product and

C. Silva (✉)
CEMUC, Department of Mechanical Engineering, University of Coimbra, Coimbra, Portugal
e-mail: cristovao.silva@dem.uc.pt

L.M.D.F. Ferreira
Economics, Management and Industrial Engineering Department, University of Aveiro, Aveiro, Portugal
e-mail: lmferreira@ua.pt

S.G. Azevedo
Department of Management and Economics, University of Beira Interior, Covilhã, Portugal
e-mail: sazevedo@ubi.pt

M. Amorim et al. (eds.), *Engineering Systems and Networks*, Lecture Notes in Management and Industrial Engineering, DOI 10.1007/978-3-319-45748-2_23

the process on both the upstream and downstream of the supply chain (Vachon and Klassen 2006).

Environmental impact of business activities has become an important issue in the last years due to the growing public awareness of environmental, and the introduction of environmental legislations and regulations mainly in developed countries (Lau 2011).

However, in recent years, more and more companies are introducing and integrating environmental issues into SCM processes by auditing and assessing suppliers on environmental performance metrics (Handfield et al. 2005). In this way they seek to ensure that they have effective tools not only for measuring environmental performance of their suppliers but also for carrying out action plans to improve their performance (Olugu et al. 2011).

The literature shows that most models for evaluating environmental performance focus on the evaluation of the organization itself (Dias-Sardinha and Reijnders 2001) and that the data used is reported by the companies in their environmental reporting (Colicchia et al. 2011). Therefore, the main objective of this article is to propose a framework for the assessment of environmental performance of an upstream supply chain.

The article is divided into five sections. This section seeks to provide an introduction to the topic in question and define the objective of the study: the definition of a model for evaluating the environmental performance of the supply chain. The second section presents a literature review. Section 3 presents a model for evaluating the environmental performance of a supply chain is proposed. Finally, the main conclusions of the study are drawn in Sect. 4.

2 Literature Review

The concept of Supply Chain Management (SCM) was born and brought a new facet to company management in the 1980s. SCM has gained a strategic relevance as a source of competitive advantage (Fine 1998).

The integration of issues related to sustainability in the legislation encourages companies to change the way they operate (Webster and Mitra 2007). These changes require not only the management of new concepts, such as the reverse supply chain, or green purchasing, but also a clear change in existing practices and concepts creating new management and production systems. It has become essential to include the management of by-products and to consider the life cycle of the product in SCM.

For any activity that has strategic implications, such as the management of the supply chain, it is essential to make performance reviews. Although many papers have been published on the topic of assessment of environmental performance within organizations, the emphasis on the evaluation of environmental performance of the supply chain has been relatively limited (Azevedo et al. 2011).

In a supply chain, a significant number of actors influence not only the costs but also the associated environmental impacts. Suppliers, producers, consumers, logistics providers, as well as suppliers of services are the main players. All these players perform most activities that impact business and the environment. Thus, it is necessary to create models that makes possible to assess the environmental performance of the supply chain, promoting also the monitoring of indicators that support decision-making and management (Olugu et al. 2011; Shaw et al. 2010).

Several attempts have been done to develop environmental supply chain performance measures. Azevedo et al. (2011) suggest a model to identify the influence of several green practices on supply chain performance. Nevertheless, the proposed model does not allow to quantitatively assess the environmental performance of a given supply chain. Braithwaite and knivett (2008) propose a model to evaluate supply chains carbon footprint. The model resulting map can be used by the supply chain parties to identify carbon emissions reduction potential and discuss SC re-design to improve it environmental performance. El Saadany et al. (2011), based on an extensive literature review, propose and categorize a set of environmental quality measures. Then, these performance measures, both quantitative and qualitative are aggregated into an environmental quality model which can be used to assess a supply chain environmental performance.

The use of composite indicators is an innovative approach to evaluate sustainable development. Computing aggregate values is a common method used for constructing indices. Indices, which can be either simple or weighted, are very useful in focusing attention and, often in simplifying the problem. Such an approach allows the evaluation of a multitude of aspects which can be deciphered into a single comparable index.

There are also some advances in measuring the environmental performance of companies and their respective upstream supply chain. Azevedo et al. (2013) suggest an "Ecosilient Index" to assess the greenness and resilience of automotive companies and the corresponding supply chain. Tsoulfas and Pappis (2008) propose a set of environmental performance indicators and multi-criteria decision-making methodologies to measure the extent to which environmental principles are fulfilled along the supply chain. Schmidt and Schwegler (2008) also study environmental performance, proposing the concept of cumulative ecointensity to help calculate a score which includes not only a company's direct effects, but also the negative indirect effects in upstream and downstream processes along its supply and waste disposal chain. Hutchins and Sutherland (2008) argue that the company's social performance can be determined by social indicators aggregated into one single weighted measure. They proposed a variety of indicators of corporate social responsibility that are aggregated into a single social sustainability metric for a company. To expand the measure to the supply chain context, value-weighted social sustainability is considered to include both the impact of the company and the social impact of its suppliers.

In the next section we propose just such a framework, describing an integrated approach to assess the level of environmental performance in both individual companies and their corresponding upstream supply chain.

3 Proposal a Framework for Environmental Assessment

In order to address the lack of structured systems for monitoring the environmental performance of the supply chains, the following framework is suggested. The proposed framework is based on the logic of the Balanced Scorecard to assess the environmental performance of the supply chain, while using ISO 14031 and the GRI guidelines to suggest the indicators.

The steps that make up the proposed framework are displayed in the Fig. 1.

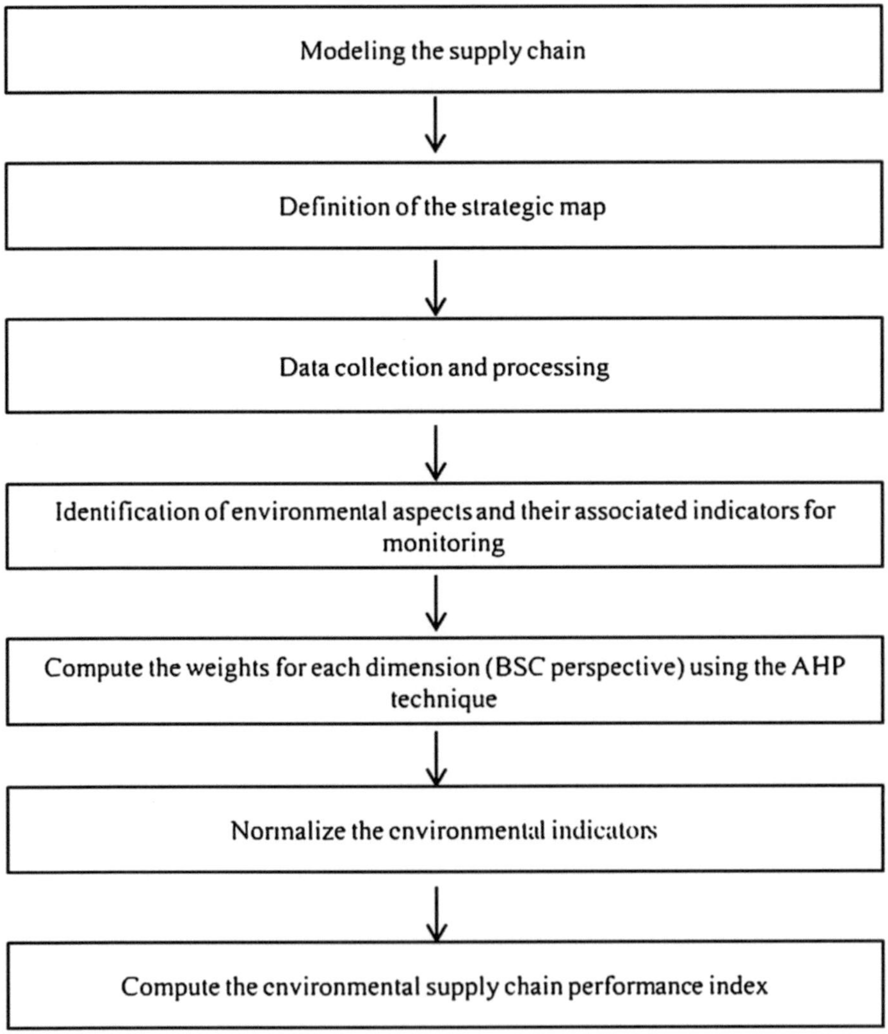

Fig. 1 Framework for evaluating the environmental performance of supply chains

There now follows a description of the different phases of the framework.

Step 1 *Modeling the supply chain.*
The project must start with the study of the supply chain in order to understand its flows, stakeholders and particularities.

Step 2 *Definition of the strategic map for the supply chain.*
The initial formulation of the BSC depicts the strategy of the company distributed over four perspectives. To develop a BSC it is advisable to draw-up a strategic map. The definition of the strategic map should take into account the strategies of the business, SCM and environmental management. In order to facilitate the management of the indicators and avoid introducing additional complexity to both the company's general performance evaluation system and the system to be created, a specific and adapted BSC to monitor the evolution of the environmental performance of the supply chain is suggested.

Step 3 *Data collection and processing.*
The instrument used for collecting the necessary data for performing this framework is a questionnaire to be sent to all first-tier suppliers. This mailing should be performed annually, to make possible that the evolution of the indicators could be monitored and compared across several years. This option represents a simple and effective way to collect the data necessary to evaluate the environmental performance of the supply chains and to incorporate it into the standard procedures associated to the supplier evaluation by most of the companies.

Step 4 *Identification of environmental aspects and their associated indicators for monitoring.*
The chosen indicators should be appropriate to each organization and should be related to the strategic objectives of the organization. The indicators should follow three criteria: measurability, data availability and the indicators should be related to the supply chain type. In this research the indicators for each perspective were selected from the Global Report Initiative (GRI) and ISO guidelines. After selecting the indicators it is necessary to create a calculation method to transform collected data into values that can be compared and analyzed.
In this study the chosen scenario for the application of the suggested framework is to analyze the environmental performance of the supply chain for a given project. Each project is associated to a particular Original Equipment Manufacturer (OEM), whose lifespan is known and where there is no sharing of components between different products that the company produces. However, in some cases the same supplier may provide components for different projects. The indicators are calculated by supplier using Eq. 1.

$$IS_i = (Ind.) \times (ShareS_i) \quad (1)$$

where

IS_i	represents the indicator under study for supplier S_i
Ind.	represents the data for the indicator under study as reported by the supplier S_i
$ShareS_i$	is the percentage of the supplier's total business volume as represented by the purchases made.

This last value is a proxy for developing the model. If for example, the volume of sales of the supplier S_i to the organization under study would be half of its total business, $ShareS_i$ will be 50 %. Ideally, the percentage would be calculated considering the contribution to the impact associated with the respective indicator of the components produced by the supplier to the company. For indicators in percentages the shares are not considered, as is the case with the following indicators: costs, other air emissions, waste water, hazardous waste, compliance with legal and customer requirements, the number of hours of training and certifications held. The indicators that record absolute numbers also do not need shares. In both cases the values of the indicator must be equal to the value reported by the supplier: IS_i = Ind.

Thus we obtain weighted indicators for different processes/business of the supplier. It is only of interest to consider the portion related to the processes associated to the manufacture/acquisition of the considered project. This suggested approach is similar to the method proposed by Hutchins and Sutherland (2008), that is based in the input-output modeling technique, to characterize the social sustainability of a given supply chain. The indicators for each project are calculated using Eq. 2:

$$IP_y = \sum_{i=1}^{m} \frac{IS_i \times SharePS_i}{m} \quad (2)$$

where

IP_y	represents the indicator for the project y
IS_i	represents the indicator under study for supplier S_i (previously calculated)
$SharePS_i$	it is the percentage of sales that the part or component represents for supplier S_i. This percentage is important because a supplier can provide more than one part/component for the same project. If supplier S_i provides only one part/component, then the $SharePS_i$ will be 100 %
m	is the total number of parts produced in the project y

Step 5 *Compute the weights for each dimension using the AHP technique.*
Analytical Hierarchy Process (AHP) is a helpful tool for dealing with complex decision making, and helps to set priorities and make the best decision possible. AHP contributes to the rationalization of the entire decision process. Additionally, the AHP integrates a practical technique for examining the consistency of the decision maker's evaluations, therefore reducing the bias in the decision making process.
An AHP hierarchy model is used to compute the weights for the four dimensions/perspectives of the BSC model. The goal is located at Level 1. Level 2 of the hierarchy contains the four dimensions/perspectives of the BSC. Level 3 contains the sub-criteria/indicators for evaluating each dimensions/perspectives (see Fig. 2). After building the hierarchy a team of evaluators is formed to assign the pair-wise comparisons to the Level 2 used in the AHP hierarchy. A nine-point scale is used. The weights of level 3 sub-criteria will not be computed using AHP pair-wise comparisons (because the possible number of pair-wise comparisons to perform would be very high). In this case we will assume that each sub-criterion will have the same weight. For example, if we have 5 indicators for one of the perspectives each will weight 20 %.

Step 6 *Normalize the sustainability indicators*
The main difficulty in aggregating indicators into the environmental performance supply chain index is because indicators may be expressed in different units. To normalize the indicators, the following procedure will be used:

$$I^{+}_{N,ij} = \frac{I^{+}_{A,ij} - I^{+}_{\min,ij}}{I^{+}_{\max,ij} - I^{+}_{\min,ij}} \tag{3}$$

$$I^{-}_{N,ij} = 1 - \frac{I^{-}_{A,ij} - I^{-}_{\min,ij}}{I^{+}_{\max,ij} - I^{+}_{\min,ij}} \tag{4}$$

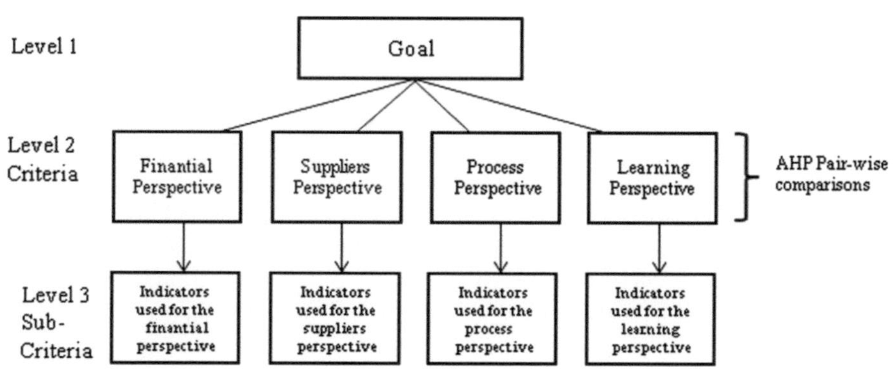

Fig. 2 AHP Model for analysis of the perspectives

where $I^{+}_{N,ij}$ is the normalized indicator *i* (with positive impact) from group of indicators j and $I^{-}_{N,ij}$ is the normalized indicator *i* (with negative impact) from the group of indicators j. In this way, it is possible to integrate different kinds of quantities with different units of measurement. One of the advantages of the proposed normalization method is a clear compatibility among different indicators, since all indicators are normalized.

Step 7 *Compute the environmental supply chain performance index*

At this stage the focus of the study is placed on the development of a methodology for deciding on which of the projects would be most advantageous in terms of the environmental performance of its upstream supply chain. Because each indicator has different units, not comparable with each other and also have a different importance an environmental performance index is proposed.

Using Eq. 5 the environmental supply chain performance index is computed:

$$Env\_Perf\_Index\_SC = \sum_{i} \sum_{j} W_i \times W_{ij} \times I_{ij} \tag{5}$$

where:

Env_Perf_Index_SC	represents the score of the environmental performance index for the supply chain of a product
W_i	represents the weight of the *i*th perspective of the BSC (calculated through the AHP judgments)
W_{ij}	it is the weight of the *j*th subcriteria of the *i*th perspective of the BSC
I_{ij}	it is the normalized score for the *j*th environmental element of the *i*th perspective of the BSC

The follow-up phase for the index is carried out jointly by the Purchasing and Environmental Management departments. In the event of deviations from the targets established, an action plan should be carry on considering the principles of the continuous improvement cycle, present in the PDCA cycle.

4 Conclusions

The starting point of this study was the need to develop a framework to support the evaluation of the environmental performance of an upstream supply chain.

The evaluation process consists of the development of a framework for the assessment of the environmental performance of an upstream supply chain, based on the four perspectives used in the BSC. A group of relevant environmental indicators for each perspective is identified considering the GRI and the ISO 14031 indicators. The model represents an effective tool for decision making support.

One of the difficulties is related to the correct application of this framework, relying on a deep understanding of the environmental impacts of the upstream supply chain. It can also be noted that the level of complexity of the supply chain can be a determining factor for the successful application of the proposed framework, due to the difficulties involved to collecting the data. As previously noted, there are several paths open to future development of the model. The next phase may include the application of this model to different industry sectors. This could contribute for enhancing the usefulness of the framework in a wider and more encompassing way.

References

Azevedo S, Carvalho H, Cruz-Machado V (2011) The influence of green practices on supply chain performance: A case study approach. Transport Res Part E Logistics Transport Rev 47(6):850–871

Azevedo S, Govindan K, Carvalho H, Cruz-Machado V (2013) Ecosilient index to assess the greenness and resilience of the upstream automotive supply chain. J Clean Prod 56:131–146

Braithwaite A, Knivett D (2008) Evaluating a supply chain carbon footprint – A methodology and case example of carbon-to-serve. In: Proceedings of the 13th logistics research network conference, Liverpool, pp 323–328, 10–12 Sept

Colicchia C, Melacini M, Perotti S (2011) Benchmarking supply chain sustainability: Insight from a field study. Benchmarking Int J 18(5):705–732

Dias-Sardinha I, Reijinders L (2001) Environmental performance evaluation and sustainability performance evaluation of organizations: an evolutionary framework. Eco-Manage Auditing 8 (2):71–79

El Saadany AMA, Jaber MY, Bonney M (2011) Environmental performance measures for supply chains. Manage Res Rev 34(11):1202–1221

Fine C (1998) Clockspeed: wining industry control in the age of temporary advantage. Basic Books, New York

Handfield R, Sroufe S, Walton S (2005) Integrating environmental management and supply chain strategies. Bus Strategy Environ 14(1):1–19

Hutchins MJ, Sutherland JW (2008) An exploration of measures of social sustainability and their application to supply chain decisions. J Clean Prod 16:1688–1698

Lau KH (2011) Benchmarking green logistics performance with a composite index. Benchmarking Int J 18(6):873–896

Olugu EU, Aliahmadi AR, Jafari-Eskandari M (2011) Development of key performance measures for the automobile green supply chain. Resour Conserv Recycl 55:567–579

Schmidt M, Schwegler R (2008) A recursive ecological indicator system for the supply chain of a company. J Clean Prod 16(15):1658–1664

Shaw S, Grant DB, Mangan J (2010) Developing environmental supply chain performance measures. Benchmarking Int J 17(3):320–339

Tsoulfas GT, Pappis CP (2008) A model for supply chains environmental performance analysis and decision making. J Clean Prod 16(15):1647–1657

Vachon S, Klassen RD (2006) Extending green practices across the supply chain: The impact of upstream and downstream integration. Int J Oper Prod Manage 26(7):795–821

Webster S, Mitra S (2007) Competitive strategy in remanufacturing and the impact of take-back laws. J Oper Manage 25(6):1123–1140

Root Cause Identification of Existing Barriers Detected by People with Disabilities in Air Transport

I. Garcia-Miranda and A. Duran Heras

Abstract EU legislation on air passenger rights try to ensure the right to free moving of people with disabilities and old persons. Despite of this, disabled people are not travelling as they could be due to existing barriers. In the European Union (EU) one in six people has some kind of disability that varies from mild to severe and due to the augmenting longevity and low levels of natality, population projections foresee that this percentage will increase in the next decades. Given the significance of the problem, research in the field has focused on identification of the barriers found by disabled air passengers so as to suggest solutions for those barriers. However, focusing on the causes that generate those barriers will generate better-suited solutions with higher effectiveness. This study analyzed existing barriers identified by disabled air passengers and applied the best methodology in order to identify root causes. Facing these root causes will contribute to a better human mobility in the Common European Space increasing welfare state and economic rewards in inclusive tourism.

Keywords Root cause determination · Disability · Air transport · Inclusive tourism

1 Introduction

A disabled passenger is defined as one who has mental, long-term physical, intellectual or sensory impairments which requires airlines to give individual attention (on enplaning and deplaning, during flight, in an emergency evacuation,

I. Garcia-Miranda (✉) · A. Duran Heras
Área de Ingeniería de Organización. Dpto. de Ingeniería Mecánica, Universidad Carlos III de Madrid, Avda. de La Universidad, 30, 28911 Leganés, Madrid, Spain
e-mail: ivan.garcia@uc3m.es

A. Duran Heras
e-mail: duran@ing.uc3m.es

M. Amorim et al. (eds.), *Engineering Systems and Networks*, Lecture Notes in Management and Industrial Engineering, DOI 10.1007/978-3-319-45748-2_24

and during ground handling at airports) that is not usually required from other passengers.

This definition is referred not only to medically ill or temporarily disabled persons, but also includes permanently handicapped persons who may not need medical authorization for air travelling although they would need exceptional assistance. In this sense, it may even refer to any person who is at a disadvantage or incapacitated, such as a pregnant woman or an obese person.

Regarding to article 2(a) of Regulation (EC) No 1107/2006, "disabled person" or "person with reduced mobility" encompasses any person whose mobility when using transport is reduced by reason of any physical or intellectual disability or impairment, or any other cause of disability, or age. Those involved in the attention, management and delivery of services should adapt their services to disabled passenger needs. For the last years, transport inaccessibility has been reduced thanks to modernization and adaptation of the infrastructure. This measures have increased passenger flows, impacting on their lives and reducing social exclusion. Air transport for disabled people is not as adequate as it should be, reducing disabled people's living standards. This accessibility should not only be limited to the assistance at the airport and in the aircraft, but also should be extended to the whole accessibility chain. The accessibility chain begins when the person chooses to travel by air and arranges his/her trip. The accessibility chain concludes in the "Post-Trip" phase. At this stage, disabled people could report a claim to the airport or airline, or simply suggest an idea improving service delivery.

European Union (EU) Regulation (EC) No. 1107/2006 supports the rights of disabled passengers in all forms of transportation. This regulation imposes legal obligations on air transport services companies regarding the service they provide to those passengers. The European Commission believes that this regulation will help disabled persons and persons with reduced mobility to meet their needs. Even with the existing EU legislation, air transport for disabled people is not as adequate as it should be. In Europe, 80 million disabled people are represented by the European Disability Forum (EDF). In 2011, EDF launched a survey whose results show that only 14.8 % of Europeans with disabilities feel pleased using public transport in another Member State and 56.4 % of interviewees recognize transportation between EU Member States uncomfortable. Besides, 68.9 % of interviewees with disabilities declare that there are barriers to freedom of movement. In addition, 62.2 % of interviewees with disabilities remark that existing barriers have prevented them from travelling. On top of that, 78 % of Europeans with disabilities state that if there were no barriers, they would consider increased their rights to Freedom of movement.

Barriers at this stage have a significant impact since they prevent disabled people and people with reduced mobility from travelling with the consequent social and economic impact. One objective of this investigation was to study and determine the root causes of barriers found by disabled air passengers. Another objective of the investigation was to develop recommended practices for dealing with the root causes as a consequence of a clear understanding of the root causes. Root cause generation methodologies will be thoroughly reviewed, in order to identify the most appropriate approach, taking into account the problems to be tackled. Thus, root causes will be identified and characterized. After cause identification is accomplished, different detailed approaches, techniques and tools can be used to define how the causes are connected to the problem. That will lead to implementing better-suited solutions with higher effectiveness to prevent the problem from happening again.

2 Accessibility for All

According to the World Report on Disability (2011), there are around 785 to 975 million people 15 years and older living with disability. Out of these, around 110–190 million experience significant difficulties in functioning due to "severe disability"—the equivalent of disability result from conditions such as quadriplegia, severe depression, or blindness. Thus, including children, the WHO estimates that over 1.000 million people (about 15 % of the world´s population) are living with a disability.

According to the European Commission's European Disability Strategy 2010–2020, one in six people in the European Union (EU) has some kind of disability that varies from mild to severe, therefore around 80 million people have less opportunities for travelling because of environmental and attitudinal barriers. Furthermore, the incidence of disability is increasing forcefully due to population ageing.

Eurostat´s population predictions foresee that the ageing process will persist in the next years. The Eurostat study "Ageing characterizes the demographic perspectives of the European societies" (Giannakouris 2008), encompassing the 2008–2060 period, shows that the EU27 population is predicted to become older with the median age calculated to scale from 40.4 years in 2008 to 47.9 years in 2060.

Given the significance of the future population problem, research in the field has to be focused on accessibility problems in air transport in order to be prepared for next years demand.

2.1 Accessibility Problems in Air Transport

Regarding Articles 45 and similar of the Lisbon Treaty, the free movement for people encompasses the right for all EU citizens to study, work, establish themselves and travel or live in another Member State.

Disabled people feel limited about their opportunities and independence. Impairments only become barriers when the environment in general, and the transportation system in particular, creates demands that the individual cannot meet.

These "barriers" to travel faced by disabled persons can be classified into three areas: economic, physical and attitudinal (Murray 1990). Burnett (2001) tested four hypotheses relating severity of disability to destination decision criteria. Difficulties

Table 1 Disabled air passengers' complaints (Chang 2012)

Airport facilities	Airline facilities and services
Distance between parking lot and terminal	User-friendly on-board restroom
Barrier-free lift	Space in on-board restroom
Barrier-free ramp	Airline wheelchair services
Waiting area at boarding lounge	Information on emergency evacuation
Barrier-free restroom	Consignment and retrieval of the wheelchair
Slip resistant doors in the airport	Cabin crew's service attitudes
Accessibility between terminals	Ground service staff's attitudes

faced during air travel by disabled people were described by Shaw (2004) and Bi et al. (2007) suggested that if disabled passengers face few barriers during their travels, they may travel even more in the future. So as to understand the disabled traveler's needs, Chang (2012) identified the most common complaints regarding airport facilities and airline facilities and services. These complaints are shown in Table 1.

An extensive analysis of all barriers found by disabled people has been conducted within an FP7-EU funded project oriented to support the European Disability Strategy 2010–2020 and its implementation plan. The general objective is to identify, characterize, justify and prioritize research and analysis approaches in those solution areas with the greatest potential towards improving access to air transportation for people with disabilities and elderly people.

Problems were identified as result of analysis of the barriers to accessibility detected by disabled people analyzing primary information captured from all stakeholders (interview recordings, questionnaires…) using social research techniques. Since the number of barriers detected exceeded reasonable figures, a cause identification analysis was proposed so as to assure if those problems could be originated by the same root causes. These barriers were used as the basic input for the cause analysis described in this paper.

2.2 *Root Causes Definition and Generation Tools*

If high impact by adopted measures is needed, solutions should go beyond the barriers perceived by the users and address the actionable causes that originate them. The definition of root cause (Rooney 2004) could encompass the following features: Root causes are specific basic and fundamental causes. Those causes that can rationally be recognized. Root causes are those over which management has control to fix.

Usually a problem is the consequence of several root causes, at different "levels" (Andersen 2006). Root causes can be classified into: Symptoms, First level causes or Higher level causes.

Symptoms are signs of existing problems.

First level causes are those that drive to the problem straight.

Higher level causes are those that conduct to the first level causes. They do not cause the problem directly but form connections in the concatenation of cause-and-effect relationships that finally constitute the problem.

The search for the root cause is a questioning process (Dew 1991) that might be ignored because root-cause analysis implies asking questions about how an organization is managed. There are several methodologies for identifying the root causes: Brainstorming; Nominal group technique; Paired comparisons; Event and causal factor diagram; The "5 Why's"; Brainwriting; Is-is not matrix; Safeguard analysis; Change analysis. Some of them are defined below (Andersen 2006):

- Brainstorming: A formal approach that can be applied throughout the root cause analysis when multiple ideas are required.
- Nominal group technique: A method used to help a group prioritize different alternatives, for instance, possible problem causes.
- Paired comparisons: A procedure used to reach consensus by allowing participants to choose between pairs of two competing alternatives.
- The event and causal factor diagram (Dew 1991) establishes a concatenation of events leading to the problem. Once the relevant events have been identified and placed in their proper sequence, the question "What allowed this to happen?" should be answered for every step. The diagram brings a visual tool for analyzing the relevant actions and for tracing the path to their roots.
- The 5 Whys is the most used tool and it is defined in the third step of the Six Sigma DMAIC (Define, Measure, Analyze, Improve, Control) methodology (Pyzdek 2003). This tool is an iterative interrogative methodology that can lead to a wrong determination of the root cause if there are not true evidences.

Although some research about comparison of root cause analysis tools and methods has been conducted (Gano 2003; Doggett 2004, 2005), none has been focused on comparison root cause determination tools. Nevertheless, brainstorming is one of the most common used tool since the publication of Osborn's (1953) influential book *Applied Imagination*, not only for cause generation but also for idea generation and as an innovation tool. Brainstorming research has targeted almost exclusively on productivity (i.e., number of ideas created) as a dependent variable, investigating possible causes for the productivity loss in interactive brainstorming groups and methods for reducing productivity loss (Rietzschel et al. 2006).

3 Methodology

As aforementioned, there are several root cause determination methodologies. In this study a brainstorming tool was used to firstly identify the causes of barriers met by disabled people. Group brainstorming was considered as a particularly effective technique for generating large numbers of creative ideas in a first approach. Brainstorming procedure was divided into cause generation and cause selection as people are thought to generate more ideas when they feel free of evaluation and criticism.

Cause generation was supported by the results of individual semi-structured interviews done with different stakeholders, experts and end users. Users were selected based on the categories established in air transport for disabled people. These categories are designed so as to classify users on the basis of their capabilities with the purpose of offering them the assistance required to set their trip. To meliorate root cause determination, nominal group technique was also applied.

4 Causes of Existing Barriers

When all the primary root causes were identified, a checklist was carried out in order to verify that experts and stakeholders could address all the problems and barriers identified with the causes shown in the checklist and in case of missing causes, new ones will be added to the checklist. 77 causes were identified thanks to this holistic methodology and some of them are shown in Table 2.

Table 2 Example of causes generated in this study

Causes				
There is no alternative channel through which information can be provided (e.g. inexistence of mobile phone applications for bidirectional communication)	There are no explanations on how to use it (aircraft controls, equipment and devices in the terminal, etc.)	The signaling is not standardized/universal; there are differences between different means of transportation or even from one airport to other	The equipment and devices in the terminal (information screens, auto check-in kiosks, call for assistance points, etc.) are not accessible for all the users	There is no specific regulation regarding the design of equipment and devices for the terminal (information screens, auto check-in kiosks, call for assistance points, etc.)
The information does not comply with universal design principles	Lack of training and awareness of the staff	The plane hold's door is too narrow	It is necessary to disassemble the wheelchairs to store them in the hold	There is no direct path to the boarding gate, it is necessary to go through the shops

Causes were classified into areas related to accessibility barrier typology. Not only were the causes classified into areas but they were also structured in “layers” according to their position in the causal chain, i.e. their level of generality.

Cause identification is usually used to accomplish causal tree/network or the appropriate root cause analysis tool. Root cause analysis tools include the ability to find causal interdependencies, factor relationships and have tools for evaluating the integrity of common conclusions.

So as to evaluate the integrity of cause identification, causal networks have been created linking the identified barriers to gradually more generic, high level causes. The high level causes would comprise top level, primary, root causes, which influence large sections of the causal network. Various types of analysis carried out (causal analysis, SWOT analysis…) have shown that above this layer there is still at least another layer of even more fundamental driving forces, e.g.: Demographic trends; Awareness and attitudes of the society at large towards diversity in general and specifically towards people with disabilities; General Technological evolution; European (and worldwide) Political, Economic and Legal-Regulatory structure and framework; Economic situation and its implications on resource availability and business priorities of involved stakeholders; Evolution of the Medical Science. All these high level causes address air transport accessibility problem.

5 Conclusions

Even though the current status of universal access to air transport is still far from achieving its desired level, the paths for improvement become clearer when root causes are identified. Given the complexity and interrelated nature of the issues involved and the heterogeneity of the target population, a holistic, comprehensive approach for the generation of the causes has been adopted, taking simultaneously into account the barriers affected, stakeholders, end users, cause integrity and the potential impact and the relationship with other root causes.

Root causes were identified so as to address problems/barriers met by disabled air passengers. Causes were classified by areas and structured by layers/level. At the highest level, causes identified in the barriers analyzed converge into common “root causes”. This convergence allows the identification of what the real causes of the problem are. Taking action on these root causes can result in effective solutions.

In the middle layers, even though each barrier is different, a certain degree of commonality and overlapping exists. This commonality/overlapping pursues defining solutions to prevent and remove barriers for disabled people and to ensure access on an equal basis with non-disabled people.

This study provides important analysis in order to raise the voice of disabled people in Europe and to seek their underlying challenges as part of a deeply committed process to provide accessibility in all daily aspects of transport for everybody.

References

Andersen B, Fagerhaug T (2006) Root cause analysis: simplified tools and techniques. ASQ Quality Press

Burnett JJ, Baker HB (2001) Assessing the travel-related behaviours of the mobility-disabled consumer. J Travel Res 40(1):4–11

Bi Y, Card JA, Cole ST (2007) Accessibility and attitudinal barriers encountered by Chinese travellers with physical disabilities. Int J Tourism Res 9(3):205–216

Chang YC, Chen CF (2012) Meeting the needs of disabled air passengers: factors that facilitate help from airlines and airports. Tour Manage 33(3):529–536

Dew JR (1991) In search of the root cause. Qual Prog 24(3):97–102

Doggett AM (2004) A statistical comparison of three root cause analysis tools. J Ind Technol 20 (2):2–9

Doggett AM (2005) Root cause analysis: a framework for tool selection. Quality Management Journal 12(4):34

Gano DL (2003) Apollo root cause analysis: a new way of thinking. Apollonian Publications, Kennewick

Giannakouris K (2008) Ageing characterises the demographic perspectives of the European societies. Stat Focus 72:2008

Murray M, Sproats J (1990) The disabled traveller: tourism and disability in Australia. J Tourism Stud 1(1):6–15

Osborn AF (1953) Applied imagination. Charles Scribner's Sons, Oxford

Pyzdek T, Keller PA (2003) The six sigma handbook, vol 486. McGraw-Hill, New York

Rietzschel EF, Nijstad BA, Stroebe W (2006) Productivity is not enough: a comparison of interactive and nominal brainstorming groups on idea generation and selection. J Exp Soc Psychol 42(2):244–251

Rooney JJ, Heuvel LNV (2004) Root cause analysis for beginners. Qual Prog 37(7):45–56

Shaw G, Coles T (2004) Disability, holiday making and the tourism industry in the UK: a preliminary survey. Tourism Manage 25(3):397–403

Spare Parts Inventory Management Using Quantitative and Qualitative Classification

Fernanda Oliveira and Clara Bento Vaz

Abstract This paper focuses on the spare parts inventory management of a maintenance provider of the health sector, where the commitment to ensure the agreed customer service level and the guarantee of maximum availability of the devices are relevant issues. Spare parts inventory management requires working with erratic and uncertain demand. As in the most cases, there is a high sporadic pattern in the demand characterized by low demand average rates, it is quite difficult to ensure that the right spare part is at the right place, at the right time and at the right quantity. This study determines an adequate inventory management policy for spare parts, specifically for unplanned maintenance operations taken into account the agreed customer service level. Considering that the criticality of a spare part has consequences regarding the availability of an equipment and service level agreement, the spare parts were classified in terms of quantity, value of usage and criticality. Based on this classification, differentiated service levels and inventory management policies were adopted for each group.

Keywords Inventory management · Spare part · ABC analysis · Criticality analysis

F. Oliveira
Polytechnic Institute of Porto, APNOR, Porto, Portugal
e-mail: fernanda.oliveira@fmc-ag.com

C.B. Vaz (✉)
Polytechnic Institute of Bragança, Bragança, Portugal
e-mail: clvaz@ipb.pt

C.B. Vaz
CGEI/INESC TEC, Porto, Portugal

M. Amorim et al. (eds.), *Engineering Systems and Networks*, Lecture Notes in Management and Industrial Engineering, DOI 10.1007/978-3-319-45748-2_25

1 Introduction

This paper is motivated by a real case study concerning the spare parts inventory management derived from unplanned maintenance activities in a maintenance provider, in the health sector. The importance of the subject in a maintenance environment, as well as the practical application of the defined method were essential elements in the development of the spare part inventory management model.

In order to support and achieve competitive advantage and create added value for the customers (Porter 1985) an efficient inventory management is essential in any maintenance department. The main objective of an inventory management system is to achieve the requested service level with the minimum inventory investment and administrative costs (Huiskonen 2001). Regarding the main maintenance types, different inventory control practices are needed for preventive maintenance and unplanned maintenance (Celebi 2008). Kennedy et al. (2002) attend that for preventive or scheduled maintenance, the demand for spare parts is predictable. The spare parts for these planned activities may arrive just in time. In the case of the unplanned maintenance, the consequences of not having the spare parts in stock may lead to a decrease of production with significant costs. In order to be more efficient and effective, a safety stock policy is necessary.

It should also be considered that the inventory planning of spare parts differ from other materials, since the service levels may be high and the demand for spare parts may be extremely sporadic and difficult to forecast (Huiskonen 2001). To support decision making, the spare parts inventory management requires a balance of the various costs related to the achievement of the agreed customer service level.

Another approach given in the literature is that spare parts are often managed by applying general inventory management principles which are independent on the specific characteristics of spare parts (Huiskonen 2001). For example, Dekker et al. (1998) define the criticality of an equipment as its importance to sustain the production in a safe and efficient way. According to De Felice et al. (2014), as the spare parts have different importance for the safe operation of a device, different inventory management policies should be adopted for each spare part group. Our approach uses this perspective and restrictions concerning the compliance with the contracted customer service level and assurance of the maximum availability of the spare parts. The high availability of spare parts improves significantly the service level, increasing the inventory costs. Insufficient stocks can lead to high downtimes of the equipment and excessive stock may increase the inventory costs and operating costs. As these factors are critical in efficient inventory management, it is important to have a detailed analysis of all the conditions that affect the logistics of spare parts in order to make the right decisions and apply the adequate inventory management policy.

Bošnjaković (2010) proposes a quantitative and qualitative classification of the spare parts in groups according to their value of usage, frequency of usage and criticality. Afterwards, the adequate inventory management policy should be defined based on the resulting combination. Following Bošnjaković (2010), this study develops a methodology based on a quantitative and qualitative classification for spare parts in groups, proposing an inventory management policy for each spare part group of the classification. This is established on classifying the spare parts in groups according to their value of usage, quantity of usage and criticality. The objective of this approach is to minimize inventory management costs and to ensure the agreed customer service level.

The remainder of this paper is organized as follows. Next section describes the methodology used and the results achieved. Finally, last section summarizes the paper findings.

2 Methodology

The case study is carried out in a maintenance department with a decentralized technical team, distributed by geographical area. In each acting area, the technician is provided with a car where the whole spare parts stock is stored. According to the agreed service level contracted with the customers, each technician has to solve any unplanned maintenance occurred within a time frame of 24 h. This issue may possibly lead to excessive stock at the technician side, since no one wants to be affected by a stock out of a spare part. This stock out situation can also lead to critical downtimes of an equipment at the customer side.

Regarding the supply of spare parts it should be mentioned that only one supplier is available and any purchase order has to be placed when needed, once a week in a specific day by the technical team. The lead time is constant and equal to 7 days.

In order to suggest a solution for the spare parts inventory stock management for each technician, an evaluation of the current inventory management costs was done and a methodology to support the spare parts supply decision was developed, by assuring the minimization of the inventory management costs and the fulfilment of the contracted customer service level.

2.1 Inventory Management Costs

The inventory management is strongly influenced by the nature of demand, lead time and costs. As these variables interact with each other, determining the efficiency and effectiveness of the inventory system, it is important to calculate the inventory management costs in the actual context, in order to compare them with the resulting costs derived from the proposed inventory management policy. The annual total relevant inventory management costs have been calculated based on the acquisition cost, the ordering costs, stock out costs and the holding costs due to regular and safety stock, by the model (1) (Ballou 2004). We estimate that the stock out costs depend on the expected number of cycles with stock out.

$$TC = d \cdot \mathrm{c} + \mathrm{F}_0 \cdot \frac{\bar{d}}{Q} + \mathrm{F}_1 \cdot \mathrm{c}\left(\frac{Q}{2} + SS\right) + \mathrm{F}_2 \cdot n_r \quad (1)$$

We shall use the following notation:

TC Total inventory cost
SS Safety stock
d Annual demand
$\bar{d}$ Average annual demand
c Item value
F_0 Ordering cost per order
Q Order quantity
F_1 Annual inventory holding cost (percent of item value per year)
F_2 Cost per cycle with stock out
n_r Number of cycles with stock out per year

2.2 *Combining the Cross ABC and Criticality Analysis*

The cross ABC analysis is combined with the criticality analysis (Bošnjaković 2010; De Felice et al. 2014) to classify the 1060 spare parts in 27 groups. The cross ABC analysis allows a more efficient management of the inventory, since it simplifies the analysis and identification of spare parts with the highest quantity of usage and value of usage, besides it constitutes an asset in the qualitative evaluation of inventory. Firstly, the spare parts are classified by the quantity of usage into A, B and C classes. Afterwards, the spare parts are classified in the three classes according to the value of usage.

According to the criticality analysis the spare parts are classified into three categories based on the impact of the consequences in the operation of the customer device: Vital, Essential and Desirable. Vital spare parts are those whose stock out causes the highest impact, the device may not work without the spare part and does not allow the medical care. Essential spare parts cause some loss, affecting the quality of service, however the machine may wait for a small period of time, as the absence of that spare part does not inhibit the operation of the medical device. Therefore, the device is still operational, but without some of its functions. Desirable spare parts are those whose the lack of items has no great interference or influence on the operation of the device. The combination of the three analysis leads to a total of 27 groups, that may be represented in a three dimensional graphic, as shown in Fig. 1.

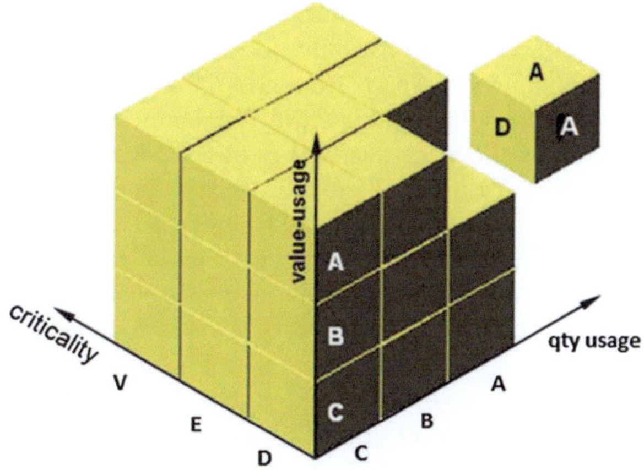

Fig. 1 3D graphic of cross ABC and criticality analysis (adapted from Bošnjaković (2010))

Although the combination of cross ABC and criticality analysis is essential for the determination of the inventory management policies, it does not identify by itself the appropriate inventory management model for each group of spare parts. The information given by cross ABC and criticality analysis supports the definition of the inventory policy and the service levels adopted for each spare parts group. Thus, the service level should increase or reduce depending on the significance given to each spare part.

2.3 Inventory Management Policies Determination

In order to clarify the inventory management policies to apply to each group, in the three dimensional model, Table 1 identifies the inventory management policies as well as the customer service level which should be adopted for spare parts in each

Table 1 Inventory management policies and service level of the 27 groups

Group	Service level (%)	Group	Service level (%)	Group	Service level (%)	Inventory management policy to be implemented
AAV	99	ABV	99	ACV	99	Adjusted continuous revision (Q*, R)
BAV	99	BBV	99	BCV	99	
CAV	95	CBV	99	CCV	99	
AAE	95	ABE	95	ACE	95	
BAE	95	BBE	95	BCE	95	
CAE	90	CBE	90	CCE	90	
AAD	90	ABD	90	ACD	90	
BAD	–	BBD	–	BCD	–	To be ordered when needed
CAD	–	CBD	–	CCD	–	

group which are constrained by the customer contract agreement. The customer service level is the probability of no stock out per order cycle.

Excluding spare parts from group CAV, a service level of 99 % was defined for all the spare parts of groups with Vital criticality, since the lack of any of these parts would not allow the availability of a device for medical operation. Regarding the spare parts from CAV group, a service level of 95 % was defined as these parts have low quantity of usage but high value of usage. The spare parts of the CAV group represent a high investment, thus it is acceptable to reduce the service level.

For the spare parts of group with Essential criticality, AAE, ABE, ACE, BAE, BBE and BCE, a service level of 95 % was defined, assuming a risk of stock out of 5 % and considering that the lack of spare parts of this group does not hinder the operation of the medical device. For the spare parts from groups CAE, CBE and CCE, a service level of 90 % was defined given their lower demand, reducing the total inventory cost.

Regarding the spare parts from Desirable group, a service level of 90 % was assigned to the spare parts of groups AAD, ABD and ACD, since these parts are characterized by a higher quantity of usage, contributing to a better technical performance. Furthermore, it avoids the need to wait for a next order to perform the maintenance intervention.

Regarding the inventory management policy, the spare parts from groups with medium and low quantity of usage (BAD, BBD, BCD, CAD, CBD and CCD) should be ordered only when needed. These spare parts can be managed through the implementation of a Kanban system. Through visual management, Kanban provides simple and intuitive indications of the stock situation, triggering purchase orders when the stock reaches a certain level or amount corresponding to the reorder point. For the remaining groups of the classification, the spare parts should be managed through the continuous review model with a small adjustment, since the purchase orders may only be placed once a week, which is equivalent to continuous review policy, discussed in the next section.

2.4 Adjusted Continuous Review Policy

Consider an inventory problem for a single installation (single-echelon) where items can be handled independently with discrete stochastic demand and constant replenishment lead time. Additionally, possible stock outs can occur and the quantity discounts are not allowed. As the demand is stochastic, it is necessary to determine the reorder points, or equivalently safety stocks. To do this we first of all need a suitable demand model during the lead time (Axsäter 2006).

According to the adjusted continuous review policy, when the inventory position drops to or below the reorder point R, an economic order quantity of Q* is ordered to replenish the inventory in the car of technician, with a small adjustment, since the purchase orders may only be placed once a week, in the established day with the single supplier. We determine Q* according to the basic EOQ formula (2), as it is a satisfactory approximation in case of stochastic demand (Axsäter 1996; Ballou 2004).

$$Q^* = \sqrt{\frac{2 \cdot F_0 \cdot \bar{d}}{F_1 \cdot c}} \tag{2}$$

The demand during the lead time (x) is a discrete stochastic variable and is derived from the total number of spare parts demanded by a set of the customers located in a given geographical area which is managed by the maintenance technician. According to the historical data, the demand during the lead time (x) has a Poisson distribution with average (*D*) score estimated on the mean of weekly demand. Effectively, the weekly demand for each spare part is very low, sporadic and unpredictable.

The reorder point R is determined according to the service level, defined in the previous section. Thus, there is a specified probability (service level) for the demand during the lead time to be lower than R, i.e., $P(x \leq R)$. The reorder point is derived from the reverse of the cumulative Poisson distribution function and the safety stock is equal to R–D.

2.5 Evaluation of the Proposed Inventory Management Policy

The proposed inventory management policy was applied to three spare parts from the AAV group, managed by a given technician, as these spare parts involve high criticality, high quantity and value of usage to calculate the total relevant inventory management costs as described in (1). These results are compared with the costs obtained from the traditional method used by the technician to manage those spare parts. Table 2 shows that the proposed policy is the most economic option in terms of ordering cost, stock out cost and total inventory management costs, fulfilling the service level contracted with the customer.

The traditional method is effectively more economic regarding the holding costs but the ordering costs and stock out cost are clearly higher than in the proposed policy. Also the service level is never fulfilled in the three spare parts.

Based on the results achieved, a decrease of the total inventory management costs is expected for all spare parts, as well as a significant increase in the service level for the global spare parts list.

Table 2 Total costs regarding the traditional and the proposed inventory policies

Item	Cost (€)	Traditional method	Proposed method
1	Acquisition	423	395
	Stock out	122	–
	Ordering	112	32
	Holding	17	50
	Total	673	476
2	Acquisition	1427	1522
	Stock out	61	–
	Ordering	207	64
	Holding	56	122
	Total	1752	1708
3	Acquisition	322	354
	Stock out	182	61
	Ordering	176	48
	Holding	6	40
	Total	686	503
	Total of three items	3111	2688

3 Conclusion

The combination of the cross ABC and criticality analysis allowed to implement an inventory management policy never before used in the maintenance department through the definition of different inventory management policies and the establishment of service levels according to the spare parts criticality, the value and quantity of usage.

By implementation the proposed continuous review policy, the three AAV group spare parts present a higher service level than in the traditional method, as well as a reduction of 14 % in the inventory management total costs due to the decrease in the ordering and stock out costs. The spare parts from groups with medium and low quantity of usage should be managed through the implementation of a Kanban system which is essential to control and maintain the inventory of these spare parts at the optimum level. With the proposed inventory management policies, improvements in the current service level and a reduction of the global inventory management costs are expected, in compliance with the strategic objective to fulfil the customer service level of the maintenance provider which decided to implement it in the near future.

For future developments, the gradual integration of each technician in the implementation of the proposed inventory management policy was suggested. This implementation should be monitored by calculating the inventory management total costs and real service level.

References

Axsäter S (1996) Using the deterministic EOQ formula in stochastic inventory control. Manage Sci 42:830–834

Axsäter S (2006) Single-echelon systems: reorder points inventory control. Springer, New York, pp 77–128

Ballou RH (2004) Business logistics/supply chain management: planning, organizing, and controlling the supply chain. Pearson Prentice Hall, Upper Saddle River

Bošnjaković M (2010) Multicriteria inventory model for spare parts. Tehnički vjesnik 17:499–504

Celebi D, Bayraktar D, Aykac DSO (2008) Multi criteria classification for spare parts inventory. Paper presented at the 38th computer and industrial engineering conference

De Felice F, Falcone D, Forcina A, Petrillo A, Silvestri A (2014) Inventory management using both quantitative and qualitative criteria in manufacturing system. Paper presented at the 19th World Congress The International Federation of Automatic Control

Dekker R, Kleijn MJ, de Rooij PJ (1998) A spare parts stocking policy based on equipment criticality. Int J Prod Econ 56:69–77

Huiskonen J (2001) Maintenance spare parts logistics: special characteristics and strategic choices. Int J Prod Econ 71:125–133

Kennedy WJ, Wayne Patterson J, Fredendall LD (2002) An overview of recent literature on spare parts inventories. Int J Prod Econ 76:201–215

Porter ME (1985) Competitive advantage: creating and sustaining superior performance. Free Press, Collier Macmillan, New York, London

Sustainable Supply Chain Management: A Case Study

Implementation the Sustainability Tools in Supply Chain Management: A Case Study

Cláudia Silva and Joaquim Borges Gouveia

Abstract The present paper aims to contribute to the development of the theory concerning the models of sustainable supply chain management. Argues that the development of a model of sustainable supply chain management, combined with a set of appropriate support tools at different levels of management (strategic and operational) will help organizations to develop integrated sustainability programs in the management of organizations. The development of a Case Study allowed sharing a set of support sustainability practices implemented in the Purchasing Process. This Case Study, combined with the Literature Review, allowed the construction of a framework for the Sustainable Supply Chain Management. The key feature differentiating of this framework its operational nature, by integrating a set of supporting sustainability practices.

Keywords Sustainability · Supply chain management · Sustainability tools · Case study

1 Introduction

The main factor that justifies the interdisciplinary of sustainability and supply chain management is the economic situation that brings new risks and challenges. Globalization has given rise to new and complex social and environmental pressures on businesses. The sustainability has been identified as the solution to this problem. However, the Sustainability Management has been rarely addressed from an operational point of view and is still closely associated with the communication and promotion the image of organizations (AEP 2011). More than knowing the

C. Silva (✉) · J.B. Gouveia
Dpto de Economia, Gestão e Engenharia Industrial, Universidade de Aveiro, 3810-193 Aveiro, Portugal
e-mail: Claudia.margarida@ua.pt

J.B. Gouveia
e-mail: bgouveia@ua.pt

M. Amorim et al. (eds.), *Engineering Systems and Networks*, Lecture Notes in Management and Industrial Engineering, DOI 10.1007/978-3-319-45748-2_26

definition of sustainability, the question arises as to reach sustainability and how it can be monitored in favour of management of organizations? (Carvalho 2010).

The main proposition of this paper argues that sustainability can and should be integrated in the management of organizations, focusing up in the supply chain. To support this proposition, were structured the following main research questions:

- Which sustainability support tools can be implemented on supply chain management?
- How sustainability support tools can be implemented in the supply chain management?

The research methodology is structured on literature review, able to support the theoretical foundations (conceptual models) used to develop a Case Study.

2 Literature Review

This section is subdivided. The first point we present the main concepts related to sustainability and the identification a set of tools that can be implemented in the sustainability program. In the second subsection, we describe the main theory contributions to sustainable supply chain management.

2.1 Sustainability and Support Tools

The major contribution of the Brundtland Report (1987) was the consensus of a general definition of sustainable development as "development that meets present needs without compromising the ability of future generations to meet their own needs."

Sustainability only makes sense when understood in a holistic, dynamic and multi-criteria conception. Holistic, because an individual approach does not lead to sustainability, an open systemic perspective is required with the identification and involvement of stakeholders (Esty and Winston 2006). Dynamic, because decisions need to support the current needs without compromising future (Brundtland 1987). Finally, multi criteria with the integration of *Triple Bottom Line*: Identification of economic, social and environmental vectors approach and understanding of the interrelationship between them (Elkington 1997).

Currently, the question arises to reach sustainability so that it can be implemented and managed in terms of the competitiveness of organizations (Carvalho 2010). From the literature, we identified several tools that support the implementation of sustainability in organizations. The authors Chalmeta and Palomero (2011) present the set of tools organized in three categories: Tools structured in government areas; Tools based on normative references; Tools supported by indicators.

Carvalho (2010) present the following tools: Environmental Impact Assessment; Environmental Management Systems; Life Cycle Analysis; Communication and Public Reporting of business commitments. Apart from these, the authors Esty and Winston (2006) add Life Cycle Product Analysis and problem identification through AUDIO methodology. Chalmeta and Palomero (2011) have been dedicated to the development of the Balanced Scorecard (BSC) with integration of sustainability concepts.

The same path is covered by performance frameworks, like EFQM[1] Framework for Corporate Social Responsibility (Pojasek and Hpllist 2011). The organization WBCSD (2000) has advocated the concept of eco-efficiency. Another tool, sustainability reports, has been developed by GRI with the publication of guidelines for sustainability reporting (GRI 2013).

Supported in the diversity of tools identified in the literature, other authors (Silva et al. 2014) present a classification tools model, built into three main levels: Strategic; Operational; and Evaluation. These authors organize the different tools according his process of implementation, since strategy management, operational management and finally the evaluation results.

2.2 *Sustainability in the Supply Chain Management*

Some professionals have launched sustainability integration projects in the supply chain just because the relief and fashionable subject, without a holistic and strategic vision to connect all the "puzzle pieces" in order to achieve beneficial results for the organization. It is therefore crucial that the research undertaken in this area can develop models to help managers create sustainable supply chains (Pagell and Wu 2009).

The authors Carter and Rogers (2008) developed a model for Sustainable Supply Chain Management (SSCM) strongly supported in the organization management. These authors demonstrated that SSCM should not be developed through social and environmental initiatives fragmented and disintegrated the organization's strategy as it will lead to the conflict with economic, environmental and social objectives.

Through a study applied to 10 companies, considered by as "exemplary" in the area of sustainability, the authors Pagell and Wu (2009) developed an integrative model of Sustainable Supply Chain Management. The study results, allowed the authors argue that the development of a sustainable supply chain requires a proactive attitude of top management, able to understand the challenges of sustainability and take it as a compromise. Sustainability is an integral part of the business and is incorporated into various aspects of the supply chain.

Others authors (Seuring and Müller 2008) present a model developed in three main components: Identification of the reasons for the Sustainable Management of

[1]http://www.efqm.org/.

the Supply Chain; Risk Management and Performance Management of suppliers; Supply Chain Management for sustainable products.

For the analysis of the models described above, the development of any sustainability program in the supply chain requires the formulation and implementation of an integrated and strategic approach addressing the economic, social and environmental vectors in the inter-relational way.

3 Research Methodology

The research methodology focused on the literature review, essential to sustain the theoretical foundations (conceptual models) used to develop a Case Study.

For selection of case study, were defined the following criteria:

- Maturity in the implementation of sustainable practices and existence of sustainable practices developed in the supply chain;
- Organization with activity in Portugal, due to accessibility.

According to these criteria, we present the case study developed in Bosch Thermotechnology, an international leader in the manufacture of heating systems and hot water. Is important to note, that the unit of analysis this study focuses only an individual operational area the supply chain management: the Purchasing Process. To increase the reliability of research (Yin 2010) was built a database to organize all the information gathered, allowing access to primary data organized, according to Table 1. This database was created on *software* WebQDA (Souza et al. 2011).

As shown by the information table, information sources are multiple, giving a valid character to study the search for various evidence and subsequent data´s triangulation (Yin 2010).

4 Results and Analysis

As described in the methodology, the work presented is the development of an empirical research through a case study, which will be the bridge between the theoretical conceptual models and practical experiences. After collecting the data,

Table 1 Sources of information

Sources of information
Sustainability report 2012 Bosch (Bosch 2012)
Five interviews: Purchasing Director (2 interviews); Purchasing Lider (2 interviews); Quality Lider (1 interview) Collection Period: From July 2013 to June 2014
Several internal documents: procedures, organization charts, codes of ethics and conduct, internal standards, internal *frameworks,* other internal documents

Source Elaborated by author

was required a consistent analysis of the information, consisting of the examination, categorization and testing the evidences (Bogdan and Biklen 2013).

The data analysis process was made with help the software WebQDA, according technique Content Analysis defined by Robert and Bouillaguet (cited in Amado 2013).

We present a set of tools implemented examples in the Procurement Process's Bosch, organized according to the classification model developed by Silva et al. (2014). We decided to present the practices/tools with direct excerpts from interviews/documents included in the data collection.

- Pro-active Tools: "Bosch Solidarity Run—a good part of the funds are collected from our suppliers";
- Strategic Tools:
  - Policies: "Bosch will not work with any supplier that proves it does not meet the basic standards set by the International Labor Organization." (Basic Principles of Social Responsibility Bosch, March 2004).
  - Standards: ISO 14001—"We have concrete examples of suppliers that have implemented the 14001 at the request of Bosch." "We do not work with suppliers that do not have environmental certification, is allowed an early stage not have the certification, but will have to develop a plan for the obtaining the same".
- Operational Tools:
  - Involvement: "*Supplier Development Learning Group*—is a meeting which join the various suppliers were subject to a *Supplier Development Program*, to share experiences" "every quarter we have a meeting at which always bring two suppliers and they make a presentation of an improvement plan"; Development the platform *online SupplyOn* to communicate and dialogue with suppliers, where designs, documents and material specifications are available.
  - Implementation: Supplier Selection—"a concept introduced at Bosch designated *Local for Local* to develop a supplier base as close to my business and a more economical solution. Failing that, develop a plan that can make this a viable and cost-effective solution".
- Assessment Tools:
  - Report: Presenting some results relating to suppliers in the Bosch Group. Report, not the specific unit of the Bosch Thermotechnology Aveiro;
  - Indicators and Indexes: Percentage billing "more than 30 % of its turnover in the Bosch division and 50 % in a group. Whenever there is a supplier that does not meet these conditions, we work with him to try to find solutions, optimize the cost structure and lower the volume of turnover, encouraging them to seek news business."

Focused on the operational side of sustainability, with the results of the case developed and the literature review, we developed the following *Framework* for sustainable supply chain management:

The model is based on an open systemic approach, with the importance of permeability required in the input and output information with stakeholders. The integration of sustainability in the supply chain management needs to be effective with a continuous improvement and permanent perspective. This challenge is possible with the implementation of the Pro-active, Strategic, Operational and Evaluation Tools (Silva et al. 2014).

For each type of tools, is recommended a classification process in Pure and Hybrid Tools. This classification leads to a self-assessment of existing practices. As experience has shown, organizations begin to have the area of sustainability implemented, however only need to be structured and aligned in a coherent program. Note that the categories of tools are joined in cycle because the strategies should be defined, operationalized and then evaluated. It is essential that the implemented tools contemplate objectives and results in the three vectors of Triple Bottom Line sustainability.

It is necessary to rethink the organizational structure, like resources to provide necessary skills and new functions. Since sustainability programs are developed along the supply chain, we identify a relation of 1 to N. The development programs will be extended to a several number of organizations, contributing to the spread of sustainability.

It is intended that this model is able to set guidelines for the implementation and structuring a program to introduce sustainability in the supply chain management.

5 Conclusion

The set of tools, identified and described, answers the problem of objectification/materialization the sustainability concept. The demonstration of its applicability in the areas of operational management, we demystify the approach to sustainability only for reasons of marketing and image. Through the examples implemented by the case study, we have shown that it is possible to develop sustainable practices respecting the Triple Bottom Line approach, touching the economic, social and environmental vectors.

The framework shown in Fig. 1 establishes a sustainability integration program in the supply chain management.

As the authors Esty and Winston (2006), in the development this work we believed that the path to sustainability is possible and tangible, it is necessary to know and implement a set of tools that enable organizations to develop a sustainable strategy. Sustainability needs to be understood, known and implemented in a structured and strategic form, supported in appropriate tools.

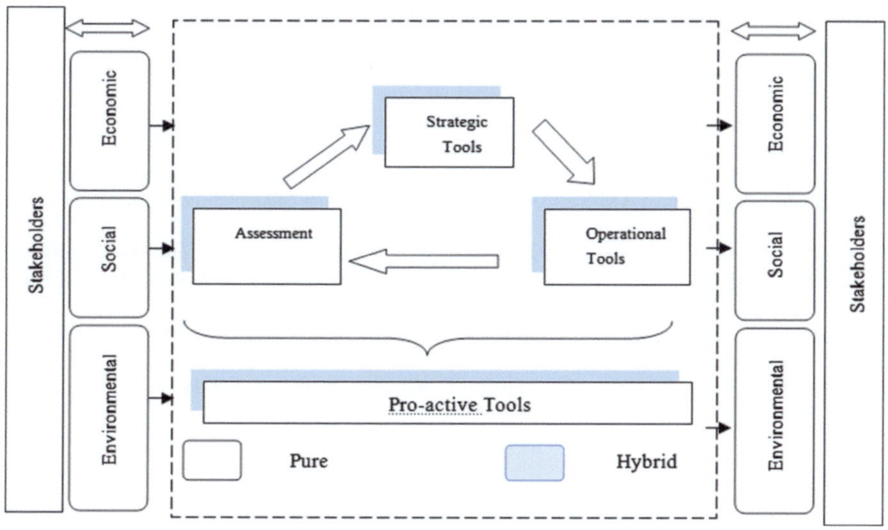

Fig. 1 Framework for implementation supply chain management sustainable. *Source* Elaborated by author

The study was conducted in a particular area of supply chain management, the purchasing process. It is pointed out as future development interest the application this model in other areas of supply chain.

References

AEP (2011) "Estado da Arte" das práticas de desenvolvimento sustentável em PME [Online]. Associação Empresarial de Portugal. http://futurcompet.aeportugal.pt/. Accessed 20 Jan 2011

Amado J (2013) Manual de Investigação Qualitativa em Educação. Imprensa da Universidade de Coimbra, Coimbra

Bogdan R, Biklen S (2013) Investigação Qualitativa em Educação Porto, Porto Editora

Bosch (2012) Sustainability Report 2012 Bosch [Online]. Bosch Group. http://www.bosch.com/en/com/sustainability. Accessed 19 Mar 2013

Brundtland GH (1987) Our common future: The World Commission on Environment and Development. Oxford University Press, Oxford

Carter CR, Rogers DS (2008) A framework of sustainable supply chain management: moving toward new theory. Int J Phys Distrib Logistics Manage 38:360–387

Carvalho JCD (2010) Logística e Gestão da Cadeia de Abastecimento. Edições Sílabo, Lisboa

Chalmeta R, Palomero S (2011) Methodological proposal for business sustainability management by means of the balanced scorecard. J Oper Res Soc 62:1344–1356

Elkington J (1997) Cannibals with forks—the triple bottom line of the 21st century. Capstone Publishing Lts, Oxford

Esty DC, Winston AS (2006) Do verde ao ouro, Casa das letras

GRI (2013) G4 Diretrizes para Relato de Sustentabilidade [Online]. Global Reporte Initiative. https://www.globalreporting.org/reporting/g4. Accessed 2 Apr 2014

Pagell M, Wu Z (2009) Building a more complete theory of sustainable supply chain management using case studies of 10 exemplars. J Supply Chain Manage 45:37–56

Pojasek RB, Hpllist JT (2011) Improving sustainability results with performance frameworks. Environ Qual Manage 81–96

Seuring S, Müller M (2008) From a literature review to a conceptual framework for sustainable supply chain management. J Clean Prod 16:1699–1710

Silva CS, Gouveia JB, Costa AP (2014) Ferramentas de Apoio à Sustentabilidade na gestão da cadeia de fornecimento: Análise Qualitativa de Relatórios de Sustentabilidade. Internet Latent Corpus J 4

Souza FND, Costa AP, Moreira A (2011). webQDA [Online]. Aveiro: Centro de Investigação Didática e Tecnologia na Formação de Formadores da Universidade de Aveiro e Esfera Crítica. www.webqda.com. Accessed 2 Apr 13

WBCSD (2000) A eco-eficiência, criar mais valor com menos impacto [Online]. http://www.wbcsd.org/. Accessed 22 Jun 2011

Yin RK (2010) Estudo de Caso: Planejamento e Métodos. Bookman, São Paulo

Transport KPIs for Supply Chain Improvement. A Literature Analysis

Pablo Domínguez-Caamaño, Jesús García-Arca, Arturo J. Fernandez-González and José Carlos Prado-Prado

Abstract Through a content analysis since 1998–2014, this article analyzes the evolution of the transport indicators in the context of managing the supply chain. Finally a set of indicators is extracted from the articles. This set of indicators can be considered a basic "pack" of KPIs (Key performance Indicators), reflecting the common needs of organizations that require freight.

Keywords Indicator · SCM · Delivery · KPI · Logistics

1 Introduction

Transport is one of the biggest environmental impact activities. Fortunately, the fact that saving CO_2 and other harmful gases to the environment are almost proportional to the fuel cost savings (Demir et al. 2014) motivates users to reduce their environmental impact. Rising fuel prices also provides an extra motivation.

Vehicles more and more efficient, better infrastructure conditions and greater awareness about environment have lowered fuel consumption per ton and kilometer (Perez-Martinez 2009), however this potential cost reduction has been generally absorbed by the aforementioned rise in fuel prices.

Key performance indicators (KPI) are measurements used to manage and control processes, organizations and projects. These systems include data collection and their analysis as a process within the company. A current model, widely used, is called Supply-chain operations reference (SCOR), proposed in 1999 by the Supply-Chain Council (SCC), which includes, inter alia, a package of more than 150 indicators that can be considered standard.

Not only in the SCM but also in business in general, KPIs are usually grouped into performance measurement systems (PMS). The latest trends seek holistic

---

P. Domínguez-Caamaño (✉) · J. García-Arca · A.J. Fernandez-González · J.C. Prado-Prado
Grupo de Ingeniería de Organización (GIO), Escola de Enxeñería Industrial,
Campus Lagoas-Marcosende, C/Maxwell, 36310 Vigo, Spain
e-mail: pablo.dominguez@gio.uvigo.es

M. Amorim et al. (eds.), *Engineering Systems and Networks*, Lecture Notes
in Management and Industrial Engineering, DOI 10.1007/978-3-319-45748-2_27

visions and alignment with business strategy (Neely et al. 1997; Medori and Steeple 2000; Vasiljevic et al. 2014).

2 Objective and Methodology

The aim of this paper is twofold, firstly illustrate and propose a set of core KPIs that can be extrapolated to any field of transport efficiency, and secondly to provide a series of recommendations and main lines of action to improve processes transport in general.

The methodology used in this work is called "content analysis", a systematic and reproducible literature review on a particular issue, to find out where and how research efforts are focused and to discover research gaps or opportunities (Meredith 1993; Guthrie et al. 2004). This methodology has been widely and reliably tested and documented, providing good results to know the state of the literature on a particular topic (Pasukeviciute and Roe 2005). To systematically execute this analysis the steps proposed by Seuring and Müller (2008) are followed: Material collection, Descriptive analysis, Categorization and Material evaluation.

The material has been delimitated by 4 criteria:

- **Content**: information has been sought about a specific topic: Transport indicators in the supply chain. The exact words used for the search are listed below.
- **Publications**: journals that represent the state-of-the-art in the selected topic have been chosen. We use as a criterion that none of them had a SJR ratio below 0,7. In alphabetical order:
  - *International Journal of Production Economics; Journal of Operations Management; Journal of Supply Chain Management; Manufacturing and Service Operations Management; Production and Operations Management; International Journal of Production Economics; Transportation Research, Part E: Logistics and Transportation Review; Journal of Business Logistics; Journal of Cleaner Production; Supply Chain Management; International Journal of Operations and Production Management; International Journal of Production Research; International Journal of Physical Distribution and Logistics Management; International Journal of Physical Distribution and Logistics Management; International Journal of Logistics Management; Journal of Purchasing and Supply Management; Journal of Advanced Transportation; Journal of Air Transport Management; Production Planning and Control; International Journal of Shipping and Transport Logistics*
- **Publication year**: all papers are subsequent to 1998.
- **Document type**: publications that were not strictly scholarly articles are excluded, such as conference papers, book chapters, articles in press, etc.

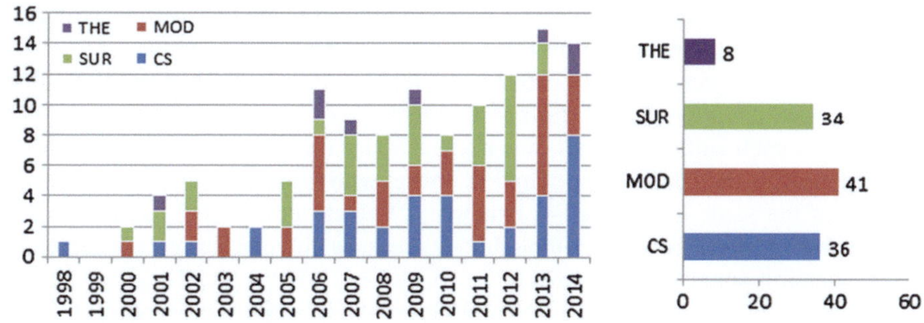

Fig. 1 Charts with the classification and quantification of content items by type and year

The database used for the collection of articles has been Scopus (www.scopus.com).

From 15,783 academic articles published in those magazines between 2000 and 2013, only 119 are about the topic under study.

The articles have been divided into 4 main groups, using a similar approach to that used by Seuring and Müller (2008): Case Studies (CS), empirical studies (SUR), mathematical modeling or simulation (MOD) and frameworks, reviews or conceptual studies (THE).

It is remarkable the low number of articles with theoretical or conceptual content, which reinforces the need covered by this article, grouping and analyzing the content developed in empirical studies, modeling and case studies This may be due to the fact that in this discipline, case studies are the predominant variant in proposing and testing new conceptual frameworks (Fig. 1).

A keywords analysis is made, as theoretically these words best describe the work content.

The term "Supply Chain Management" (SCM) is most often used as a keyword. Probably, the reason for this is that "SCM" is the name of the discipline in which it is framed the analyzed topic. The same applies to "supply chains".

The fact that "customer satisfaction" is above the "costs" or "sales" reflects a growing tendency to include the client as center of business strategy.

3 Material Analysis

After an individual analysis of the documents under consideration, the indicators proposed by different authors have been tabulated. Thereby which transport indicators are most used in the literature has been determined. Transportation or distribution indicators have been exclusively analyzed; other indicators not closely related to transportation management are discarded.

Only articles that mention at least 3 different indicators are considered (with one exception, which is detailed below). Finally, 22 articles are selected.

We have also rejected indicators that have not been clearly defined and those that have not been cited by at least 2 different papers.

Although each author used his own nomenclature for indicators, they have been grouped within the same denomination those who present authors considered variants of the same indicator. For example, "Vehicle utilization" has joined "% empty space in truck" because, although they measure different things, conceptually the information sought is the same.

Indicators with the same meaning but different units (years, days, hours, etc.) are also grouped. While some authors only get the average value, others perform comprehensive statistical analysis (variance, standard deviation, percentile, etc.).

The collection of KPIs by Author is detailed in Fig. 2 in the Appendix.

The 12 KPIs obtained, sorted by the number of papers that are cited:

Indicators/authors	1	2	3	4	5	6	7	8	9	10	11	12	TOTAL
Packdil & Moustafa 2014	x	x				x	x	x	x	x	x		8
Ala-Harja & Helo, 2014	x	x	x										3
Ding et al, 2014										x			1
Böhme et al, 2014	x		x					x					3
Bansal & Dyer, 2014				x									2
Thunberg & Persson, 2014	x				x								2
Packdil & Moustafa 2013	x				x		x					x	5
Appelqvist et al, 2012		x	x	x	x								4
Wan et al, 2012			x	x		x							3
Jung, 2012		x		x	x								3
Riane et al, 2011		x	x	x								x	4
Mondragón et al, 2011	x		x			x	x		x				5
Van Kampen et al, 2010	x												1
Bongsug, 2009	x	x		x									3
Whitten et al, 2009		x				x							2
Sezen, 2008	x	x	x	x					x	x			6
Skipworth & Harrison, 2006	x				x		x	x					4
Wacker & Sheu, 2006	x	x				x		x					4
Cheng et al, 2002	x	x			x		x						4
Gunasekaran et al 2001	x	x	x	x	x	x	x		x		x		11
Wacker & Sheu, 2000	x	x						x			x		4
Mapes et al, 2000	x	x	x			x				x			5
TOTAL	15	13	9	8	7	7	6	5	4	4	3	2	
%	68 %	59 %	41 %	36 %	32 %	32 %	27 %	23 %	18 %	18 %	14 %	9 %	

Fig. 2 Indicators classified by article

1. **Order cycle time or Lead time**: two ways of interpreting the cycle time were identified. On the one hand, under customer perspective it measures time elapsed since the moment the customer orders until the order is served. On the other hand, under the company perspective it measures the time since the department in charge of the expeditions is ordered by the sales department. This indicator is one of the most important elements of customer service.
2. **On-time delivery (%)**: all articles agree on-time delivery is not the same that in-time delivery, the latter can be considered nonpoint as a delivery before the agreement can cause undesirable consequences like excess of stock or misalignment of a time window. In addition, an author (Bongsug 2009) also proposed two different indicators for on-time arrivals and on-time departures, so it would be easier to find the root of delays in deliveries.
3. **Average transport cost**: it allows an approximate but direct allocation of transportation cost to products or orders. Can be calculated per kg, by reference, per pallet, full truck, etc.
4. **Vehicle utilization (fill rate)**: the use of resources is a key issue in SCM in general and in transport in particular, thus the fill rate in trucks, shipping containers, and others receptacles is one of the most important indicators. Note that the vehicle fill rate should always be calculated with respect to the limitation that is more restrictive: weight or volume.
5. **Perfect order fulfillment**: This indicator is a combination of on-time delivery and errors in orders; it determines the reliability of the transport and dispatcher service.
6. **% of products returned**: this indicator focuses on the amount of product that is returned. Generally includes transportation errors, but also returns for lack of quality, after-sales service, warranties, etc. Therefore its aim is to evaluate the performance of reverse flow of goods, not customer service, as it brings together all kinds of returns.
7. **Average processing time for orders**: it is a part of the lead time, but is often considered separately to evaluate the level of bureaucracy in the process.
8. **Average delay**: it provides complementary information to the on-time delivery, measuring how late they have been. This indicator can vary from hours to days depending on the sector and type of transport.
9. **% errors in orders**: it refers to errors in orders, either by billing issues, picking or errors in the order of the seller. All mismatched items whether to excess or lack thereof count as errors.
10. **Customer satisfaction (Surveys)**: the fact that the customer is the center of business strategy is something already widely assumed in academic and business circles, however the measure of this satisfaction is complex, so surveys with varying degrees of reliability are commonly used to measure customer perception of service.
11. **Number of complaints**: a quantitative way to measure the level of customer satisfaction is the evaluation and analysis of complaints received. The comparison of this indicator with perfect order fulfillment, on-time delivery and

errors in orders shows the tolerance and sensitivity of the customer to the service received.

12. **Average transport distance**: This indicator measures the distance to customers and it helps establish zones of influence and geographical segmentation. Its importance is noticeable, as most of the transport costs are directly proportional to the distance traveled.

Indicators 1, 2, 5, 7, 8 and 9 measure directly and objectively the service provided to the customer, while indicators 11 and 12 measure customer perception of that service. Concern for logistics processes is reflected in the indicators 4 and 6. They take into account the efficiency of both direct and reverse logistics system. Only indicator 3 refers to a cost directly, but eventually all other indicators can be converted to financial indicators.

4 Discussion and Future Research

Nowadays organizations have had to harness logistical improvements in order to get a reduction of their transportation costs. Such improvements are basically directed to 3 issues:

- **Distance reduction**: different methods have been used since the middle of last century. Some are relatively simple (Clarke and Wright 1964; Segerstedt 2013) but others use complex algorithms with high computational requirements (Zhu 2000; Glover 1991). These methods have been widely used, especially in road transport.
- **Time reduction**: specifically decreasing waiting times, which directly affect labor costs. Following the premises of the Lean philosophy these times should be minimized. The combined use of time windows with positioning and planning software allow real-time restructuring of schedules, tending to reduce these waiting times (Mogre et al. 2014; Lasserre 2004).
- **Load unit rationalization**: to improve the occupation ratio and lower empty travels (McKinnon and Ge 2006), routes can be optimized and reverse logistics must be taken into account (Turrisi et al. 2013). Improvements in packaging also let increase the utilization of trucks (García-Arca et al. 2014).

To quantify these improvements and to know the performance of the transport process is necessary to use indicators as the basic set we propose.

In order to provide more information on the origin of this set of indicators the objective of proposing indicators has been qualitatively analyzed.

The work of Gunasekaran et al. (2001) and Bongsug (2009), with a more theoretical or conceptual content, propose sets of indicators based on literature review. The first focuses exclusively on transport and it is the article that most transport indicators suggests. The second proposes indicators for the whole supply chain, so it is less complete in the specific part of the carriage. Both papers propose indicators to serve to guide process improvement and problem solving.

In articles about simulation and mathematical modeling, indicators have been used as variables to optimize or to verify proper system operation. Punctuality and fill rate are variables used by the most of them.

In the remainder, case studies and empirical studies, indicators are used primarily as a method of comparing performance between sectors, businesses or product lines. They are also used to check the impact, positive or negative, of certain practices in the transport process; examples of this are "form postponement" or using time windows.

Only one article (Ding et al. 2014) refers directly to a systematic analysis of indicators (actually, only one indicator, this article is the only admitted that propose only one) linked to improvement actions. Its methodology is to: Defining KPIs, data capture and measurement, data and deviations analysis and corrective actions and improvements. This absence suggests that there is no consensus in the literature on how to deal with process improvement through indicators, at least in terms of methodology and standardization.

Further studies could increase the level of detail of each indicator, look for indicators in other fields with the same methodology or try to make a similar set of indicators with greater bibliographic analysis (which may include journals from other fields), to provide a holistic view on a specific topic such as measuring the performance of transport.

Another line of research would be an empirical confirmation based on responses from companies dedicated to freight, using a survey. This would make it possible to see the differences between the academic and business world in choosing KPIs.

As is common in the investigation of the supply chain, case studies can serve as implementation experience or merely to verify the benefits of these and other indicators to assess the performance of the transport process.

5 Conclusion

The present work is intended to serve as a basic framework, providing a generic package of KPIs easily measurable, interpretable and adaptable to particular cases indicators.

This content analysis has revealed a progressive increase of publications about transport indicators in the most prestigious journals in SCM. It has also demonstrated a concern for the customer service provided and the degree of satisfaction.

Appendix

See Fig. 2.

References

Bongsug C (2009) Developing key performance indicators for supply chain: an industry perspective. Supply Chain Manage: Int J 14(6):422–428
Clarke GU, Wright JW (1964) Scheduling of vehicles from a central depot to a number of delivery points. Oper Res
Demir E, Bektaş T, Laporte G (2014) A review of recent research on green road freight transportation. Eur J Operat Res 237(3):775–793
Ding MJ, Jie F, Parton KA., Matanda MJ (2014) Relationships between quality of information sharing and supply chain food quality in the Australian beef processing industry. Int J Logistics Manage 25(1):85–108
García-Arca J, Prado-Prado JC, González-Portela AT (2014) "Packaging logistics": promoting sustainable efficiency in supply chains. Int J Phys Distrib Logist Manag 44(4):325–346
Glover F (1991) Multilevel tabu search and embedded search neighborhoods for the traveling salesman problem. In: Conference on artificial intelligence, University of Colorado
Gunasekaran A, Patel C, Tirtiroglu E (2001) Performance measures and metrics in a supply chainenvironment. Int J Oper Prod Manage 21(1/2):71–87
Guthrie J, Petty R, Yongvanich K, Ricceri F (2004) Using content analysis as a research method to inquire into intellectual capital reporting. J Intellect Cap 5(2):282–293
Lasserre F (2004) Logistics and the Internet: transportation and location issues are crucial in the logistics chain. J Transp Geogr 12(1):73–84
McKinnon AC, Ge Y (2006) The potential for reducing empty running by trucks: a retrospective analysis. Int J Phys Distrib Logistics Manage 36(5):391–410
Medori D, Steeple D (2000) A framework for auditing and enhancing performance measurement systems. Int J Oper Prod Manage 20(5):520–533
Meredith J (1993) Theory building through conceptual methods. Int J Oper Prod Manage 13(5): 3–11
Mogre R, Wong CY, Lalwani CS (2014) Mitigating supply and production uncertainties with dynamic scheduling using real-time transport information. Int J Prod Res 52(17):5223–5235
Neely A, Richards H, Mills J, Platts K, Bourne M (1997) Designing performance measures: a structured approach. Int J Oper Prod Manage 17(11):1131–1152
Pasukeviciute I, Roe M (2005) Strategic policy and the logistics of crude oil transit in Lithuania. Energy Policy 33(7):857–866
Perez-Martinez PJ (2009) The vehicle approach for freight road transport energy and environmental analysis in Spain. Eur Transp Res Rev 1(2):75–85
Segerstedt A (2013) A simple heuristic for vehicle routing—a variant of Clarke and Wright's saving method. Int J Prod Econ 157:74–79
Seuring S, Müller M (2008) From a literature review to a conceptual framework for sustainable supply chain management. J Clean Prod 16(15):1699–1710
Turrisi M, Bruccoleri M, Cannella S (2013) Impact of reverse logistics on supply chain performance. Int J Phys Distrib Logistics Manage 43(7):564–585
Vasiljevic D, Trkulja Z, Danilovic M (2014) Towards an extended set of production line performance indicators. Total Qual Manage Bus Excellence 25(5–6):618–634
Zhu KQ (2000) A new genetic algorithm for VRPTW. In: Proceedings of the international conference on artificial intelligence

Using Big Data for Competitive Dimensions Improvement in a Telco Company

Rafael Fernandes Novo and José Manoel Souza das Neves

Abstract This is a case study on how a telecommunications company is improving its competitive dimensions using Big Data. Even though the company were already using traditional data analytic tools, they are building Big Data systems to analyze larger amounts of data, in a faster and more efficient way. They are achieving improvements in some competitive dimensions, such as reducing operational costs, increase revenue through more accurate measurement of their service usage and improved quality of service delivered to customers.

Keywords Big data · Competitive dimensions · Business intelligence · Technological innovation

1 Introduction

The term Big Data defines systems capable of storing large amounts of information and process such information more efficiently. However, as explained by Davenport et al. (2012), Big Data is much more than that, as it can deliver to companies the ability to understand their business environment in a much more granular level, enabling those companies to create new products and services, respond to changes more efficiently and better explore the relationship between companies and their customers.

R.F. Novo (✉) · J.M.S. das Neves
Centro Paula Souza—Unidade de Pós-Graduação, Extensão e Pesquisa, Rua Bandeirantes 169, São Paulo, SP, Brazil
e-mail: rfnovo@hotmail.com

J.M.S. das Neves
e-mail: jmneves.fatec@gmail.com

M. Amorim et al. (eds.), *Engineering Systems and Networks*, Lecture Notes in Management and Industrial Engineering, DOI 10.1007/978-3-319-45748-2_28

Brynjolfsson and McAfee (2012) state that Big Data have a greater potential than traditional analytical solutions to bring businesses benefits as it will enable company executives to have a more holistic understanding of theirs business.

2 Big Data Overview

There is a comparison—and sometimes a lack of a clear and objective definition—between Big Data and traditional business intelligence solutions, as both have the ability to analyze data and convert them into information to generate improvements to the competitive dimensions of companies. However, there are three major improvements in Big Data, as listed by Brynjolfsson and McAfee (2012) and corroborated by Casonato et al. (2013): volume, velocity and variety. Note that the existence of only one of these characteristics may already consist of an opportunity for data analytics improvement, using Big Data systems.

2.1 *Volume*

The amount of data available is growing rapidly. According Brynjolfsson and McAfee (2012) in 2012, we created about 2.5 exabytes[1] of data daily, and the data volume accessed every second on the Internet was higher than all the information stored in the network 20 years ago. This creates a new opportunity for companies to analyze this huge volume of data that may contain relevant information to generate a more accurate view of the market, improving the relationship model with customers and other help develop other business aspects that allow these companies to improve their competitive dimensions.

2.2 *Velocity*

The second aspect of Big Data differentiation is the information processing speed, which allows for almost instantaneous data analysis, enabling the company to be more agile. Gallant (2011) explains the evolution of traditional data analysis model to this new model, comparing it to the human behavior itself. The human being can process a lot of information and make decisions in real time. When a person is driving, he or she must pay attention to various information sources, processing everything in real time to make the best decision. If he or she spots an obstacle in the way, needs to respond immediately because there is no point to break after

[1] 1 exabyte = 10^9 gigabytes or 10^{18} bytes (considering decimal base).

reaching the obstacle. This metaphor can be applied to the corporate use of Big Data solutions, which can analyze real-time massive amounts of information that come from different sources. It gives to companies the ability to do predictive analysis, testing hypotheses and scenarios.

2.3 Variety

The third listed item, variety, refers to various data sources and types, such as messages; sensor readings; GPS devices; mobile phones, among others, which can be used for analysis and subsequent extraction of information relevant to business. Davenport et al. (2012) stated that sources for Big Data may include everything from completely unstructured data such as call center transcripts to traditional structured data such as application databases, and may come from various sources, as simple as a copying machine or complex such as a jet engine, accordingly to Brown et al. (2011). Note that some of the most relevant data sources for Big Data analysis are relatively new, such as social networks, among which stand out Facebook, launched in 2004, and Twitter in 2006. Social Networks allied with the also recent proliferation of smartphones, tablets, ultrabooks, wireless networks and high speed mobile data networks are allowing people to be always connected, generating content, ideas and opinions.

3 Methods

The case study was conducted at the Brazilian subsidiary of a multinational company in the telecommunications industry. The company provides mobile phone broadband connection, both for corporate market and retail. With national coverage, the company is among the three largest telecommunications company in Brazil.

Company is in constant research for technological innovations, considered a pioneer in adopting new technologies and providing innovative products to its customers.

A semi-structured questionnaire was used as it allows participants to respond more comprehensively, better capturing the real life experience of these and the complexity of the scenarios for this study.

The interviews and collection of documents described in Table 1 were performed during the year of 2014.

Table 1 Case study: interviews and documents

Interviews	Documents
Project manager—IT business intelligence and analytics	Project documentation
IT director	Company web site
Architect—IT business intelligence and analytics	Business presentations

4 Results

Telecommunications is a highly competitive market, and the company is under pressure by the internal competition with other major players and by new entrants and substitute products, for example, WhatsApp and Skype, both solutions that use data network to replace, respectively, SMS messages and voice calls.

The company has grown over the past years through the acquisition of new customers, taking advantage of a growing market in Brazil. However, this model is now saturated, with a low emergence of new customers, therefore forcing the company to compete directly with other players for existing customers.

During the last years, the company has invested in data analysis solutions such as Teradata, ultra-high capacity storage systems, ETL[2] software and analytical software such as MicroStrategy and SAS.

The company currently has more than one petabyte[3] of data stored in data warehouse systems and therefore possible to be analyzed. However, due to restrictions of traditional data analysis systems to include such a high volume of data, the company has limitations in this area, making it impossible to consult all available data in a single analysis.

Due to the large data volume, analysis had become very slow; with this, the company ended up restricting the analysis to only a fraction of the available data. Additionally, any change in the structure of these queries, for example inclusion of a new field or modify existing fields, becomes very expensive and time consuming to perform. These factors forced the company to create analytical islands, which have a subset of the total data available. Each analytical island belongs to separate department (marketing, engineering, quality control, financial) containing the data that this department deems appropriate. Each island was built on a different technology and do not communicate with other islands. Using this model, analysis becomes faster, but creates problems such as:

- Data duplication between the islands;
- Increase in the cost of infrastructure;
- Decreased assertiveness in the analysis result;
- Disparate views among areas for the same object or business counter, for example, engineering and quality assurance departments had different values for services availability meter.

---

[2]ETL is the process of *extracting* data from transactional databases, usually used to support company's systems, *transforming* such data through conversion or filtering of the data and *loading* the transformed data into the data warehouse used by analytical systems. This process is generally time consuming and complex to modify.

[3]1 petabyte = 10^{15} bytes (considering decimal base).

Analysis of CDR[4] is another challenge for companies in this sector, as CDR is a text-based file generated in high quantity (millions of records per day). Do the ETL process in all CDR records to insert it in a traditional analytical system requires large computational power, exceeding the limits of traditional analytical solutions and forcing the company to limit the fields and amount of CDRs analyzed.

In order to solve these problems, the company acquired a Big Data solution that uses a single data repository—called Data Lake—based on EMC Isilon technology, with the ability to store data from various sources and in various formats, such as structured data from corporate databases and semi-structured data such as CDRs.

Within this architecture, data as CDRs, even being semi or completely unstructured, can be stored directly on that date lake, without the need to go through an ETL. Data analysis is performed using a distributed data processing system based on Hadoop. This technology allows an increased storage capacity and low cost processing using parallel nodes, creating an architecture with a relatively low cost but with high storage and processing capacity.

The Big Data project is transformation of the analytical model of the company, aiming to improving the following competitive dimensions:

- Reduce IT costs by eliminating or reducing analytical islands, thus reducing the duplicated copies of data, resulting in reduction in the amount of hardware, software, power, cooling and operational costs;
- Increase agility searching for information in order to comply with legal determinations in tracking calls and monitoring users, thus avoiding fines of the regulatory agency;
- Reduce operational costs through increased flexibility in utilization of data set, without the requirement of ETL process, allowing a more accurate analysis to identify network faults, enabling a more precisely identification of affected users. Company can provide a better repair service in the network and, if necessary, compensate users for any downtime experience, thereby preventing heavy fines by the regulatory agency;
- Measure more accurately the use of antennas by roaming users, allowing the correct charging of the service. During the project pilot, company identified a twenty million dollar gap to be received from roaming users, that had not been identified by traditional analysis of CDRs;
- Increase the speed and processing power, getting close to real-time processing, allowing, in this first stage of the project, provide the customer service department accurate information on the operations of services and users impacted, thus allowing a faster response to the end user, increasing his satisfaction, which will result in a higher retention rate.

In order to be more competitive, be able to analyze the information as soon as possible, enabling company management to take better, well-informed decision is

---

[4]Call Detail Record is the record containing a cellular connection details. It is generated and stored by the telephony companies.

paramount. Figure 1 shows the company understanding on the competitive gain brought by Big Data. The red line shows the delay between an event and the reaction of the company reducing their earning potential over time. A solution such as Big Data, making possible to react more quickly to an event, can bring greater benefit to the company, as shown by the blue line.

Figure 2 summarizes the main differences between the use of traditional data analysis and Big Data, as well as improvements in competitive dimensions for the company.

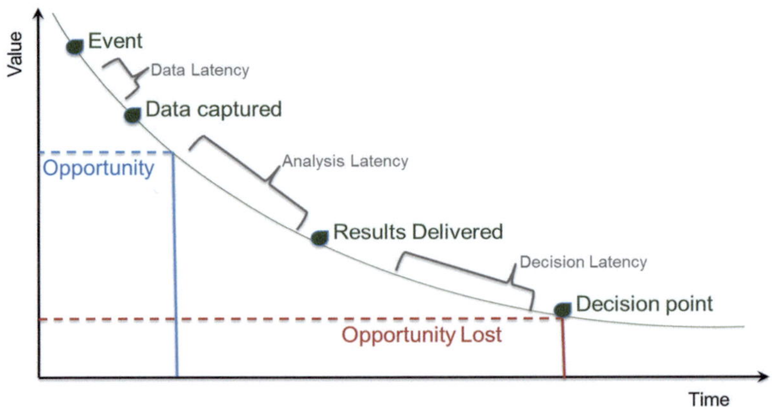

Fig. 1 Opportunity curve

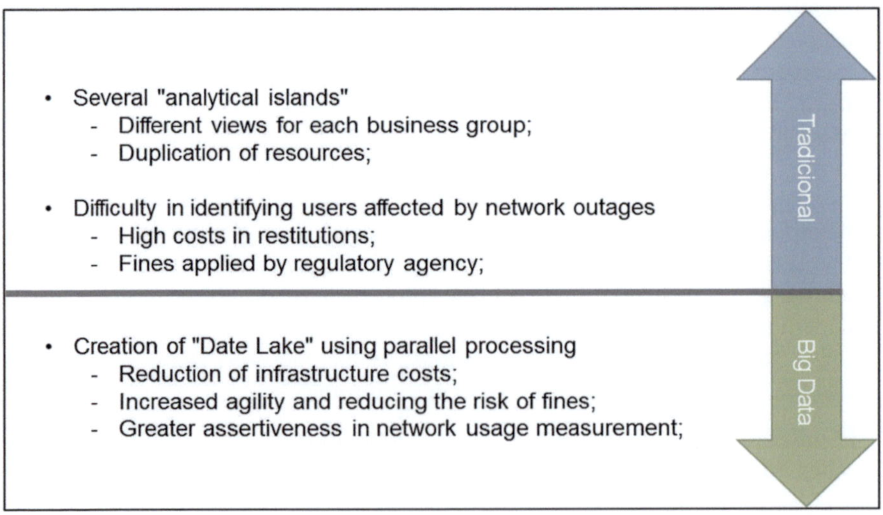

Fig. 2 Case study overview

5 Conclusion

Big Data sets a new paradigm as a more efficient method/technology for processing large data sets. It seems clear that the concepts embedded within the term, including the analysis at high speed of large unstructured data volumes, are creating a paradigm shift in data analysis.

This change occurs because companies are swimming in a sea of data, which are too bulky or too complex for traditional analytics tools, and the analysis of this huge mass of data is becoming the new technological frontier for the creation and implementation of a corporate strategy that allows the company to compete more efficiently.

The telecommunications company studied is using Big Data solutions with these main objectives:

- Process data much faster than traditional data analytics solutions, allowing the company to generate near real-time answers to certain conditions or business transactions;
- Analyze large volumes of data that could not be analyzed in traditional systems due to its limited capacity at an affordable cost;
- Include in the analysis unstructured data, such as CDRs and voice transcripts of call center, allowing insights that are more accurate or generating new analytical studies.

Although Big Data implementation is still in the early stages, the company is already reporting some gains in the competitive dimensions:

- Reduced costs of IT infrastructure;
- Reduction of costs of fines for the local regulatory entity;
- Reduction of restitutions to customer due to service unavailability;
- Increased revenue through more accurate measurement of the use of services;
- Improved customer service quality.

References

Brown B, Chui M, Manyika J (2011) Are you ready for the era of 'Big Data'? McKinsey Global Institute. Available from http://www.mckinsey.com/insights/strategy/ are_you_ready_for_the_era_of_big_data. Accessed 10 June 2013

Brynjolfsson E, Mcafee A (2012) Big data—A Revolução da Gestão. Harvard Bus Rev. Available from http://www.hbrbr.com.br/materia/big-data-revolucao-da-gestao Accessed 15 May 2013

Casonato R et al (2013) Top 10 technology trends impacting information infrastructure. Gartner. Available from https://www.gartner.com/doc/2340315. Accessed 10 June 2013

Cearley D, Claunch C (2013) The top 10 strategic technology trends for 2013. Gartner. Available from https://www.gartner.com/doc/2335015/top–strategic-technology-trends. Accessed 10 June 2013

Davenport T, Barth P, Bean R (2012) How 'Big data' is different. MIT Sloan Manage Rev 54 (1):22–24

Davenport T, Harris J (2007) Competição Analítica: Vencendo Através da Nova Ciência. Campus, Rio de Janeiro

Gallant J (2011) TIBCO CEO: how real-time computing will change the landscape. ComputerWorld. Available from http://www.computerworld.com/article/2511353/business-intelligence/tibco-ceo–how-real-time-computing-will-change-the-landscape.html. Accessed 10 June 2013

Gantz J, Reinsel D (2012) The digital universe in 2020: Big Data, nigger digital shadows, and biggest growth in the far east. IDC. Available from http://www.emc.com/collateral/analyst-reports/idc-the-digital-universe-in-2020.pdf. Accessed 2 June 2013

Porter M Estratégia competitiva: Técnicas para análise de indústrias e da concorrência. Rio de Janeiro, Campus

The Economist (2010) New rules for big data. Available from http://www.economist.com/node/15557487. Accessed 10 Oct 2013

Reduction of Drying Process Time of Natural Cork Stoppers Process in Lean Improvement Efforts

Tatiana M. Pinho, Daniel Campos, J. Boaventura-Cunha, Américo Azevedo and A. Paulo Moreira

Abstract Cork is a material with a significant economic, social and environmental impact. Due to its characteristic properties, this material exhibits a diversified applicability, incorporating several economic sectors. Among these, the natural cork stoppers industry reveals the greatest potential, being its production higher than 50 % of the total cork products. This work is encompassed in the Pilot Case IV of the FOCUS (Advances in Forestry Control and aUtomation Systems in Europe) project. The aim is to develop lean improvement suggestions for the cork-stoppers value stream which if implemented could lead to shorter production lead time and increased efficiency. The lean method of Value Stream Mapping (VSM) was used, since this provides an overview of the entire production process, rather than having process-specific focus, and offers a systematic way of finding the sources of problems and solving them. Based on this, it will be possible to propose and develop solutions, to improve or reformulate the necessary processes, in order to make the production line more efficient. Through the developed VSM and analysis of thermal images was identified as critical the cork stoppers drying process. A conceptual proposal of a new drying machine is also presented.

Keywords Drying · Natural cork stoppers industry · Model predictive control · VSM

T.M. Pinho (✉) · J. Boaventura-Cunha
Escola de Ciências e Tecnologia, Universidade de Trás-os-Montes e Alto Douro, UTAD, Quinta de Prados, 5000-801 Vila Real, Portugal
e-mail: tatiana.m.pinho@inesctec.pt

T.M. Pinho · D. Campos · J. Boaventura-Cunha
INESC TEC Technology and Science, Porto, Portugal

A. Azevedo · A. Paulo Moreira
INESC TEC Technology and Science and FEUP—Faculty of Engineering, University of Porto, Porto, Portugal

M. Amorim et al. (eds.), *Engineering Systems and Networks*, Lecture Notes in Management and Industrial Engineering, DOI 10.1007/978-3-319-45748-2_29

1 Introduction

Portugal and Spain are the countries with higher area of cork oak (*Quercus suber* L.) forest, with an occupation of 54.5 % from total area of 2275 thousands hectares existent in Europe (63 %) and Africa (37 %). Besides these, other Mediterranean climate countries, such as Italy, France, Morocco, Algeria and Tunisia, are also cork oak producers (Campos et al. 2008; Nunes et al. 2013). "Montados" (Portugal's designation for cork oak forests) reveal high economic, social and environmental impact, covering an area of 0.74 M ha, 23 % of the national forest area, and employing about 15,000 people, also representing 0.7 % of Gross Domestic Product (GDP) (Nunes et al. 2013; Besson et al. 2014).

Cork exhibits several interesting properties, whereby this raw material is currently used in a vast range of products (Abenojar et al. 2014; Bullitta et al. 2011). Despite this fact, in order to maximize raw material use, the first transformation objective is to develop natural cork products. Among these, cork stoppers represent the relevant product of the cork industry. In fact, its production percentage is higher than 50 % of the total cork products production. Overages of this production line and non-adequate raw material are used in other products development (Fortes et al. 2006). The ever-increasing demand for quality and process efficiency has been an important issue in the cork sector, also pro-actively accompanied by the wine industry. In addition to initiatives aimed specifically at increasing the quality of the raw material, a strategy to increase market competitiveness of this industry sector is through waste elimination along the supply and manufacturing chain process, in order to achieve higher levels of efficiency, responsiveness and dependability.

A methodological approach suitable to address in a structured way the identification of opportunities of improvement is grounded in Lean techniques and tools. Basically, the lean approach to eliminate wastes is to capture non-value added activities and work to reduce or totally eliminate them. The typical tool used to capture value added and non-value added activities is the value stream mapping (VSM).

VSM is a mapping tool that is used to map the production process, representing 3 things: flow, information and time. That important tool can be applied to almost every industry, to enhance its competitiveness at cost, service and quality levels (Forno et al. 2014; Bertolini et al. 2013).

This work proposes the analysis of natural cork stoppers production processes, through VSM elaboration of the production line. Its objective is to identify the process with large wastes in terms of processing time and energy consumption and propose solutions. Besides this introductory section, this article is organized as follows: Sect. 2 presents the natural cork stoppers industrial processes; Sect. 3 details the case study and methods; Sect. 4 presents the results, identifying the processes improving points; Sect. 5 presents the machine prototype proposal to reduce the drying process time; and, at last, Sect. 6 contains the main conclusions.

2 Industrial Process of Natural Cork Stoppers Production

Since natural cork stopper represents the key-factor in the cork industrial sector, it is important to understand the whole process behind its production. Therefore, after the cork extraction from the tree, its processing begins in preparer activity, involving boiling, gauge (i.e., thickness) selection, and waste separation, among others. It should be noted that the most relevant stages are regulated in the International Code of Cork Stopper Manufacturing Practice (Celiege 2011). Namely, as shown in Fig. 1, in a first step, are performed: stabilisation (cork bark stacking in the exterior during about 6 months to dry the cork); packing or stacking (place cork bark in piles); boiling (pile submersion in boiling water for approximately one hour in order to release internal tensions and clean the cork, with thickness increasing and tannins extraction); stabilization (cork bark stacking during two to three weeks to tree shape elimination and humidity reduction); sorting (cork bark selection according gauge and waste and defects separation, allowing quality inquiry to natural cork stoppers production); second boiling (optional boiling during 30 min to improve the characteristics of cork planks); snip (corners and edges cut to aspect improvement and quality enhancement); sorting (planks separation in qualities and gauges); and again stacking.

After preparation, the performed processes are: slicing (longitudinal cut of planks in slices with length and thickness adequate to the cork stoppers production); punching (cork slices drilling to obtain the cork stoppers); pre-sorting (separation of cork stoppers with defects or chunks); drying (cork stoppers drying in the outdoor or in kilns in order to reduce humidity); tops rectification (length rectification through sanding); bodies rectification (diameter rectification through sanding); top's rounding operation (cork stoppers top rounding through sanding); conical corks production (cork stoppers diameter modification); chamfering cork stoppers (shape

Fig. 1 Natural cork stoppers production industrial processes

finishing); sorting (manual or automatic separation of cork stoppers with defects as cracks, crevices or bad finishes, based on visual criteria like appearance and porosity); washing (cork stoppers washing to disinfection or bleaching); draining/spinning (water excess reduction); drying; sorting; colmation (fill inferior quality porous surfaces with a cork powder and aliment glue mixture to pore blockage); marking or printing (logo printing by ink or fire); treatment (cork stoppers surface lubrication); sterilization (cork stoppers with surface treatment packaging in waterproof bags, in a sulphur dioxide atmosphere, to reduce fungi contamination); and packaging (placement of cork stoppers in bags) (Ineti 2001).

3 Case Study and Methods

This work was developed within FOCUS (Advances in Forestry Control and aUtomation Systems in Europe) project. FOCUS is a 7 FP SME-target collaborative RTD project started at 1st January 2014 with duration of 30 months. According to project website (www.focusnet.eu) its goal is to "improve sustainability, productivity, and product marketability of forest-based value chains through an innovative technological platform for integrated planning and control of the whole tree-to-product operations, used by forest-producers to industry players". This project comprises four Pilot Cases covering the value chains in Europe of lumber, pulpwood, biomass and cork transformation. In particular the Pilot Case IV involves the study of natural cork stoppers production, stated in this article. One of the objectives of this Pilot Case is to identify the processes in the industrial supply chain that have more wastes in terms of time and energy. Based on this it becomes possible to propose solutions to overcome these weaknesses. The method chosen for this analysis was the Value Stream Mapping (VSM).

The VSM analysis was developed concerning a Portuguese cork stopper industry. This industrial partner, certified by several entities, supplies numerous markets in the Europe and North America. Furthermore, an industrial equipment developer company is involved regarding the design, implementation and testing of new technological solutions to improve the efficiency of the production processes.

4 Analysis

Concerning the processes fulfilled in the cork industrial partner, it was developed the VSM diagram presented in Fig. 2. The blue boxes indicate process activity, triangular shapes represent waiting times between processes (i.e., inventories), and the arrows show material transport to different locations. The time line allows the identification of unproductive (in which no task is being fulfilled) and productive processes (in which are done raw material operations). The presented times were provided by the processes' responsible and operators. It is important to refer that the represented processes are the main ones performed by this company. Thus, some of the processes described in Sect. 2 may not be mentioned in this VSM diagram. Through VSM analysis it is possible to verify that the productive time corresponds to a total of 26.1 h, and the unproductive time to 84 h, being only analysed the processes after the beginning of cork processing at mills and until the final packaging. In this way the processes before the 1st boiling, corresponding the pre-processing phase, are not considered here. It should also be noted that there is a significant unproductive time in this pre-processing phase of about 6 months. However, the industry does not have significant interest in reduce it. This fact is related with cork seasonality, so this period guarantees the raw material availability for all year.

VSM provides a clear perspective regarding processing time. A critical process is the first cork stoppers drying process used to achieve an intended humidity content level. The drying lasts about 12 h which creates a delay in the production and impedes the batch finish in the same day. To complement the information provided by the VSM it was used a thermal camera TiR 110 from Fluke to identify processes with exceeding energy consumption and heat losses. Drying process is also one of those with large wastes.

5 Machine Prototype Proposal to Reduce Drying Process Time

The current drying solution of the industrial partner is composed by a conventional kiln with hot air closed circulation and a gas heater. Each production cycle dries about 200,000 cork stoppers arranged in nets and takes approximately 12 h. This is a long time-consuming process, with large energy consumption and a non-uniform temperature distribution within the nets. Although this is one of the most commonly used methods, several research works and practical experiments were conducted with the aim of improving this drying process. Firstly, to understand cork drying kinetics and provide theoretical support to drying equipment development, in Belghit and Bennis (2009), the authors developed a summary of adsorption isothermal curves of cork. Still in this context, Martins (1990) elaborated a study about cork stoppers characteristic drying curves, through the execution of drying experiments with laboratorial material modified to present thermal and hydrodynamic conditions similar to an industrial dryer.

In Magalhães and Pinho (2008) the performance of a laboratorial scale spouted bed dryer was analysed concerning Porto wine cork stoppers drying in several operating conditions. Based on drying characteristic curves of cork, Julião (2008) performed the design of a spouted bed dryer with a capacity of 5000 cork stoppers/h. In Oliveira (2010) work, was performed the design of a continuous drying system. This system, based on rolls carpets, intended to have a capacity of 40,000 cork stoppers/h. Furthermore other drying methods were found at commercial level

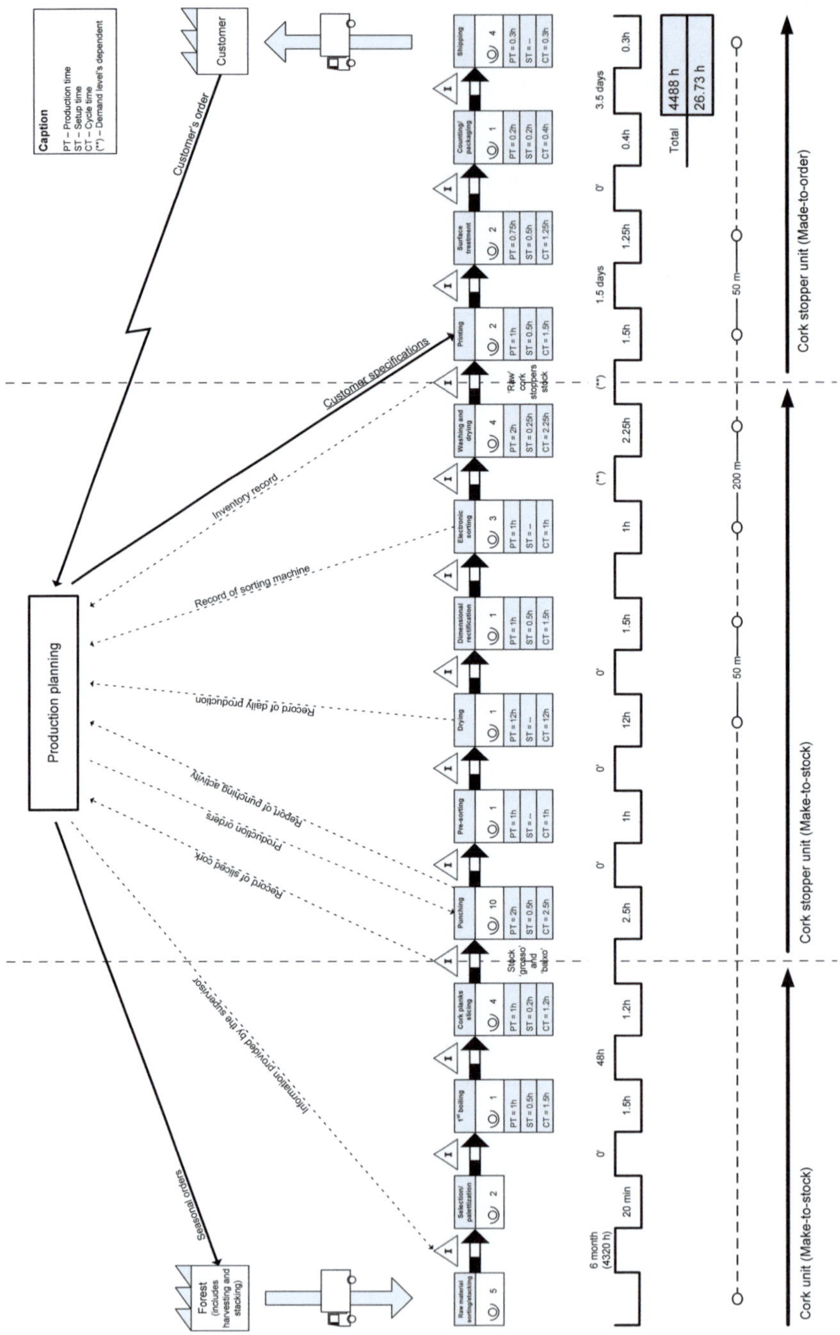

Fig. 2 VSM of natural cork stoppers production industrial process

concerning conventional kilns (CCL 2014), dehumidification kilns (Hainox 2014), continuous linear conveyors with hot air (A Eléctrica 2014), and continuous linear conveyors with microwaves (Adasen 2014). However, the several methods available have one or more of the following drawbacks: large energy consumption and processing time, cork deformation, contamination, among others.

In this way, based on the VSM analysis and the characterization of the drying process performed at the cork stopper company, one of the objectives of the FOCUS project is to develop a new drying system to reduce the process time and the energy consumption, ensuring the quality of cork stoppers. Figure 3 shows a scheme for this prototype, consisting on a multiple layer conveyor belt dryer with hot air injection. The system working principle is based on introducing the cork stoppers in the dryer, which has a fan and an extractor on each conveyor belt layer injecting hot air and extracting the contaminated air respectively. In this way, it is expected the reduction of the contamination factors, namely the 2,4,6-trichloroanisole (TCA) extracted from the cork stoppers in the bottom layers, towards the ones in the top levels. Also are defined multiple drying cycles if the moisture content of the exiting cork stoppers is above the required specification of 7 ± 1 %. At the dryer entrance is decided the cork stoppers feeding source, which can be from a new batch or from a previous cycle needing a new drying cycle. After choosing the source, cork stoppers are subjected to hot air drying.

The air entering the dryer is extracted from the environment, being filtered for moisture and other contaminants. Afterwards, the dryer input air passes through a heat exchanger, which uses the dryer output air as a heat source, transferring the temperature to the input air. Also, another heating system, based on electrical resistances is used, using fans as a pump system. This system pumps the air to each layer of the dryer, increasing the air temperature to improve the efficiency of the drying process. The recycling of the dryer output air as a heat source increases the energy efficiency, since it pre-heats the air reducing the energy consumption needed by the air heating system. The dryer is designed to control the air temperature and

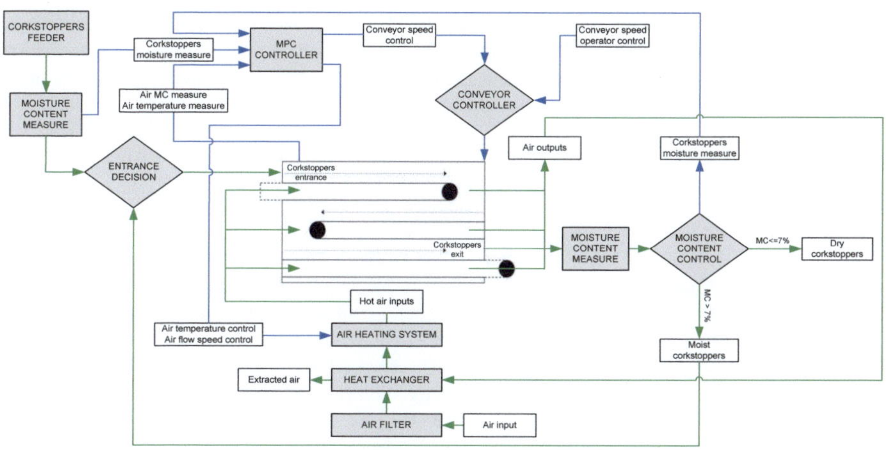

Fig. 3 Cork stoppers drying machine block diagram

flow speed and also the conveyor speed using a Model Predictive Controller (MPC) using the cork stoppers moisture measurement, the air moisture and temperature measures as inputs. This system will allow an improvement of the drying process, reducing the energy used and the time required, and ensuring the desired cork moisture content. Furthermore, the use of a multiple layer conveyor dryer allows a more uniform drying, since the stoppers stir when changing conveyor layers. This solution is considered capable of reducing the drying time, enabling a more homogeneous drying and can be adapted to use hot vapor, like in ROSA Evolution system (Sarkar 2012), as a way of not only drying but also sterilizing the stoppers, reducing TCA contamination. This prototype is currently being assembled by an industrial equipment developer company in order to start the tests and the system improvements during 2015.

6 Conclusions

Since cork has a relevant economic, social and environmental context, particularly natural cork stoppers industrial process, it is important to analyse it in order to improve its efficiency and competitiveness. In this work, this analysis was made through VSM lean tool, concerning an industrial enterprise, within the scope of the FOCUS project. The objective was to identify the process that presents large unproductive time, more energy consumption and heat losses. To fulfil the analysis relatively to energy consumption and heat losses, a thermal camera was used to acquire images of several cork mill processes. With this analysis, one process was identified as critical, namely cork stoppers drying. The current adopted solution of drying implies large time and energy consumption, and presents a non-uniform temperature distribution. So, as presented, one of the objectives is to develop an industrial prototype for a new drying process equipment, which reduces the drying process time and the energy consumption.

Acknowledgments This work was supported by the FCT—Fundação para a Ciência e Tecnologia through the PhD Studentship SFRH/BD/98032/2013, program POPH—Programa Operacional Potencial Humano and FSE—Fundo Social Europeu, and by the FOCUS project—Advances in the Forestry Control and aUtomation Systems in Europe, grant agreement no: 604286.

References

A Eléctrica (2014) Máquina de secagem de rolhas. A Eléctrica, Lda. Available from: http://www.aeletrica.com/pt/produtos.html?page=shop.product_details&flypage=flypage.tpl&product_id=60&category_id=10. Accessed 2 Oct 2014

Abenojar J, Barbosa AQ, Ballesteros Y, Real JC, Silva LFM, Martínez MA (2014) Effect of surface treatments on natural cork: surface energy, adhesion, and acoustic insulation. Wood Sci Technol 48:207–224. doi:10.1007/s00226-013-0599-7

Adasen (2014) Continuous tunnel type microwave equipment for drying wood cork. Shandong Adasen Trade Co., Ltd. Available from: http://www.jnadasen.cn/product/652874744-21042 3917/continuous_tunnel_type_microwave_equipment_for_drying_wood_cork.html. Accessed 2 Oct 2014

Belghit A, Bennis A (2009) Experimental analysis of the drying kinetics of cork. Energy Convers Manag 50:618–625. doi:10.1016/j.enconman.2008.10.014

Bertolini M, Braglia M, Romagnoli G, Zammori F (2013) Extending value stream mapping: the synchro-MRP case. Int J Prod Res 51(18):5499–5519. doi:10.1080/00207543.2013.784415

Besson CK, Lobo-do-Vale R, Rodrigues ML, Almeida P, Herd A, Grant OM, David TS, Schmidt M, Otieno D, Keenan TF, Gouveia C, Mériaux C, Chaves MM, Pereira JS (2014) Cork oak physiological responses to manipulated water availability in a Mediterranean woodland. Agric For Meteorol 184:230–242. doi:10.1016/j.agrformet.2013.10.004

Bullitta S, Dettori S, Manchinu M, Filigheddu MR, Piluzza G (2011) Characterization of Sardinian cork oak (*Quercus suber* L.) genetic resources for economically important traits. Genet Resour Crop Evol 58:1007–1020. doi:10.1007/s10722-010-9636-7

Campos P, Daly-Hassen H, Oviedo JL, Ovando P, Chebil A (2008) Accounting for single and aggregated forest incomes: Application to public cork oak forests in Jerez (Spain) and Iteimia (Tunisia). Ecol Econ 65:76–86. doi:10.1016/j.ecolecon.2007.06.001

CCL (2014) Secagem. Corticeira Cardoso. Available from: http://corticeiracardoso.com/site/produtos.php?menuId=produtos&id=8. Accessed 2 Oct 2014

CELIEGE (2011) Código Internacional das Práticas Rolheiras, Versão 6.03. CELIEGE European Cork Federation

Forno AJD, Pereira FA, Forcellini FA, Kipper LM (2014) Value Stream Mapping: a study about the problems and challenges found in the literature from the past 15 years about application of Lean tools. Int J Adv Manuf Technol 72:779–790. doi:10.1007/s00170-014-5712-z

Fortes MA, Rosa ME, Pereira H (2006) A Cortiça—2a edição. IST Press, Lisboa

Hainox (2014) Estufa de secagem com sistema de desumidificação. Hainox - Inoxidáveis para Indústria da Cortiça, Lda. Cardoso. Available from: http://www.sabemais.pt/mais_infor_foto.asp?tipo=2&cod=23704&pic=101. Accessed 2 Oct 2014

INETI (2001) Guia Técnico Sectorial—Indústria da Cortiça. Instituto Nacional de Engenharia e Tecnologia Industrial

Julião RJMS (2008) Dimensionamento de um secador de rolhas de leito em jorro para 5000 rolhas/hora. Faculdade de Engenharia da Universidade do Porto, Dissertação de Mestrado Integrado em Engenharia Mecânica

Magalhães A, Pinho C (2008) Spouted bed drying of cork stoppers. Chem Eng Process 47:2395–2401. doi:10.1016/j.cep.2007.11.009

Martins LASB (1990) Determinação de curvas características de secagem de rolhas de cortiça. Faculdade de Engenharia da Universidade do Porto, Dissertação de Mestrado em Engenharia Térmica

Nunes LJR, Matias JCO, Catalão JPS (2013) Energy recovery from cork industrial waste: Production and characterisation of cork pellets. Fuel 113:24–30. doi:10.1016/j.fuel.2013.05.052

Oliveira JCN (2010) Dimensionamento de um Secador Linear de Tapetes para 40.000 Rolhas de Cortiça por Hora. Dissertação de Mestrado Integrado em Engenharia Mecânica, Faculdade de Engenharia da Universidade do Porto

Sarkar S (2012) Message in a bottle: process innovations in the cork stopper fightback. CEFAGE-UE working papers

Waste Types in People Processing Services

José Dinis-Carvalho, Rui M. Lima, Andromeda Menezes and Marlene Amorim

Abstract The globalization needs and the increased competition between all types of companies put an increased focus on service improvement. Services are mostly customized, intangible, knowledge-based, and one type of services, called here as People Processing Services (PPS), can have the direct participation of customers both as object being processed and as a co-producer. These characteristics of PPS make it difficult to standardize processes, which contribute for generating non-value added activities, classified as wastes in the Lean Thinking knowledge area. This work aims to contribute to the waste classification, proposing a clarification for the types of waste associated with People Processing Services, and further applying this classification to a particular PPS case, the Hip Surgery Production Process of a Portuguese Public Hospital. It was observed that a large number of activities do not add value to the service and waiting is the most common waste.

Keywords Lean services · Waste classification · Hospital processes · People processing services

J. Dinis-Carvalho (✉) · R.M. Lima · A. Menezes
Department of Production and Systems Engineering, University of Minho, Guimarães, Portugal
e-mail: dinis@dps.uminho.pt

R.M. Lima
e-mail: rml@dps.uminho.pt

A. Menezes
e-mail: agcmenezes@dps.uminho.pt

M. Amorim
Department of Economics, Management and Industrial Engineering, University of Aveiro, Aveiro, Portugal
e-mail: mamorim@ua.pt

M. Amorim et al. (eds.), *Engineering Systems and Networks*, Lecture Notes in Management and Industrial Engineering, DOI 10.1007/978-3-319-45748-2_30

1 Introduction

Technological advances and globalization has been changing the way operations managers deal with the production of goods and services; including the way factories, suppliers and customers have interacted to perform their business (Womack and Jones 2005; Wemmerlöv 1990). This worldwide context changed the way companies operate, so organizations that were once local, have become national and global with all its benefits and challenges. The business competitiveness has leveraged organizations in overcoming the competition, and therefore the need for continuous improvement of the production process; productivity; and the quality of its products and services, so that has sought to innovate by adding value to them (Zysman 2006; Chesbrough and Spohrer 2006; Heizer and Render 2004).

This change applies to all kind of companies, manufacturing or service focused companies. Considering that, a service can be seen as a change in the condition of a customer as the result of the activity of some economic entity (Chesbrough and Spohrer 2006), even manufacturing focused companies have an increased concern with services they offer to their customers. Other service definitions emphasize an exchange between two or more parties and a transformation (potentially intangible) received by a customer. Services differ from goods because they are usually unique, intangible, produced and consumed simultaneously, with a high level of interaction with the customer (Bitran and Lojo 1993; Heizer and Render 2004). The services characteristics make it remarkable difficult in measuring the output of a service, because it does not depend solely on delivering it efficiently, but also in creating an atmosphere and a rewarding experience (Heizer and Render 2004).

Customer participation is a defining characteristic of service production systems. Customers provide key inputs for service production, including personal information, personal items to be handled by the service provider, etc.), for which their contact and presence is often required in the providers' premises (Sampson and Froehle 2006); but they have also been increasingly invited to assume other active roles in service operations, co-producing the service, and often fully self-serving themselves (Bowers and Martin 2007; Eichentopf et al. 2011; Prahalad and Ramaswamy 2004). According to these perspectives services can be addressed as processes, i.e. sequences of operations for the transformation and value adding to customer-provided items, involving both customers and providers in a co-production experience. Wemmerlöv (1990) proposed a useful classification of service processes that distinguished materials processing services, i.e. those in which the main transformation is done on a material item provided by the customer (e.g. car repair services, delivery services, etc.), information processing services (e.g. telecommunications, financial services, etc.) and people processing services (PPS) (e.g. healthcare, education, etc.). In PPS the value adding operations in service production focus on customers' themselves, i.e. on their minds, bodies, etc. For this reason, in these services the presence of the customer is required and they often have an important contribution in the process.

By assuming an active role in service production processes, customers contribute for the adding value activities that he or she will benefit from. As Chesbrough and Spohrer (2006: 40) referred, a service can be seen as creation of value as a result of a provider-client interaction. These characteristics make the link with the Lean thinking that is focused on the value as perceived by the client, and the reduction of all activities that do not contribute to this value. In other words, the Lean approach can be seen as the "way to do more using less and less equipment, time and space, always meeting the customer expectations" (Womack and Jones 2003).

The concept of Lean Thinking arises in the context of improvement of production processes, with a commitment to the reduction of wastes, i.e. reduction of activity that do not add value (Womack and Jones 2003). Known by the Japanese word "Muda", waste refers to all activity, which absorbs resources but creates no value. Ohno (1988) identified and classified waste in seven categories: Over-production; Inventory; Transportation; Motion; Waiting; Over-processing; and Defects.

Whereas the application of Lean principles in service industries has been taking place for several years, this practice has been mostly limited to a few service contexts, with a particular emphasis on those involving the handling of physical items (e.g. logistics and retail services) (Atkinson 2004), as well as in the context of handling patients in health care services (Waring and Bishop 2010; Ng et al. 2010), i.e. addressing patients as items flowing in a transformation process. The applicability of Lean principles to other service contexts has been explored with other service domains, such as in administrative services and operations (e.g. order-receipt, quotation, sales processing, accounting, etc.) (Holmes 2007), therefore supporting the possibility of transferring Lean tools and practices from manufacturing to service contexts in general. Womack and Jones (2005) introduced the concept of Lean Consumption as way to focus operations management on maximizing the service satisfaction of the customer, delivering exactly what he/she wants.

Driven both by a cost saving rationale, as well as by the belief that participation in service operations could improve customers' perceptions of service quality (Bendapudi and Leone 2003) the prevalent logic has been to increase the role of customers in service production systems (Bowers and Martin 2007). However, the quality of service experiences can be significantly affected by the heterogeneity of customers in what concern their capabilities, as well as their motivation to enrol, in an effective manner, in the tasks that are required to obtain a service (Frei 2006; Metters and Marucheck 2007). Nevertheless, service providers still lack the tools to assess the impacts of customer participation for service quality in a systematic manner, in order to inform the specification of adequate levels for customer involvement in service operations. Notably, the prevalent service quality models don't include explicit dimensions to assess the impacts of customer performance for the quality of service outcomes and the overall service experience (Amorim et al. 2014).

The application of lean approaches in the service sector has been underway for several years (Abdi et al. 2006; Bowen and Youngdahl 1998; Atkinson 2004). The clarification of a terminology of types of wastes for service systems could lead to the development of improved methods of identification of wastes. Thus, the main objective of this article is the development of a classification framework for the types of waste associated to PPS, by definition (Wemmerlöv 1990) the type of service processes where the presence and participation of the customer is more vital for value creation, and the application of this classification to a particular PPS case.

The authors made an analysis of the existent knowledge about Lean concepts and waste classification, and crossed that information with the characteristics of PPS systems in order to create a proposal of a classification terminology. Furthermore, that classification terminology was used as a framework applied within a specific case study: the Hip Surgery Process of a Portuguese Public Hospital. The data collection techniques used in this case study were process observation, log book and interview with managers.

2 Proposal for Classification of Waste Types in PPS

According to the principles of the Toyota Production System (TPS) all activities that do not add value to a product are considered as waste. Ohno (1988) published the 7 waste types classified in the TPS which were also adopted by Womack and Jones (2003) when they coined the TPS approach as Lean Manufacturing. The TPS classification of waste was created in industrial context and therefore its terminology may not be completely obvious in PPS. It is important to notice that in manufacturing context the object being processed is most of the times material based, and sometimes also information. This nature of the object that is being processed is enlarged, in service production systems, to include people (Wemmerlöv 1990). In this work, a terminology is proposed simultaneously related with the processing provider point of view and the user point of view. The different terminology proposed in Table 1 although following the original concepts intends to be more understandable in service context.

Too much information or material: this waste is analogous to over-production waste in industrial environments and in short means that more information is

Table 1 A terminology proposal for service waste types

Traditional waste types (Ohno 1988)	Services provider/user
Over-production	Too much information or material
Inventory	Information or material waiting
Transportation	Transport of Information or material
Motion	People motion
Waiting	People waiting
Over-processing	Complex and redundant processes
Defects	Defects

generated than the required or needed by the costumer. The act of generating more information than the necessary uses time and resources without adding value and also results in higher quantity of information stored. Examples of this type of waste are: deliver more copies than needed; input of data in several locations and several times; printing electronic forms to fill by hand; deliver information to people that does not need it; deliver reports that no one reads.

Information or material waiting: being equivalent to Inventory this type of waste represents storage of data, information waiting to be processed, material to be processed, and storage of material. Storage of data is referred to all of that data stored in memory of hard disks that is never used. As for information waiting to be processed can be exemplified by administrative processes waiting for the next processing step. Finally, an example of storage of material is visible in piles of paper or large quantities of office supplies. In PPS when customers are waiting to be processed (e.g. in a hairdresser) is here classified as material waiting. Although people should not be classified as material, for the logics of the processing flow, they can be seen as being waiting to be processed.

Transport of Information or material: when information is transferred between processes or material is transported from one location to another no value is added to the products and therefor is considered as waste. Examples of this type of waste are: people transporting documents between processes, transference of information between processes, people carrying office items. In PPS the service user also performs this type of non-value adding activities.

People motion: when people need to move from one location to another location they are not adding value to anything. In offices people often need to walk to the copy machine or to collect information from distant files. One of the main reason for this type of waste is poor layouts. Service users are also in many cases forced to move through different processes.

People waiting: it is a very common waste especially from service user's point of view. In many services such as dentists, hairdressers, or shops, most waiting is actually assigned to service users. People waiting is also an issue within service provides since many times people must wait for the printer, wait for documents, wait for decisions, wait for authorization, and so on.

Complex and redundant processes: this type of waste includes absent or improper process instructions (e.g. use of outdates forms), faulty data storage, obsolete process steps, and process steps that are no longer needed. One example sometimes occurring in office work is the act of printing an electronic form, then fill in the form with a pen and then digitalize the form with a scanner and then sending it by email. This type of waste lead to inconsistency in results and service quality, difficulty in finding files in the computer, inefficient use of resources, additional work, and higher probability of errors and mistakes.

Defects: this waste type includes wrong services, wrong information, errors, and damaged materials. This waste type leads to rework, corrections, additional transport and additional cost. Defects in people processing services are very common and some examples are: use of wrong document, unreadable faxes, sending mail to wrong addresses, sending emails with unclear purpose, sending a document with missing information, and so on.

3 Applying the PPS Waste Classification in a Real Case

A case study of the Hip Surgery process was conducted on Guimarães' Public Hospital, Portugal. The basic process of hip surgery production is composed of a set of necessary sub-processes, with the objective of submitting the user to an intervention. Hip Surgery therefore offers a clear case of PPS, i.e. a setting where customer presence and participation is fully embedded in the service production process. For this reason, the application of Lean principles to the improvement of such a service needs to address both the operations handled by the provider, and the activities performed by the customer himself.

Using a short description of the National Health System in Portugal, it is composed, among others, by the health centers and the hospitals. In the health center there are family doctors for each user, which make the monitoring and control of the evolution of the health of those patients. When necessary they redirect their patients' processes for hospitals for more complex procedures such as a surgery.

The hospital's response time is organized according to the following priority levels: (1) Normal—9 months; (2) Priority—two months; (3) High Priority—15 days; (4) Emergency Upheld—3 days. In this context, users who are coming from the Health Center can only be contemplated in the first two levels. These processes begin at the time the hospital receives, via a specific system (Alert P1 System), data on the imminent need of the user. The process may close in different ways, but in general, it is expected to be closed after post-surgical consultation.

The complete process was studied, described and modeled in order to carry out the analysis of the type of existing wastes, from both points of view of the service: supplier and user. Due to the limits for manuscript length this article focuses only on the study of the first sub-process: "Schedule of Medical Appointment", whose activities are described in Table 2. The choice was made because it is the initial part of the process by which all users must pass, and furthermore, is the one having the greatest evidence of different types of waste. This sub-process determines all the necessary steps for scheduling the first appointment with the orthopedist.

The waste classification for each sub-process was performed according to the classification proposed for waste in services defined in the previous section. Thus, the following table contains the sub-process steps and the waste classification for each of these steps. The sub-process "Schedule of Medical Appointment", as can be

Table 2 "Schedule of medical appointment" sub-process activities from the hip surgery process

Description activities of on "schedule of medical appointment"	Service provider	Service user
The orthopedic center receives the user process via alert system P1, from the health center, and the user automatically enters a queue of orthopedics screening	Information or material waiting	People waiting
The user waits for the response of screening to know whether or not forwarded to an orthopedic consultation and the date thereof	Information or material waiting	People waiting
A triage orthopedist studies the process of the user who is in Alert P1 system	Value adding	People waiting
The triage doctor takes one of the decisions on the case of the user: reject the application; return process; or mark consultant	Value adding	People waiting
Reject the request: the triage doctor rejects the request via alert P1 system and the user process leaves the waiting list for triage and exits	Transport of information or material	People waiting
Return process: the triage doctor returns the process to the family doctor, of the health center, via alert P1 system, to request additional information due to the same being: incorrect, incomplete or raising new questions	Transport of information or material	People waiting
Schedule appointment: the triage doctor includes the user in an orthopedics appointment waiting list	Value adding	People waiting
The user receives written notification by letter post, in his address, indicating the date of the 1st consultation	Transport of Information or material	Value adding

verified in Table 2, is composed of a set of activities classified mostly as non-adding value, where the user is most of times waiting.

From the BPMN model, (Fig. 1) it is possible to understand the process sequence and observe the detected waste. From the interviews and the observation of the process it is possible to identify two activities that add no value, which could be resolved promptly: (1) The triage doctor should consult the system at a lower frequency, checking the users who are in the screening waiting list. This would reduce wastes, both for the user and the hospital, because the hospital would reduce the inventory of information, and the customer would reduce the time on hold. (2) The triage doctor could enter himself the user on the system of waiting list for appointment. In this case, it is possible to verify from the Hospital point of view unnecessary people movement, too much information or material generated, people motion, redundant processes; and from the user point of view, people waiting.

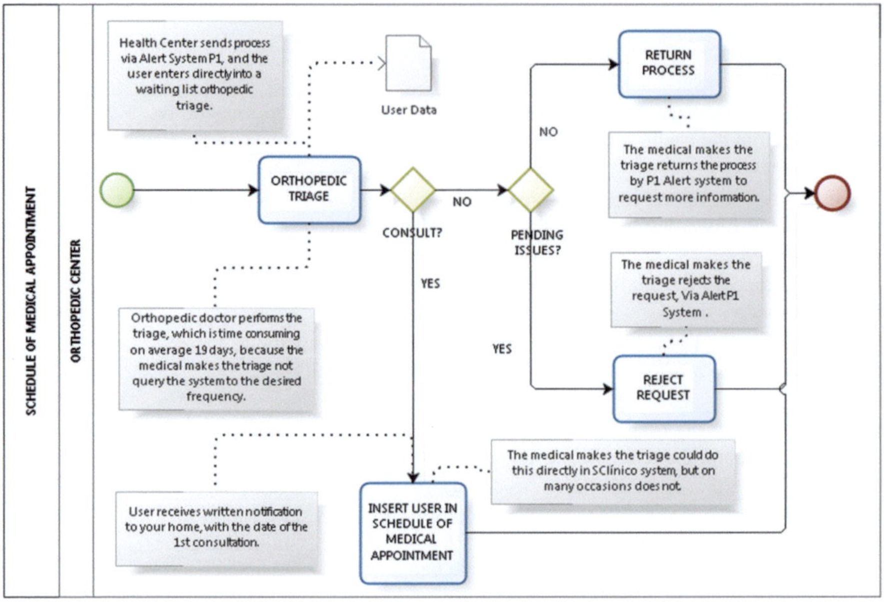

Fig. 1 BPMN model of the "schedule of medical appointment" process

4 Final Remarks

The reduction of waste in the production of a service process is critical to ensure high performances and the satisfaction for the customer (Womack and Jones 2005), contributing for delivering the right service and effectively. When there is direct participation of customers, either as an object being processed or as a co-producer, the need for identification and reduction of wastes becomes even more complex. In which case, the service consumer is a special type of partner user that needs to have the right conditions, and all the necessary infrastructure and guidelines in order to co-produce properly and/or to have a satisfactory experience as a processed object.

The perception of waste on the process and result of a PPS from the point of view of the user makes more evident the number of steps that do not add value. Therefore, the development of a classification framework for the types of waste associated to People Processing Services (PPS), and the application of this classification to a particular PPS case are important, in the search for continuous improvement.

This study had some limitations on data collection because there was no possibility of access to hospital users, and a reduced availability of time from orthopedists participants in the process. Further interviews and observation with more participants are expected to improve the identification of types of wastes and even contribute to its quantification.

References

Abdi F, Shavarini SK, Seyed Hoseini SM (2006) GLean lean: how to use lean approach in service industries? J Serv Res 6:191–206
Amorim M, Rosa MJ, Santos S (2014) Managing customer participation and customer interactions in service delivery: the case of museums and educational services. Organizacija 47:166–175
Atkinson P (2004) Creating and implementing lean strategies. Manage Serv 48:18–33
Bendapudi N, Leone RP (2003) Psychological implications of customer participation in co-production. J Mark 67:14–28
Bitran G, Lojo M (1993) A framework for analyzing the quality of the customer interface. Eur Manage J 11:385–396
Bowen DE, Youngdahl WE (1998) "Lean" service: in defense of a production-line approach. Int J Serv Ind Manage 9:207–225
Bowers MR, Martin CL (2007) Trading places redux: employees as customers, customers as employees. J Serv Mark 21:88–98
Chesbrough H, Spohrer J (2006) A research manifesto for services science. Commun ACM 49: 35–40
Eichentopf T, Kleinaltenkamp M, Jv Stiphout (2011) Modelling customer process activities in interactive value creation. J Serv Manage 22:650–663
Frei FX (2006) Breaking the trade-off between efficiency and service. Harvard Bus Rev 84(11):92
Heizer J, Render B (2004) Operations management. Prentice Hall, NJ
Holmes F (2007) Is your office as lean as your production line? Manufact Eng 139:20–21
Metters R, Marucheck A (2007) Service management—academic issues and scholarly reflections from operations management researchers. Decis Sci 38:195–214
Ng D, Vail G, Thomas S et al (2010) Applying the lean principles of the Toyota production system to reduce wait times in the emergency department. Can J Emerg Med 12:50–57
Ohno T (1988) Toyota production system: beyond large-scale production. Productivity Press
Prahalad CK, Ramaswamy V (2004) Co-creating unique value with customers. Strateg Leadersh 32:4–9
Sampson SE, Froehle CM (2006) Foundations and implications of a proposed unified services theory. Prod Oper Manage 15:329–343
Waring JJ, Bishop S (2010) Lean healthcare: Rhetoric, ritual and resistance. Soc Sci Med 71:1332–1340
Wemmerlöv U (1990) A taxonomy for service processes and its implications for system design. Int J Serv Ind Manage 1:20–40
Womack JP, Jones DT (2003) Lean thinking: Banish waste and create wealth in your corporation. Free Press, Revised and Updated
Womack JP, Jones DT (2005) Lean consumption. Harvard Bus Rev 83:58–68
Zysman J (2006) The algorithmic revolution—the fourth service transformation. Commun ACM 49:48

Part III
Quality and Management in Engineering Systems and Networks

Characterising Knowledge Workers' Job Positions

Ana Moreno-Romero, Eva Ponce-Cueto and Ruth Carrasco-Gallego

Abstract This research proposes a model for identifying job positions that correspond to professional knowledge workers profile. To this end, the job positions are classified according to three variables: the extent to which the activity is dematerialised (working with atoms or bits), the flexibility allowed by the tasks for managing time and location, and the level of complexity of the knowledge associated with the job. Some studies suggest that the always-on model associated with ICTs can disperse attention, lead to a drop in productivity and increase stress. This categorization of knowledge positions is useful for studying productivity as it enables, according to job profile, to differentiate the measures for improvement productivity. The objective of this work is to apply the model and test its usefulness for characterizing net profile job positions. To test the validity of the model, this research presents case studies of companies with knowledge workers in which the proposed variables are studied, together with the way the work is organized and the techniques used to allocate times.

Keywords Knowledge workers · Job positions · Networked organization · Productivity · Work organization

A. Moreno-Romero (✉) · E. Ponce-Cueto · R. Carrasco-Gallego
GIOS, Grupo de Investigación de Organizaciones Sostenibles,
School of Industrial Engineering, Madrid Polytechnic University (UPM),
C/José Gutiérrez Abascal, 2, 28006 Madrid, Spain
e-mail: ana.moreno.romero@upm.es

E. Ponce-Cueto
e-mail: eva.ponce@upm.es

R. Carrasco-Gallego
e-mail: ruth.carrasco@upm.es

M. Amorim et al. (eds.), *Engineering Systems and Networks*, Lecture Notes in Management and Industrial Engineering, DOI 10.1007/978-3-319-45748-2_31

1 The Study of Work from the Beginning of Its Scientific Organisation up to Innovative Organisations

The publication of Taylor's (1911) "Principles of Scientific Management" was followed by the theories that extended the activities of a system to include psychological, social and technological processes, to finally arrive at the present-day changes towards networked models. Throughout all this time, the organisation of work has been a discipline that has attempted to conduct a systematic examination of the methods adopted to carry out activities, with the purpose of improving the effective use of resources and setting performance standards regarding the tasks being carried out.

Traditional work study and organisation schemes for improving productivity have been mainly restricted to material transformation operations (assembly lines) and are difficult to apply to knowledge organisations. In the new context of networks, the ways to measure productivity are changing as a result of the changes in work organisation. If measuring productivity was already difficult in traditional organisations, it is even more complex when we are dealing with knowledge workers, for whom job positions are less standardised.

In his book, "The Design of Efficient Organisations" (1992) Minzberg discussed five possible coordination systems (mutual adjustment, direct supervision, process standardisation, production standardisation and knowledge standardisation), which, if used properly, lead to the consistency and harmony of the parts of an organisation (strategic apex, techno structure, support staff, middle line and operating core). Of all the possible structures, (simple structure, machine bureaucracy, professional bureaucracy, divisional form, adhocracy or innovative, missionary and political organisation) the one that seems to be best adapted to the networked organisation is adhocracy, which is designed for dynamic, complex systems (Mintzberg 1989). However, in Mintzberg (1989) said they do have problems of inefficiency because they are not oriented to doing "everyday things" and because of the high cost of being always-on and of meetings. The solution was to evolve towards other less efficient and more bureaucratic structures. Twenty years later, Mintzberg (2009) spoke about reconstructing companies as communities, assuming that the adhocratic (or innovative) structures must be accepted as permanent in networked societies. The new concept is "community ship". Community means "taking care of our job, our colleagues and our geographic place in the world and, at the same time, feeling inspired by that caring attitude". From an organisational point of view this requires: changing from top-to bottom schemes to schemes that go from the centre outwards. It also means generating an environment that promotes trust, possessing a strong culture and placing leadership at the centre.

In the new development mode described as "informational" by Castells (1998), the source of productivity stems from the technology of the knowledge generation, from information processing and the communication of symbols. This, together with Drucker's conceptualisation, "Working more efficiently means working more productively without working harder for longer" (Drucker 1999), are the pillars on which to begin modelling knowledge workers' productivity. The organisation of labour in organisations now becomes more complex. IBM's study on leadership trends "Capitalizing on Complexity" (2010) now identifies the "complexity gap" as being the major concern of the leaders interviewed. Qualified organisation is based on four principles: providing project teams, dynamic project logics, cooperation and friendly working, and forethought (Zarifian 2005).

Inside an organisation, it is important to identify what job positions affect network models and which do not. Given the complexity and singularity of the subject under study, it would be a mistake right from the very start to apply it to all types of workers as some could fit in with traditional work organisation models, even if they are part of networked organisations. In job positions that have a high degree of specialisation, it is those that are related to data and particularly to people where the greatest number of so-called knowledge workers is to be found. These are characterised by making an intensive use of Information and Communication Technologies (ICTs) and by information and knowledge being the basis of their work. They also often have a high degree of flexibility when organising their activities.

The opportunities opened up by the incorporation of ICTs are clearly consolidated. However, the expected improvement in performance that should have resulted in ICT-intensive work environments is not happening. Moreover, there are studies that are beginning to show that the always-on model disperses attention to the point that productivity falls. The problem is being analysed from the standpoint of key industrial management positions. In the McKensey Quarterly Journal, Davenport (2011) discusses re-thinking knowledge work, "There's little evidence that massive spending on personal computing, productivity software, knowledge-management systems, and much else has moved the needle. What's more, a wide variety of recent research has begun suggesting that always-on, multitasking work environments are so distracting that they are sapping productivity". The only thing on which the knowledge workers' productivity study is based is measuring the working hour. The ability to break down the number of hours required to perform a certain activity is the key to job planning. However, allocating specific times, identifying unnecessary activities or controlling one's own time in a hyper connected environment do not appear to be easily attainable goals.

This research proposes a model for characterising knowledge positions. This would allow traditional job studies to be completed with time study techniques and methods that have been specifically tailored to these kinds of jobs. Since the measurements for improving productivity and worker comfort are different, partly due to the job profile, clearly identifying what we might call a networked knowledge position is valuable for the productivity of organisations that are adopting networking models.

2 KUBJO Model for Characterising Knowledge Positions

In order to identify knowledge positions, this research classifies job positions according to three variables: the extent to which the activity is dematerialised (working with atoms and bits), the flexibility allowed by the tasks for managing time and location, and the level of complexity of the knowledge associated with the job (KUBJO model: Knowledge, Ubiquity and Bits in Job Organisation) (Moreno 2009) (Fig. 1; Table 1).

Going into detail concerning the characterisation of these job profiles is complex and goes beyond the aims of this work. The objective of this stage of the research is to explore the job positions that most intrinsically reflect the knowledge society, the KUBs, (the task consists of working with bits under a mobile scheme and with a high level of knowledge of the task: e.g. company executive) and to characterise the opportunities, challenges and skills required by networkers.

3 Methodology for Testing the Model

Having put forward the KUBJO model, the objective of this work is to apply it and test its usefulness for characterising net profile job positions (KUB), since it is these positions that are the greatest challenge for taking standard measurements to

Fig. 1 KUBJO model

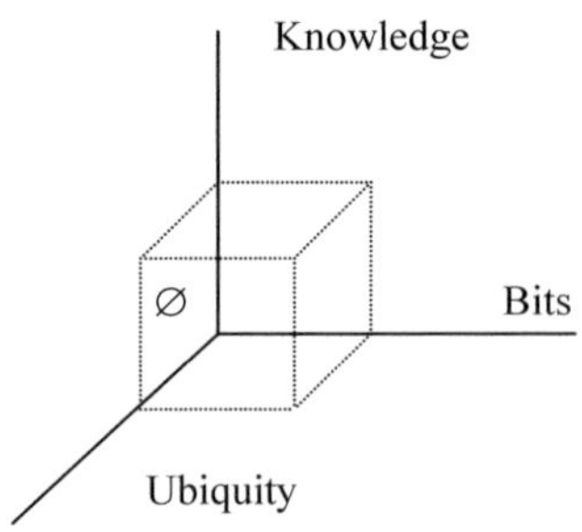

Table 1 Classification and definition of the variables comprising the KUBJO model (Moreno 2011)

KUBJO base	Definition of bases	Level type	Level definition
Knowledge base	Task knowledge level	High knowledge versus low knowledge	Task monitoring versus task creativity
Bits base (Dematerialisation)	Level of task dematerialisation	Atoms versus bits	Productive processes versus knowledge processes
Ubiquity base	Level of flexibility in time and location to perform the task	Fixed scheme versus mobile scheme	Presence in the job versus flexibility (time and location) in the position

improve productivity. To this end, a **questionnaire** was prepared that characterises each axis of the model as well as the worker's perception of their job in the organisation and of time control (Table 2).

To test the usefulness of the KUBJO model for characterising knowledge-based job positions, five pilot **case studies** were developed using the above questionnaire.

Table 2 KUBJO model variables

Variable	Questions	Scale
Ubiquity	1a. My job allows me the flexibility to work out of the office	Likert 1–5
	1b. I can organise my timetable with the freedom to start and finish work when I like	
Bits (Dematerialisation)	2a. What proportion of your work is linked to information and knowledge that is in a digital format?	%
	2b. What percentage of your working time do you devote to interacting with other people in a digital format? (e.g.: email, video conferences, WebEx, net meetings, etc.)	
	2c. What percentage do you think should necessarily be face-to-face?	
Knowledge	3a. I believe the structure and organisation of my job gives me the autonomy to plan, control and regulate my context	Likert 1–5
	3b. I believe my job lets me learn, develop myself and adapt to the circumstances surrounding it	
	3c. I believe my job offers a variety of experiences or activities, so I don't consider it to be repetitive	
	3d. I feel I take an active part when it comes to designing or planning what my job consists of or the changes that are made	
	3e. My work is important and has repercussions on the organisation as a whole	
Work organisation	4a. I plan my work on a yearly basis	Likert 1–5
	4b. I plan my work on a monthly basis	
	4c. I plan my work on a daily basis	
	4d. I manage to make what I do fit in with what was planned	
	4e. Unscheduled meetings are a distorting element in my planning	
	4f. The telephone has a negative impact on my planning	
	4g. Email has a negative impact on my planning	
	4h. The amount of work is more than I can do in my working day	
	4i. I think that working with certain levels of stress is inevitable	
Time control	5a. I can estimate the time I spend on each of my actions	Likert 1–5
	5b. It would be useful to have a corporate tool for knowing the time I spend on each task	
	5c. I personally think it would be a good idea to devote part of my time to improving individual and group planning systems	

The case studies were chosen to cover different knowledge worker profiles. They ranged from people in consultancy projects, and therefore used to controlling their working hours as a billing unit, up to people who had mainly procedural tasks bordering on industrial processes but with a significant knowledge work component.

For each of the case studies, the information set out in the results section is:

- Description of the organisation
- Description of the pilot scheme (sample, timeframe, research scenario)
- Net profile analysis (KUB) of the sample workers.

Having analysed the cases individually, all the data were analysed as a whole to see if any patterns could be observed in the KUB worker sub-types.

4 Results

Only one of the 5 cases is included as an example.

Case B

Description of the Organisation
Company B is a domestic company from the energy sector with 1500 workers. It has implemented systems of management by objectives, management by competences, internal processes defined according to EFQM standards, top level information and knowledge management systems, and leading edge corporate social responsibility policies, among other strategic and organisational resources.

Pilot Scheme Description
The Human Resources department has two service headquarters, one of which is responsible for developing professionals in a very wide sense. Five of the ten people in this unit took part in a pilot scheme in the last quarter of 2009 and the first quarter of 2010. The purpose was to help workers improve the unit's productivity.

Net Profile Analysis
From the KUBJO questionnaire it was observed that the average evaluated job type could be defined as a job position that allows a considerable flexibility of timetable, lets work be done out of the office and is well-designed and organised. This means that the employee is motivated and feels committed to their work. The workers plan their activity but find it difficult to make their plans fit in with the actual circumstances. This is due to elements like the telephone, email and unscheduled meetings, which have a negative impact. There is also a certain work overload for the working day. According to the means of the questions in block B, these are therefore net profiles.

Table 3 Means and deviations for case B

Variable	Question	Mean	Deviation
Ubiquity	1a	3.25	0.96
	1b	3.50	0.58
Bits (Dematerialisation)	2a	50.00 %	40.82 %
	2b	67.50 %	17.08 %
	2c	58.75%	26.58 %
Knowledge	3a	4.00	0.00
	3b	4.75	0.50
	3c	4.00	0.82
	3d	4.00	0.82
	3e	4.50	0.58
Work organisation	4a	3.75	1.89
	4b	4.00	0.82
	4c	3.50	1.29
	4d	2.50	0.58
	4e	1.25	0.50
	4f	2.50	1.00
	4g	2.75	0.50
	4h	2.00	0.82
	4i	3.75	0.50
Time control	5a	2.75	0.96
	5b	4.00	0.82
	5c	4.00	0.82

The results for case B are shown in Table 3.

For the responses in blocks 3 and 4, it should be pointed out that the high standard deviation together with the low number of responses prevents the means being taken by way of guidance, specifically, questions 4a and 4c. It appears that each worker plans their work according to their own criteria without any common guidelines.

Aggregate Quantitative Analysis

This analysis was performed to obtain additional information on the validity of the KUBJO questionnaire and hence on the validity of the work organisation model in knowledge positions for identifying worker' profiles according to three network model variables (knowledge, use of bits in the job and ubiquity). The data of 77 of the 79 workers, from the 5 cases, who responded to the questionnaire were examined. The analyses were done as a whole and comparing the results of the case studies with one another and with the total.

The methodology has three steps:

– An analysis by an expert in networked organisation to predict the expected responses in the model variables according to the profile of each organisation, using a relative scale (1st, 2nd, 3rd, 4th, 5th in intensity of that variable).

- A comparative analysis of the questionnaire results by organisation, including both the individual questions and the questions as a whole by variable. The results of these two steps confirm that the expert estimation broadly is aligned with questionnaire results.
- A conceptual evaluation of the above results in order to evaluate the behaviour of the questions and the variables.

For the aggregate analysis of the variables, a factorial analysis was previously performed that would let the significance of the questions as an aggregate variable be confirmed. The factors that are extracted and the aggregate description that could be attributed to this set of questions are:

1. 3b, 3c, 3d (professional development),
2. 1a, 1b, 2a, 2b, 2c (description of the mobility of the job)
3. 3a, 3e (autonomy and relevance)
4. 4a, 4b, 5a (macro planning)
5. 4c, 4d (micro planning)
6. 4e, 4f, 4g, 4h (time management)
7. 5b, 5c (need for planning tools)

Therefore, questions 1a and 1b are part of a single factor together with those in block 2, while blocks 3 and 4 are broken down into 2 and 3 variables respectively.

5 Conclusions

In view of the case study data, the first conclusions of interest are related to the characterisation obtained for each of the companies studied. There is a set of results that appears to be linked to the circumstances of the organisation and the job, while others are more linked to the worker's profile and skills.

From a knowledge positions research point of view, the objective of the case studies presented was three-fold. On the one hand, it was wished to test the usefulness of a job analysis methodology for identifying the positions closest to network models. The wide involvement of the companies and participants in the pilot schemes together with the evaluation they made of the experiences would clearly seem to indicate the usefulness of the methodology. On the other hand, it was also wished to prove whether or not the knowledge, ubiquity and bits variables were significant for positioning the studied groups. Finally, it was wished to prove if the questions for each variable and analysis block were right. It has been seen that some questions require a better explanation and maybe need revising. However, generally speaking, the factors comprising the questions lead to an explanation of the sought variables.

References

Castells M (1998) La era de la información. Economía sociedad y cultura, vol 3. Fin del Milenio, Alianza Editorial, Madrid
Davenport T (2011) Rethinking knowledge work: a strategic approach. McKinsey Q 1:88–99
Drucker P (1999) Management challenges for the 21st century. Butterworth-Heinemann, Oxford
IBM (2010) Capitalizing on complexity. Insights from the Global Chief Executive Officer Study
Mintzberg H (1989) Mintzberg on management: inside our strange world of organizations. The Free Press, New York. ISBN 0-02-921371-1
Mintzberg H (2009) Rebuilding companies as communities. Harvard Bus Rev 87:140–143
Moreno A (2009) Adaptación a las nuevas tecnologías organizativas del trabajo en entornos de red. Telos. Cuadernos de comunicación e Innovación, 81, ISSN 0213-084X
Moreno A, Mataix C, Mahou A (2011) Claves para la gestión de puestos de trabajo en las organizaciones en red. Revista de Dirección y Organización, Cepade
Taylor F (1911) Factory management. McGraw-Hill
Zarifian P (2005) Compétences et stratégies d'entreprise. Les démarches compétences á l'épreuve de la stratégie des grandes enterprises. Editions Liaisons, Paris

Information Quality in Companies Committed to TQM

Marta Zárraga-Rodríguez, Manuel Francisco Suárez-Barraza, M. Jesús Álvarez, Elisabeth Viles and Carmen Jaca

Abstract In the turbulent global environment in which organizations now compete, having reliable, accurate and readily available information is a key factor. Information is a strategic company resource. To the extent that TQM is an information-intensive management model, it is not unreasonable to assume that companies committed to TQM models should be efficient when managing information. This study has two main objectives. First, we aim to explore whether the information managed by companies committed to TQM models (in particular a quality management system based on the ISO 9000 international quality standards series) meets the dimensions required for information quality. Second, we aim to explore how information quality dimensions influence information consumers' overall feeling of being well informed in those companies. Using a quantitative approach, we used a questionnaire survey to capture the perceptions of managers in companies committed to a quality management system based on the ISO 9000 international quality standards series with respect to the information they managed. More specifically we applied a nonparametric test and performed a multiple regression analysis to address the objectives of the study. The analysis shows

M. Zárraga-Rodríguez (✉)
ISSA-School of Management Assistants, Universidad de Navarra,
Edificio Amigos, Campus Universitario, 31009 Pamplona, Spain
e-mail: mzarraga@unav.es

M.F. Suárez-Barraza
EGADE Business School, Instituto Tecnológico y de Estudios Superiores
de Monterrey, Monterrey, Mexico
e-mail: manuelfrancisco.suarez@itesm.mx

M.J. Álvarez · E. Viles · C. Jaca
TECNUN, Universidad de Navarra, P Manuel Lardizabal,
13, 20018 San Sebastian, Spain
e-mail: mjalvarez@tecnun.es

E. Viles
e-mail: eviles@tecnun.es

C. Jaca
e-mail: cjaca@tecnun.es

M. Amorim et al. (eds.), *Engineering Systems and Networks*, Lecture Notes
in Management and Industrial Engineering, DOI 10.1007/978-3-319-45748-2_32

evidence that the information managed already meets many dimensions of information quality among the companies analysed.

Keywords Information quality · TQM · ISO

1 Introduction

In the turbulent global environment in which organizations now compete, information has become a key factor (Sen and Taylor 2007; Tee et al. 2007). Poor quality information can lead companies to adopt decisions which cause adverse effects at operational, tactical and strategic levels, and also on customer and employee dissatisfaction, lower revenues, conflicts among organizational units, etc. (Gorla et al. 2010; DeLone and McLean 1992). In particular, when a company is committed to TQM, information is indispensable (Doeleman et al. 2012) because TQM is an information-intensive management system (Matta et al. 1998); quality information is especially critical in companies that implement TQM practices because providing quality information is critical to the success of efforts to achieve quality.

Organizations are committed to quality management mainly within two frameworks (Martínez-Costa et al. 2008; Bayo-Moriones et al. 2011; Heras-Saizarbitoria et al. 2011): a quality management system based on the ISO 9000 international quality standards series or a total quality management (TQM) model.[1] In this study we aim to explore the quality of the information managed when a company has a quality management system based on the ISO 9000 international quality standards series and to clarify what impact the information quality dimensions have on employees' feeling of being well informed.

2 Framework and Objectives of the Study

Information quality is a multidimensional construct. Numerous attempts have been made to define information quality and to identify its dimensions (studies cited by DeLone and McLean 2003; Lee et al. 2002; Gorla et al. 2010; Nelson et al. 2005; among others). In this exploratory study we selected the set of information quality dimensions that have been the most extensively studied and we used them when we defined the statements in the questionnaire survey. Table 1 presents this set of dimensions and the statements included in the questionnaire.

[1]Total Quality is embodied in several models, including: the EFQM Excellence Model, the Deming Management Model, the Latin American model for Excellence and the Malcolm Baldrige Model.

Table 1 Information quality dimensions

Information quality dimensions	Statements in the questionnaire
Accessibility (IQD1)	I can easily locate and access the information I need
Consistency (IQD2)	The information is integrated and coordinated (although I make use of different sources the information does not change)
Relevance (IQD3)	Information is relevant, clear and concise, and it is processed (not just data)
Completeness (IQD4)	I get the information I need to perform my job
Accuracy (IQD5)	The information I receive is accurate
Believability (IQD6)	I can trust the information I receive
Timely for use (IQD7)	I receive the information in a timely manner

This study has two main objectives. First, we aim to explore whether the information managed by companies committed to TQM models (in particular a quality management system based on the ISO 9000 international quality standards series) meets the dimensions required for information quality. Second we aim to explore how information quality dimensions influence information consumers' overall feeling of being well informed in those companies. We defined the following research questions:

Q1. Does the information managed by companies committed to a quality management system based on the ISO 9000 international quality standards series meet the dimensions of information quality?
Q2. Which are the information quality dimensions that influence the feeling of being well informed among the employees in those companies?

In order to answer these questions we followed the methodology described in the next section.

3 Research Methodology

A survey was administered to companies in Mexico that have a quality management system based on the ISO 9000. The sample consisted of 47 companies whose managers reported their agreement with each statement in the questionnaire.

The items on the questionnaire were presented as statements, and respondents had to indicate their agreement on scale of 1 (strongly disagree) to 10 (strongly agree). The statements aimed to measure the perceived quality of the information managed and the feeling of being well informed. Statements were developed from a literature review and the final instrument contained 8 items, 7 of which addressed information quality and one that measured the feeling of being well informed. We used a nonparametric test to explore the quality of the information and we used multiple regression analysis to examine the relationships among the collected data.

The instrument was pre-tested in order to avoid misunderstanding effects, i.e. to make sure the items' meanings were clear and that the questionnaire was easy to answer.

Because a random sample was selected, the data analysis process involved the use of statistical techniques. The research questions were evaluated using statistical hypothesis tests, and the Minitab 16® software package was used to analyze the results. A nonparametric test was applied because the ordinal responses are not normally distributed.

In order to check whether the proposed information quality dimensions are commonly adopted by companies, a one-sample Wilcoxon signed rank test was applied. This technique is a nonparametric alternative method to the one-sample t-test. This is a test for the median of a single population, which compares a measurement on a single sample in order to assess whether its population mean rank differs and thus it can be used to test whether the location (median) of the measurement is equal to a specified value.

In the tests applied in this study, the criteria established in order to reject the null hypothesis was p-value < 0.05.

The relationship between information quality dimensions and the feeling of being well informed was evaluated using multiple regression analysis.

4 Results and Discussion

Table 2 presents a summary of the scores given to each statement by the respondents. All the means, except for one quality dimension, are higher than 7.5, which means that a high level of information quality is commonly perceived.

This is consistent with the idea that the TQM is an information-intensive management system that usually has a good information system in place. In order to answer question Q1, we decided that an information dimension can be considered to be a high quality dimension by companies if more than 50 % of them evaluate it at 7.5 or above on a 1–10 scale. Therefore, the following statistical hypothesis was defined:

H0: The median is 7.5
H1: The median is greater than 7.5

Table 3 shows the median and mean value of the scores given by the companies to each information dimension studied. It also shows the results obtained when applying the Wilcoxon test to those dimensions. According to the results all the

Table 2 Quantitative results: mean ($\bar{x}$) and deviation (σ) of the questionnaire scores

	IQD1	IQD2	IQD3	IQD4	IQD5	IQD6	IQD7
($\bar{x}$)	7.94	7.85	7.57	7.79	7.57	7.70	6.70
σ	1.7	2.0	2.04	1.77	1.69	2.04	2.48

Table 3 Results of Wilcoxon tests for scores of the companies' dimensions of information quality (significance is indicated by asterisks next to each value)

Practices	Median	Mean	*P*-value (Wilcoxon test)
IQD1	7.94	8.0	0.026*
IQD2	7.85	8.0	0.045*
IQD3	7.57	8.0	0.101
IQD4	7.79	8.0	0.031*
IQD5	7.57	8.0	0.225
IQD6	7.70	8.0	0.049*
IQD7	6.70	7.0	0.942

dimensions except relevance, accuracy and timeliness are considered to have achieved a high quality level by companies.

We used multiple regression analysis to answer research question Q2; in particular we used the best subsets regression method to determine which predictor variables should be included in the multiple regression model. We computed this method with a statistical software program (Minitab).

Results are shown in Table 4. Different models are presented and for each one R^2 shows the % of the dependent variable that can be explained by the predictor variables in the model.

It can be observed that a model with three variables has the highest Adj.R^2 value. Moreover, the lower the value of S is, the more accurate the predictions made with the regression line are, and the lowest value of S corresponds to that model with three variables in Table 4. Therefore, the best subset of variables would be given by these variables: IQD1, IQD4 and IQD7.

Table 4 Best subsets analysis

Vars	R^2	Adj. R^2	Mallows' Cp	S	IQD1	IQD2	IQD3	IQD4	IQD5	IQD6	IQD7
1	61.3	60.4	8.7	1.3328		*					
1	56.9	56.0	14.5	1.4059	*						
2	66.1	64.5	4.3	1.2623		*					*
2	65.2	63.6	5.5	1.2784	*			*			
3	69.5	67.4	1.7	1.2100	*			*			*
3	68.8	66.7	2.6	1.2235	*	*					*
4	70.1	67.3	2.8	1.2116	*	*		*			*
4	69.7	66.8	3.4	1.2207	*			*		*	*
5	70.5	66.9	4.4	1.2187	*	*	*	*			*
5	70.4	66.8	4.5	1.2212	*	*		*		*	*
6	70.8	66.4	6.0	1.2284	*	*	*	*	*		*
6	70.6	66.2	6.3	1.2326	*	*	*	*		*	*
7	70.8	65.5	8.0	1.2439	*	*	*	*	*	*	*

Note: *Vars* No. of predictors in the model; R^2 Model fit; *S* Standard error of the regression; * Predictors that are present in the model

We consider that model to be the best one considering the values of R^2, Adj.R^2 and S, although the value of Mallows' Cp is not adequate.

4.1 Multiple Regression Analysis

Once we have selected the best subset, a three-predictor multiple linear regression model is proposed with three predictor variables: accessibility (IQD1), completeness (IQD4) and timeliness (IQD7). The model allows us to explain how the predictor variables influence the dependent variable. The equation for the proposed multiple linear regression model is illustrated as follows:

$$\mathrm{FWI} = \mathrm{b0} + \mathrm{b1(IQD1)} + \mathrm{b2(IQD4)} + \mathrm{b3(IQD7)} + \mathrm{e}$$

where

FWI feeling well-informed (dependent variable),

b0 constant,

e error.

We used Minitab statistical software for the regression analysis, and Table 5 shows the output.

According to Hair et al. (1987), variation inflation factor (VIF) values should be lower than 10 to ensure the non-existence of multicollinearity, and this is satisfied. According to the value of Adjusted R2 (0.695), the model has a very reasonable fit. Moreover, the proposed model is adequate as the F-statistic = 32.689 is significant at the 1 % level (p-value < 0.001).

Three information quality dimensions were found to have a positive and significant relationship with the feeling of being well-informed: accessibility (IQD1;

Table 5 Multiple regression analysis of information quality dimensions (3) on perception of feeling well-informed

Regression analysis results					
R^2			0.695		
F-Statistic			32.689 $P < 0.001$		
Adj. R^2			0.674		
S			1.2100		
N			47.000		
Predictor	Coeff.	SE Coeff.	T	P	VIF
(Constant)	−0.7645	0.8888	−0.8601	0.395	
IQD1	0.4853	0.1582	3.0679	0.004	2.2696
IQD4	0.4071	0.1556	2.6098	0.012	2.3911
IQD7	0.2092	0.0846	2.4731	0.017	1.3790

b = 0.4853; $p = 0.004$), completeness (IQD4; b = 0.4071; $p < 0.05$) and timeliness (IQD7; b = 0.2092; $p < 0.05$).

We used residual plots to examine the goodness of model fit and to determine if the regression assumptions are being met.

5 Conclusions

As a general conclusion, we see that the information managed in the companies analyzed seems to meet many dimensions of information quality. The mean of the score given to each dimension by company managers is usually above 7.5; this was to be expected since those companies are committed to a quality management system based on the ISO 9000 international quality standards series. In particular, all the dimensions except relevance, accuracy and timeliness are considered to have achieved a high quality level by companies.

We have found a positive and significant relationship between some information quality dimensions and the feeling of being well informed. To the extent that the employees perceive that they can easily locate and access the information needed, they get the information needed to perform their job and they receive the information in a timely manner the feeling of being well informed will be higher.

The information gathered provides interesting information but, taking into account the size of the sample, this study can be seen just as a first qualitative approach to the subject.

References

Bayo-Moriones A, Merino-Diaz-de-Cerio J, Escamilla-de-León SA, Selvam RM (2011) The impact of ISO 9000 and EFQM on the use of flexible work practices. Int J Prod Econ 130: 33–42

DeLone WH, McLean ER (1992) Information systems success: the quest for the dependent variable. Inf Syst Res 3(1):60–95

DeLone WH, McLean ER (2003) The DeLone and McLean model of information system success: a ten-year update. J Manage Inf Syst 19(4):9–30

Doeleman HJ, Have S ten, Ahaus K (2012) The moderating role of leadership in the relationship between management control and business excellence. Total Qual Manage Bus Excellence 23 (5–6):591–611

Gorla N, Somers TM, Wong B (2010) Organizational impact of system quality, information quality, and service quality. J Strateg Inf Syst 19:207–228

Hair JF Jr, Anderson RE, Tatham RL (1987) Multivariate analysis, 2nd edn. Macmillan, New York

Heras-Saizarbitoria I, Casadesús M, Marimón F (2011) The impact of ISO 9001 standard and the EFQM model: the view of the assessors. Total Qual Manage 22(2):197–218

Lee YW, Strong DM, Kahn BK, Wang RY (2002) AIMQ: a methodology for information quality assessment. Inf Manage 40:133–146

Martínez-Costa M, Martínez-Lorente AR, Choi TY (2008) Simultaneous consideration of TQM and ISO 9000 on performance and motivation: an empirical study of Spanish companies. Int J Prod Econ 113:23–39

Matta K, Chen HG, Tama J (1998) The information requirements of total quality management. Total Qual Manage Bus Excellence 9(6):445–461

Nelson RR, Todd PA, Wixom BH (2005) Antecedents of information and system quality: an empirical examination within the context of data warehousing. J Manage Inf Syst 21(4):199–235

Sen BA, Taylor R (2007) Determining the information needs of small and medium-sized enterprises: a critical success factor analysis. Inf Res 12(4). http://informationr.net/ir/12-4/paper329.html

Tee SW, Bowen PL, Doyle P, Rohde FH (2007) Factors influencing organizations to improve data quality in their information systems. Acc Finan 47:335–355

Performance Measurement Systems for Designing and Managing Interoperability Performance Measures: A Literature Analysis

Alexandre da Costa Castro, José Marcelo Almeida Cestari, Eduardo Rocha Loures, Edson Pinheiro de Lima and Eduardo Alves Portela Santos

Abstract In a globalized and networked market, organizational interoperability has a major role, defining how companies rethink their organizational processes to assess the collaboration and cooperation level their partner companies exert over their productive processes. Classical approaches on Enterprise Interoperability Assessment (EIA) do not identify (clearly) structural elements regarding the composition of their performance measures. Not all frameworks assess a set of specific measures for their respective models regarding the interoperability perspectives and also the organizational performance. The Performance Measurement Systems (PMS) have reached a higher level and maturity, and the implementation of such recommendations in EIA measurement systems may contribute to a better qualification, foreseeability, and standardization of the measures' composition.

A. da Costa Castro (✉) · J.M.A. Cestari · E.R. Loures ·
E.P. de Lima · E.A.P. Santos
Industrial and Systems Engineering Postgraduate Program, Polytechnic School,
Pontifical Catholic University of Parana, Rua Imaculada Conceição,
Prado Velho, 1155, Curitiba-PR, Brazil
e-mail: costa.castro@pucpr.br

J.M.A. Cestari
e-mail: jose.cestari@pucpr.br

E.R. Loures
e-mail: eduardo.loures@pucpr.br

E.P. de Lima
e-mail: e.pinheiro@pucpr.br

E.A.P. Santos
e-mail: eduardo.portela@pucpr.br

M. Amorim et al. (eds.), *Engineering Systems and Networks*, Lecture Notes
in Management and Industrial Engineering, DOI 10.1007/978-3-319-45748-2_33

The objective of this paper is propose a PMS-based model, in which a set of recommendations, applied to the EIA, enables the *superqualification* of these performance measures, contributing to a better relationship with the business layers and maintaining the original structures prescribed in the EIA literature.

Keywords Performance · Measurement · Systems · Interoperability · Assessment · Strategic management

1 Introduction

Interoperability is a term used in the most varied intra- and inter-enterprise collaboration contexts to present aspects related to functional integration and coordination among departments and other companies. Its understanding, however, is commonly limited to technology-related aspects, without considering the involved business processes and the strategic concerns. Aspects such as collaboration, compatibility, performance, cost, and availability are recurring issues in the interoperability context. In turn, once these aspects lie beyond the technological level, they require principles that define their characteristics as well as methods associated with its measurement, dimensioning, and performance assessment. These principles assumes a definition of performance measures, and once qualified with attributes and properties correlated to the business specific objectives, such measures allow an interoperability performance assessment. This paper presents a systematic review of the existing literature on EIA, and makes use of primary studies and secondary literature reviews related to performance measurement processes in interoperability assessment and current recommendations for PMS. Not only the identification of relevant aspects for the investigation of the EIA literature is considered, but also their specific approach regarding interoperability measures. The context preconized by the maturity models, as well as recommendations in use, determine EIA measures as required components in the identification and understanding of specific organizational maturity levels related to interoperability. In the literature review, however, no unified criteria or objectives were identified in its composition, entitled "formation law" of its measures. Such scenario motivates a better understanding and review of the organizational interoperability performance dimensions through the diagnostic investigation of EIA specific literature and exploration of the literature related to performance measurement in the industry. The present study focuses on evincing that the measurement system design in EIA can emerge from literature contributions that deals with PMS, leading to better qualification in interoperability assessment context, highlighting the business correlation, specific metrics identification, measurement, record and maintenance as central issues.

2 Scientific Scenario and Research Strategy

Considering the classical EIA approaches, no structural elements are identified in the composition of the performance measurements. Each of the evaluated frameworks present a specific set of metrics for its respective models, and likewise, in EIA there is no clarity with regard to the measures formation law, which is a central and open issue in EIA domain. In the research, main online sources were listed, such as Scopus, IEEE Xplore, ScienceDirect, Cross Ref, and ISI Web of Science. Figure 1 illustrates the macro stages of definition, selection, and management of the related sources in a research strategy proposal. Regarding the taxonomic definitions, the interoperability problematic demanded more attention during the documents search and selection process, once since the year 2000, the number of articles and publications on the topic has significantly increased, even though with predominance of technological perspective. The taxonomic issue led to the initial definition of: (i) a comprehensive research criteria considering the database adopted and (ii) the application of selection methods and restrictive search criteria managing the selected documents. The selection criterion based in the string "Interoperability Measurement" was considered more comprehensive, providing a wider range of results. Each searching tool applies different methods and engines and, eventually, unified storage formats for the content.

The R platform (R Core Team 2014) was chosen for the manipulation of the documents. From the 1775 papers composing the off-line database, around 316 were declined due to duplicity. Considering the remaining documents, 559 resulting papers were chosen applying an automatic contextual evaluation. As a result of the thematic selection process, 54 papers were taken into account—papers related to the identification of the terms Enterprise, Interoperability, and Assessment. Such selection was analyzed with a detailed interpretation of the Abstract section. From the 54 papers, only 19 were selected through contextual analysis, thus forming a referential basis for the literature review in EIA. A detailed study of the whole content was conducted, evincing PMS elements related to EIA. The idea of this section was to present general information regarding the construction of the rational around the EIA model based on PMS guidelines, discussed in Sect. 4.

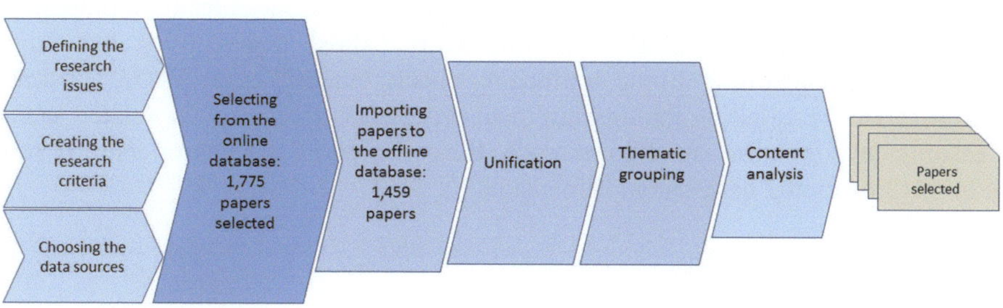

Fig. 1 Research strategy

3 Enterprise Interoperability and Maturity Models

An important aspect of interoperability is the assessment of the adherence regarding some specific model or maturity degree. That is, the evaluation of how adherent (or how mature) is an organization in comparison to a baseline model and/or in comparison with other organizations. Enterprise Interoperability Assessment (EIA) provides an organization the opportunity to know their strengths, weaknesses and prioritize actions to improve its performance and maturity (Cestari et al. 2013). The characteristic of a PMS for collaborative environments are related to such perspectives and must be considered by such systems in order to hand an integrated view of the performance of the networked enterprises (Alfaro et al. 2009). Some authors define "non-interoperability" as "heterogeneity". In fact, in order to achieve plain interoperability, the systems needs to get rid of this heterogeneity (Blanc et al. 2007). Alfaro et al. (2009) advocate that the interoperability improvement of business process is achieved by finding out up to what extent a process is efficient and effective with regard to the established criteria of interoperability. Measuring the performance of these processes from a global (inter-enterprise) and individual (intra- enterprise) perspective is, thus, made necessary. The authors in Blanc et al. (2007) deal with this question by means of specific ontologies in the organizations and through the suggestion of conceptual assessment among the knowledge sources (ontologies) for the identification of semantic relations oriented to performance. The objective consists in assuring their communication within their mutual business, bearing in mind the aspects present in the ontological models and the knowledge sharing with regard to the performance measures involved in the interoperability processes.

Considering this global collaborative scenario, different maturity models applied to interoperability have been proposed, not only in the technology field, but also regarding the different layers of the company (Rezaei et al. 2014). In most cases, the existing maturity models only define the interoperability business levels; some maturity models also include a set of best-practices, advising companies on how to become interoperable (Daclin et al. 2006; Pardo et al. 2011). In general terms, however, those proposals do not go beyond the aspects assessed, lacking further details on methodological aspects, assessment methods, measurement elements, among others. Existing interoperability maturity models commonly focus on a simple interoperability concern (data, technology, concept, business modeling, etc.), listing countless attributes that characterize those perspectives and, commonly, being object to a simple conformity check. Guédria et al. (2009) illustrate such scenario by targeting the measurement process in which aspects related to the measurement formation law are unknown, thus weakening the interoperability role as an organizational assessment dimension.

4 Measurement Concepts and Approaches in EIA

The fact that interoperability can be improved means that its measurement parameters can be defined. Interoperability measurement provides an enterprise with the understanding of its strengths and weaknesses in order to interoperate and prioritize actions focused on improving interoperability (Chen 2008; Chen et al. 2008). As a result, the potentiality of an enterprise to interoperate with an unknown partner can be evaluated through aspects such as use of standards, organizational flexibility, open ICT infrastructure, and existence of business models. Camara et al. (2014) defined interoperability measures related to costs: the interoperation cost, the information exchange cost, and the costs required for validating the exchanged information. Additionally, they defined interoperation quality by considering three dimensions: the exchange quality, the use quality, and the conformity quality.

The way to deal with interoperability measures has been focused in the literature on technological aspects the last decade. Process aspects (coordination), on the other hand, are greatly valuable for establishing an appropriate operation between organizations as well as permeating internal processes. The execution and monitoring of these business processes provide valuable information about the operation, the health of departmental activities, and relations with the market, suppliers, and customers. Measuring the potential of a company to interoperate with an unknown partner can be assessed through its use of standards, flexibility, open ICT infrastructure, existence of business models, etc. (Chen and Daclin 2007). All these aspects are related to the enterprise strategic direction, dealing with internal and environmental (market) relations. Given that the business strategy is developed in accordance with the type of competition environment intended, they strongly impact on the operation of the supply chain, targets, and performance assessment. In the EIA field we have the presence of frameworks that address the interoperability on quantitative and qualitative perspectives. The LCIM, the SoSi Morris et al. (2004), and the Kingston et al. (2005) are examples of qualitative models. The limitation of qualitative approaches, when compared to quantitative approaches, resides in the fact that they provide a less accurate assessment of interoperability (Camara et al. 2014). In their turn, quantitative models allow the numerical measurement of interoperability features. In many of these models (Blanc et al. 2006; Camara et al. 2014; Huynh and Osmundson 2006), the main disadvantage consists of the existence of a large number of criteria during the interoperability assessment process (Camara et al. 2014).

5 Literature Contributions on Performance Measurement

Literature review on performance measurement recognizes the tendencies to inter-organizational work, besides regularly recommending the research on performance measurement in supply chains and collaborative organizations

(Bititci et al. 2012). Furthermore, literature on performance measurement for inter-organizational collaboration identifies an additional complexity level associated with collaborative organizations (Busi and Bititci 2006; Folan and Browne 2005). Two properties seems to be particularly relevant in a functional context: reliability on the performance measurement and functional specificity (Artz et al. 2012). The former refers to discrepancy level in performance indicators (Banker and Datar 1989; Merchant et al. 2012), or the accuracy level to ensure that information is reasonably error-free and faithfully represents what is intended (Christensen and Demski 2003). Functional specificity concerns the concept of functional direct correlation, in which measures must be associated with strategic guidelines, business processes, sector and department aspects, and other contributions to the continuous process of performance measurement. Such aspects can be present in its attributes. Kaplan and Norton (1992) say that this problem can be overcome when the organization adopts a balanced set of measures, meeting both finance, business, customer, innovation, and learning perspectives; the core of the BSC model. The traditional view consists of the performance measures as a component of the planning and control cycle. Supposedly, the measure provides a way to capture performance data that can be used to inform the decision-making (Bourne et al. 2000). Consequently, it is intended that such measures carry (in their structural formation law) data or complete information about their origin, processing, and destination, thus supporting the decision process in an interoperable environment.

Globerson in (1985) adopts an approach oriented to exploring the formula issue (how the measure is calculated) as well as its utilization and formation of its structure in a wide context. According to him, a measure must: (i) have strategy derivatives; (ii) provide fast and accurate feedback; (iii) be specific and possibly achievable within the process; (iv) be founded on quantities that can be either influenced or controlled by the user or by the collaboration among users; (v) be clearly defined; (vi) be part of a defined management cycle; (vii) have an explicit purpose; (viii) be based on an explicitly defined formula as well as its respective data source; (ix) have numerical proportions instead of absolute numbers; (x) collect data, whenever possible, through automated tools as part of a process. According to recommendations present in the literature related to performance measurement aspects and its role in the organization, Neely et al. (1997) defined a group of twenty-two features recommended as appropriate performance measures. Neely et al. (1997) also suggest a specific and standardized record, based on the required characteristics to meet the twenty-two guidelines that define a good measure as well as the Globerson (1985) directives. Additionally, keeping a single record contributes to the standardization and uniformity of the record process, potentializing its perception and utilization by all parts involved in the processes in which the measure is used. These facts lead to an easy understanding of performance measurement for new users involved and new sectors/departments of the organization, customers, suppliers, and interoperation partners.

6 Discussion

Based on the scientific scenario presented, there is no structural elements identified in the composition of the performance measurements in classical EIA approaches as well there is a lack of clarity with regard to the measures formation law. These issues can be formalized by the following research questions in order to guide the discussions on the EIA model proposal based on PMS guidelines.

Q1: *Is it possible a structural model for interoperability measures considering the organization performance concerns*?

Neither the maturity models explored nor the specific literature in EIA presents, in a structured way, recommendations to performance measures. The inter-organizational cooperation has been one of the most evident strategies to compete and get adapted to the global market, so the internal processes of the cooperative businesses must interact in order to achieve common objectives for all parts (Verdecho et al. 2012). The approaches typically related to the supply chain of goods and services suggest that specific approaches in EIA are meant to conceptually explain and qualify the characteristics of the interoperability measures. This fact may possibly explain the reduced availability of papers dealing with performance in interoperability context. An additional basic element that helps improve the interoperability of business processes in the long term resides in finding out up to what extent the process is efficient and effective with regard to the established interoperability criteria. As a result, it is necessary to measure the performance of these business processes from a global (inter-enterprise) and individual (intra-enterprise) perspective, typically addressing the inter-organizational relation issues. The core question, however, lies in the efficiency measurement processes through indicators defined both as feedback to the market and as seasonal initiatives, or through performance measures related to the strategy. In this way, a lot has been developed in measurement systems related to finance measures. Models such as the BSC were originally based on the evaluation of accounting and finance aspects, aspects still accepted and established in the market and the academy. Regardless the model or framework, measuring the aspects related to interoperability allows a company to recognize its strengths and weaknesses, and then, interoperate and prioritize actions.

Q2: *What is the contribution of an approach deriving from the PMS literature to EIA measurement concerns*?

Suggestions for documentation, present in Neely et al. (1997), indicate a specific and standardized record, based on the required characteristics to meet the guidelines of a good measure. Although such recommendations and the ones presented in the previous section shows themselves widely applicable within various approaches and models, they manage the measure qualitative aspect instead of the performance itself. The observation of such recommendations suggests that the result of their application contributes to the measure quality, instead of the performance quality. Given that the core topic of this paper is the aspects related to measure and not to the measure result, the suggestion to apply the fulfillment recommendations of specific criteria may ensure a complementary qualification (or *superqualification*) of the performance measures typically assessed in EIA, e.g. *Performance*, *Potentiality*, and *Compatibility*. Keeping a single record contributes to the standardization and uniformity of the record process, potentializing its perception and utilization by all parts involved in the processes in which the measure is used, besides providing an easy understanding for new users involved and new sectors/departments of the organization, customers, suppliers, and interoperation partners.

7 Conclusion

Specific criteria for the qualification of performance measures were not observed in the literature restricted to *Enterprise Interoperability Assessment*. Literature on Performance Measurement Systems, however, presents a higher maturity level in the timeline, both for users and academy. Papers that suggest specific attributes and properties of the performance measures, conceptually contribute to *superqualifying* measures in EIA, possibly approximating them to strategic and managerial aspects, besides keeping qualifiers linked to their manipulation, record, and history. Evidences from the present paper will support future researches, which may take advantage both of a performance measure method that considers more qualified measures according to the formation laws proposed by the literature in PMS and decision-making processes for supporting measurement attributes and measure properties in interoperability.

References

Alfaro JJ et al (2009) Business process interoperability and collaborative performance measurement. Int J Comput Integr Manuf 22:877–889

Artz M, Homburg C, Rajab T (2012) Performance-measurement system design and functional strategic decision influence: the role of performance-measure properties. Acc Organ Soc 37:445–460

Athena (2007) The ATHENA Framework—deliverable number : D.A4.2 Specification of interoperability framework and profiles, guidelines and best practices. Integrating and strengthening the European research, March 2007 (pp 1–215). Retrieved from http://interop-vlab.eu/ei_public_deliverables/athena-deliverablesATHENA

Banker RD, Datar SM (1989) Sensitivity, precision, and linear aggregation of signals for performance evaluation. J Acc Res 21–39

Bititci U et al (2012) Performance measurement: challenges for tomorrow. Int J Manage Rev 14:305–327

Blanc S, Ducq Y, Vallespir B (2006) Interoperability problems in supply chains context. In: 12th IFAC symposium on information control problems in manufacturing, INCOM 2006, and associated industrial meetings: EMM'2006, BPM'2006, JT'2006. Saint—Etienne
Blanc S, Ducq Y, Vallespir B (2007) Evolution management towards interoperable supply chains using performance measurement. Comput Ind 58:720–732
Bourne M et al (2000) Designing, implementing and updating performance measurement systems. Int J Oper Prod Manage 20:754–771
Busi M, Bititci US (2006) Collaborative performance management: present gaps and future research. Int J Prod Perf Manag 55(1):7–25
Camara MS, Ducq Y, DUPAS R (2014) A methodology for the evaluation of interoperability improvements in inter-enterprises collaboration based on causal performance measurement models. Int J Comput Integr Manuf 27:103–119
Cestari JP, Loures E, Santos E (2013) Interoperability assessment approaches for enterprise and public administration. In: Demey Y, Panetto H (eds) On the move to meaningful internet systems: OTM 2013 workshops SE—13. Lecture notes in computer science [s.l.], vol 8186. Springer, Berlin, Heidelberg, pp 78–85
Chen D (2008) Enterprise interoperability framework. LAPS/GRAI, University Bordeaux 1, 351, Cours de la liberation
Chen D, Daclin N (2007) Barriers driven methodology for enterprise interoperability
Chen D, Doumeingts G, Vernadat F (2008) Architectures for enterprise integration and interoperability: past, present and future. Comput Ind 59(7):647–659
Christensen J, Demski JS (2003) Accounting theory: an information content perspective [s.l.]. McGraw-Hill Companies
Daclin N, Chen D, Vallespir B (2006) Enterprise interoperability measurement—Basic concepts: 3rd open interop workshop on enterprise modeling and ontologies for interoperability, EMOI—INTEROP 2006—Co-located with the 18th conference on advanced information systems engineering, CAiSE 2006, Luxembourg
Folan P, Browne J (2005) A review of performance measurement: towards performance management. Comput Ind 56:663–680
Globerson S (1985) Issues in developing a performance criteria system for an organization. Int J Prod Res 23(4):639–646
Guédria W et al (2009) A maturity model for enterprise interoperability, vol 15504, pp 216–225
Huynh TV, Osmundson JS (2006) A systems engineering methodology for analyzing systems of systems using the systems modeling language (SysML). In: Proceedings from the 2nd annual system of systems engineering conference, Ft. Belvoir VA, sponsored by the National Defense Industrial Association (NDIA) and OUSD AT&L. Anais
Kaplan R, Norton D (1992) BSC: measures that drives performance. Harvard Bus Rev
Kingston G, Fewell S, Richer W (2005) An organizational interoperability agility model
Merchant K, Stringer C, Shantapriyan PT (2012) Relationships between performance evaluation elements. In: PMA 2012 conference. Anais
Morris E et al (2004) System of systems interoperability (SOSI): final report
Neely A et al (1997) Designing performance measures: a structured approach. Int J Oper Prod Manage 17:1131
Pardo TA, Nam T, Burke GB (2011) E-Government interoperability: interaction of policy, management, and technology dimensions. Soc Sci Comput Rev 30(1):7–23
R Core Team (2014) R: a language and environment for statistical computing. Vienna, Austria. http://www.r-project.org [s.n.]
Rezaei R et al (2014) Interoperability evaluation models: a systematic review. Comput Ind 65:1–23
Verdecho M-J et al (2012) A multi-criteria approach for managing inter-enterprise collaborative relationships. Omega 40:249–263

Systematic Analysis of Economic Viability with Stochastic Approach: A Proposal for Investment

José Donizetti de Lima, Marcelo Gonçalves Trentin, Gilson Adamczuk Oliveira, Dayse Regina Batistus and Dalmarino Setti

Abstract This paper aims to expand a Systematic for the Analysis of the Economic Viability of Investment Projects (SAEVIP) proposed by Lima et al. (Joint conference CIO-ICIEOM-IIIE 2014—XX international conference on industrial engineering and operations management 2014, p. 87, 2014), (Int J Eng Manage Econ, 2015) using methods of Engineering Economics. The SAEVIP received the "Research Award 2014" of the joint conference CIO-ICIEOM-IIIE. For this purpose, adopts a stochastic treatment on principal variables that affect the expected economic performance of an investment project. The method of Monte Carlo simulation along with some distributions of probabilities, such as: uniform, triangular and normal are used to achieve the proposed objective. An example is used to illustrate the use of this approach in SAEVIP. The results extend the quality and reliability of the information generated and better substantiate the decision-making process on investment in fixed assets.

Keywords Economic analysis · Risk and return · Sensitivity analysis · SAEVIP · Stochastic approach · Method of Monte Carlo simulation

J.D. de Lima (✉) · D.R. Batistus
Department of Mathematics, Federal Technological University of Paraná, Pato Branco Campus, Via Do Conhecimento. Km 01, Curitiba-PR CEP 85503-390, Brazil
e-mail: donizetti@utfpr.edu.br

D.R. Batistus
e-mail: batistus@utfpr.edu.br

M.G. Trentin
Department of Electrical Engineering, Federal Technological University of Paraná, Pato Branco Campus, Via Do Conhecimento. Km 01, Curitiba-PR CEP 85503-390, Brazil
e-mail: marcelo@utfpr.edu.br

G.A. Oliveira · D. Setti
Department of Mechanical Engineering, Federal Technological University of Paraná, Pato Branco Campus, Via do Conhecimento. Km 01, Curitiba-PR CEP 85503-390, Brazil
e-mail: gilson@utfpr.edu.br

D. Setti
e-mail: dalmarino@utfpr.edu.br

M. Amorim et al. (eds.), *Engineering Systems and Networks*, Lecture Notes in Management and Industrial Engineering, DOI 10.1007/978-3-319-45748-2_34

1 Introduction

The usual way to incorporate the risk in the analysis of economic viability of Investment Projects (IP) consists in the sensitivity analysis (SA), which, in general, involves the simulation of results for several levels of capital cost (MRA) and/or growth rates in sales, prices, costs (operational and/or of maintenance) and revenues (Bruni et al. 1998; Souza and Clemente 2008; Casarotto and Kopittke 2010; Rasoto et al. 2012; Lima et al. 2014, 2015).

In SA it is studied the effect that the variation of a given input may cause in results. When a small variation in a parameter dramatically changes the profitability of an IP, it is said that the IP is very sensitive to this parameter and may be interesting concentrate efforts to obtain less uncertain data (Casarotto and Kopittke 2010). Nowadays, the spreadsheets of calculation are one of the best instruments to draw up an SA study.

According to Souza and Clemente (2008), the technique of SA has been used for the case in which only a few components of the Cash Flow (CF) are subject to a small degree of randomness. It is the case of small variations in Minimum Rate of Attractiveness (MRA), the initial investment (CF_0), in periodic net benefits (CFj) or in duration of the project (N). To apply the SA is necessary to vary the input parameters, one at a time, solve the problem of investment in study and record the results. Thus, rather than having a single result, we will have a summary of results as a function of the values of problem parameters. For each discount rate used (MRA), there will be a Net Present Value (NPV). The same will happen for each growth rate of sales or duration of projects, for example.

Lima et al. (2014, 2015) contributed to this theme on organizing, supported by literature, the main indicators of return and risk associated with IP and proposed indexes for SA, which determine the maximum variation for MRA, to the estimated costs and for the expected revenues for the IP in analysis, prior to the economic unfeasibility occurrence. On the other hand, if the IP is economically unfeasible, the proposed indexes for the sensitivity analysis will show what needs to make the IP viable. On the other hand, if the IP is economically unfeasible, the proposed indexes for the sensitivity analysis will show what needs to make the IP viable.

According to Lima et al. (2014, 2015) in order to adequately assess the economic viability of an IP we need to examine in depth the risk and return dimensions associated with the expected performance of the IP. Furthermore, it is of fundamental importance to promote an SA in the main factors involved in economic performance of IP in study (MRA, costs and revenues). The purpose of SA is to deepen the perception of risk that the IP is subjected. In addition, according to these authors it is necessary to build the CF diagram, to build the two-dimensional graph: NPVs versus MRAs, to draw up a report evaluating the risks and returns associated with IP and issue a conclusive opinion.

However, it must be considered in a formal way the risks that involve the future CF of an IP, which represent projections. In this sense, this article aims to discuss the analysis of the economic viability of an IP with the use of probability distributions associated to the main intervening factors and of Method of Monte Carlo Simulation (MMCS) through of an illustrative example implemented in MS-Excel®.

2 Theoretical Framework

According Lima et al. (2014, 2015), the Systematic of Analysis of the Economic Viability of Investment Projects (SAEVIP) was established based on index for return and risk existing in the literature. The SAEVIP consists of indicators in three dimensions, which are: return, risk and a SA that aims to improve the perception of risk associated with the IP in study. The latter is complemented by verification of the spectrum of the decision validity. Figure 1 shows the dimensions considered with their respective indicators and parameters of inputs.

The uncertainty introduces a new dimension in the analysis of an IP. This is due to the impossibility of controlling the future events. We can make predictions about future events, but we cannot determine exactly when and how intensively they will occur. Classic examples of such events are the future behavior of the economy, the future sales, wear and maintenance costs of equipment. Despite the uncertainty, the decision-making has to decide, sustained on the information available, what is the best action to be taken (Souza and Clemente 2008).

SAEVIP proposed by Lima et al. (2014, 2015) is based on the premise of the deterministic knowledge of the components that make up the CF, i.e. it is known with certainty the discount rate (MRA) to be used, the duration of the project (N) and what would be the revenue (Rj) and costs (Cj) provided for each period (j). However, estimating the values (cash inflows and outflows) and compose the CF representative of an IP with the resulting values means use approximations or averages. To avoid the weakness of this approach, we can make use of analysis techniques that take into account the randomness of the elements that make up the CF of an IP.

The most known techniques for dealing with risk and uncertainty are Sensitivity Analysis (SA), analytical generation of probability distribution of Net Present Value (NPV) of IP and numerical generation of the probability distribution of the Net Present Value (NPV) of IP or Simulation (Souza and Clemente 2008). According to Bruni et al. (1998), the treatment mathematical and/or statistical traditional is complex and outside the reality of the majority of decision-making and/or analysts of projects, by the level of knowledge required.

According to Gitman and Zutter (2014), a similar approach to SA, but of broader scope, can be used to evaluate simultaneously the impact of various parameters in the return of the IP. Thus, instead of isolating the effect of the change in a single variable, the scenario analysis is used to assess the impact, in the return of the IP, concurrent changes in several variables, such as: cash inflows (revenue), cash outflows (costs) and the cost of capital or cost of opportunity (MRA), resulting from different assumptions about the economic and competitive conditions.

Dimension	Indicator and Formula	Input
Return	$PV = \sum_{j=1}^{N} \frac{CF_j}{(1+MRA)^j}$	MRA, N, CF_0, CF_j and N
	$NPV = -\lvert CF_0 \rvert + PV$	PV and CF_0
	$NPVA = \frac{NPV \times [MRA \times (1+MRA)^N]}{[(1+MRA)^N - 1]}$	NPV, MRA and N
	$IBC = \frac{PV}{\lvert CF_0 \rvert}$ or $IBC = \frac{VP(R)}{VP(C)}$	PV and CF_0 or PV(R) and PV(C)
	$ROIA = \sqrt[N]{IBC} - 1$	IBC and N
	$Index\ ROIA/MRA = \frac{ROIA}{MRA}$	MRA and ROIA
	$ROI = (1+MRA) \times (1+ROIA) - 1$	MRA and ROIA
Risk	$-\lvert CF_0 \rvert + \sum_{j=1}^{N} \frac{CF_j}{(1+IRR)^j} = 0$	N, CF_0 and CF_j
	Payback = Minimum {j} such that: $\sum_{j=1}^{N} \frac{CF_j}{(1+MRA)^j} \geq \lvert CF_0 \rvert$	N, MRA, CF_0 and CF_j
	$Index\ MRA/IRR = \frac{MRA}{IRR}$	MRA and IRR
	$Index\ Payback/N = \frac{Payback}{N}$	Payback and N
Sensitivity Analysis	$\Delta\%MRA_{máx\uparrow} = \gamma = \frac{IRR}{MRA} - 1$	IRR and MRA
	$\Delta\%C_{máx\uparrow} = \varphi = IBC - 1$	IBC
	$\Delta\%R_{máx\downarrow} = \lambda = 1 - \frac{1}{IBC}$	IBC
	$\Delta\%(MRA_{máx\uparrow} \cap C_{máx\uparrow}) = \phi = \frac{\gamma \cdot \varphi}{\gamma + \varphi}$	γ and φ
	$\Delta\%(MRA_{máx\uparrow} \cap R_{máx\downarrow}) = \theta = \frac{\gamma \cdot \lambda}{\gamma + \lambda}$	γ and λ
	$\Delta\%(C_{máx\uparrow} \cap R_{máx\downarrow}) = \alpha = \frac{\lambda \cdot \varphi}{\lambda + \varphi}$	λ and φ
	$\Delta\%(MRA_{máx\uparrow} \cap C_{máx\uparrow} \cap R_{máx\downarrow}) = \beta = \frac{\lambda \cdot \varphi \cdot \gamma}{\lambda \cdot \varphi + \lambda \cdot \gamma + \varphi \cdot \gamma}$	λ and φ and γ

Fig. 1 Frame SAEVIP. *Source* Adapted from Lima et al. (2015)

In this context, a company could assess both the impact of a high inflation scenario (scenario 1) and another of low inflation (scenario 2) in NPV of an IP. Each scenario will affect the cash inflows of the company, the outflows of cash and the cost of capital (MRA), resulting, in this way, in different levels of NPV. The decision-making can use this estimate of NPV to assess superficially the risk related to the level of inflation (Gitman and Zutter 2014). The wide availability of spreadsheets on computers has strongly increased the ease and expanded the use of this technique and of SA.

However, the average value used is the result of other averages weighting of each CF (sales, revenue and costs, for example) (Bruni et al. 1998). According to these authors, the main set of risk factors of elements of cash flow is: prices, quantities sold, costs and expenses, including default rates of distribution channels.

According to Corrar (1993), the MMCS was developed in 1940 by researchers Stan Ulam and John Von Neumann, to solve problems of shielding in nuclear reactors. MMCS was discovered during World War II, in the research for the development of the atomic bomb. The adaptation of MMCS for the area of investment analysis was developed by David Bendel Hertz in 1964 (Hertz 1964).

The MMCS is a technique that involves the use of pseudo-random number and probability distribution for resolution of problems of various research areas (Gujarati 2002). According to Moore and Weatherford (2005), MMCS is one of several methods for the analysis of uncertainty propagation. Its main advantage is to determine how a random variation, which is already known, or error, affect the performance or the viability of the system that is being modelled.

For Moore and Weatherford (2005), MMCS can be used in the evaluation of IP, in which the risks involved can be expressed in a simple and easy to read, and the simulations help the decision. Thus, the indicators are no longer deterministic and become stochastic.

An alternative to the solution of this type of problem is to use MMCS in calculation of the variability of NPV of an IP (Bruni et al. 1998). MMCS involves the use of random numbers, which are chosen to generate results and distributions of corresponding probabilities. This procedure facilitates the calculation of IP risk, mainly as the aid of an electronic spreadsheet for calculation, such as MS-Excel® or the use of a specific software for this purpose, such as the Oracle Crystal Ball® and @risk®. This allows the automatic generation of results.

The simulation is an approach based on statistics. It is used in the capital budget for that perception of risk through the application of probability distributions pre-determined and random numbers to estimate the outcomes. Bringing together the various components of the CF in a mathematical model and repeating the process several times, the analyst can obtain the probability distribution of the returns of an IP (Gitman and Zutter 2014). For Casarotto and Kopittke (2010), the simulation is a powerful tool, but we must be very careful to define types and parameters of distribution of each variable, under penalty to obtain useless results.

This methodology, incorporated into the models of finance, provides as a result approximations for the probability distributions of the parameters that are being studied. Various simulations are performed, in which each one of them, are generated random values for the set of input variables of the model that are subject to uncertainty. Such random values generated follow probability distributions specific to be identified or estimated previously (Costa and Azevedo 1996).

3 Research Methodology

The modeling and simulation methods were used to achieve the proposed objective to adopt a stochastic approach in SAEVIP (Martins et al. 2014). According to these authors, the MMCS involves the use of random numbers and probability to analyze and solve problems.

The main steps for implementation of MMCS in SAEVIP are: (i) set the IP; (ii) collect, and validate data; (iii) define the variables to be treated as random variables; (iv) identify the best distribution of probability for each variable defined in the previous step; (v) define the parameters of each probability distribution identified in the previous step; (vi) establish the mathematical function for each indicator of SAVEPI; (vii) determine the number of simulations for each variable; (viii) computational implementation; (ix) to generate the simulation results; and (x) analysis of the results and issue of the conclusive opinion on the IP in study.

4 Results and Discuss

In this section, an example illustrates SAEVIP with stochastic approach by MMCS. For this purpose, we will consider the analysis of the economic viability of an IP in fixed asset with the following characteristics.

The initial investment or cash flow zero (CF_0) is a random variable that follows a Uniform probability distribution: **U** (a; b), described by:

$$\mathbf{CF_0} \sim \mathrm{U}(-\mathrm{R\$}\,900{,}000.00;\ -\mathrm{R\$}\,1{,}100{,}000.00) \tag{1}$$

The Minimum Rate of Attractiveness (MRA) is a random variable that follows a Triangular probability distribution: **T** (min; mp; max), described by:

$$\mathbf{MRA} \sim \mathrm{T}(7.25\ \%;\ 11.25\ \%;\ 20.00\ \%) \tag{2}$$

The amount of estimated sales per period (Q_j) is a random variable that follows the Normal probability distribution: **N** ($\boldsymbol{\mu}$; $\boldsymbol{\sigma}^2$), described by:

$$\mathbf{Qj} \sim \mathrm{N}\left(15{,}000;\ 1{,}000^2\right) \tag{3}$$

The Unit Contribution Margin (sales price minus variable costs) per period, UCM_j, it is a random variable that follows a Triangular probability distribution: **T** (min; mp; max), described by:

$$\mathbf{UCM_j} \sim \mathrm{T}(\mathrm{R\$}\,50.00;\ \mathrm{R\$}\,70.00;\ \mathrm{R\$}\,100.00) \tag{4}$$

The fixed cost (FC) for the activity levels evaluated is constant (deterministic), **FC** = R$ 1,000,000.00. On the other hand, the planning horizon or of analysis or the useful life of IP is fixed, **N** = 13.

In this context, the CF for each period is given by:

$$CF_j = Q_j \times UCM_j - FC \tag{5}$$

Table 1 Confidence interval for the average NPV—sample size equals 10,000

Confidence level	95 %	99 %
$Z\gamma$	1.9600	2.5758
Lower limit	R$ 813,837.86	R$ 811,127.43
Upper limit	R$ 831,089.57	R$ 833,800.01

The residual value (RV) is a random variable that follows a Triangular distribution probability: **T** (min; mp; max), described by:

$$\mathbf{RV} \sim \mathrm{T}(\mathrm{R\$}\,50{,}000.00;\ \mathrm{R\$}\,100{,}000.00;\ \mathrm{R\$}\,200{,}000.00) \quad (6)$$

Thus, we can establish the function of NPV by:

$$\mathrm{NPV}(\mathrm{CF_0}, \mathrm{CF_j}, \mathrm{N}, \mathrm{MRA}) = -|CF_0| + \sum_{j=1}^{N} \frac{CF_j}{(1 + MRA)^j} \quad (7)$$

In an analogous way we can establish a function for each indicator of SAEVIP.

The calculations were performed by using spreadsheet calculations on MS-Excel®. The results found for 10,000 simulations are presented in Tables 1 (confidence interval for the average NPV) and 2 (frame SAEVIP) and Fig. 2 (probability distribution for NPV).

According to the values obtained at the time of analysis, it is expected that the investment manages average benefits in the order of R$ 822,463.72 (NPV). NPV indicates the wealth provided by IP, this is equivalent to R$ 127,680.38 per year

Table 2 SAEVIP with stochastic approach

Indicator	Mean	Standard deviation	Coefficient of variation (%)
PV	R$ 1,822,412.07	R$ 435,995.63	23.92
NPV	R$ 822,463.72	R$ 440,102.60	53.51
NPVA	R$ 127,680.38	R$ 61,504.68	48.17
IBC	1.8288	0.4518	24.70
ROIA	4.52 %	2.05 %	45.44
Index ROIA/MRA	38.99 %	22.71 %	58.24
ROI	17.88 %	2.16 %	12.10
IRR	23.18 %	4.74 %	20.44
Index IRR/MRA	57.53 %	23.03 %	40.04
Var. MRA	88.59 %	55.33 %	62.46
Var. C	82.88 %	45.18 %	54.51
Var. R	41.59 %	16.32 %	39.25
Var. MRA and C	42.61 %	24.64 %	57.83
Var. MRA and R	27.79 %	12.43 %	44.73
Var. C and R	27.45 %	11.83 %	43.11
Var. MRA and C and R	38.42 %	33.89 %	88.20

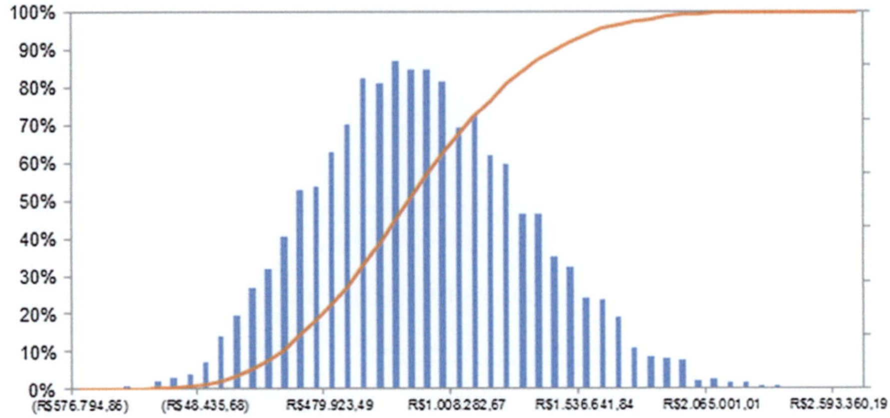

Fig. 2 Probability distribution of NPV

(medium NPVA). Every monetary unit invested in the project is estimated the average return of 1.8288 unit. This implies a ROIA of 4.52 % in addition to MRA, that is, 38.99 % above the gain of the market. Other indicators follow the conventional interpretation and, therefore, are omitted.

The probability of IP be economically unfeasible can be measured through the NPV is negative. For the problem under study, we have:

$$\mathrm{Prob}(\mathrm{NPV} < \mathrm{R\$}\,0.00) = 3.08\ \% \tag{8}$$

Thus, the probability that the IP present losses is of 3.08 %. Similarly, we have:

$$\mathrm{Prob}(\mathrm{NPV} < \mathrm{R\$}\,1{,}000{,}000.00) = 65.67\ \% \tag{9}$$

This approach can be adopted for the other parameters of SAEVIP. Figure 2 shows the probability distribution for NPV of IP in study.

Given the above, we recommend the IP implementation in study. A special care for studies of future cases is to identify the best probability distribution for each variable involved in the performance of IP.

5 Conclusion

The central objective of this paper was to propose a stochastic approach for SAEVIP proposed by Lima et al. (2014, 2015). For this reason, it was considered the main intervening variables as stochastic, that is, they are treated as random variables that follow some distribution of known probability. An illustrative example was explored to clarify the extension proposed in SAEVIP. The results showed that the probabilistic approach gives greater credibility to the information generated, which better substantiate the decision-making process on the IP in fixed assets.

This article presented main contribution to the theme to extend SAEVIP proposed by Lima et al. (2014, 2015) to assign a stochastic approach. For further work, we suggest using the real options theory (Brealey and Myers 1992), the investment theory under uncertainty (Dixit and Pindyck 1994) and the incorporation of a heterogeneous inflation (Souza and Kliemann Neto 2012) to complement the SAEVIP with stochastic approach.

Acknowledgments We are especially grateful to the National Council for Scientific and Technological Development (CNPq, n. 457.473/2014-2) and to the Coordination for the Improvement of Higher Education Personnel (CAPES, AEX 402.915/1) for their financial support, which enabled this study to be conducted and to presented.

References

Brealey R, Myers SC (1992) Princípios de finanças empresariais. McGraw-Hill, Portugal

Bruni AL, Famá R, Siqueira JO (1998) Análise do risco na avaliação de projetos de investimento: uma aplicação do método de Monte Carlo. Cadernos de Pesq. em Administração 1(6):62–74

Casarotto Filho NC, Kopittke BH (2010) Análise de Investimentos: Matemática Financeira, Engenharia Econômica, Tomada de Decisão, Estratégia Empresarial, 11th edn. Atlas, São Paulo

Corrar LJ (1993) O modelo econômico da empresa em condições de incerteza—aplicação do método de simulação de Monte Carlo. Caderno de Estudos. n.8. FIPECAFI, São Paulo

Costa LGTA, Azevedo MCL (1996) Análise fundamentalista. FGV/EPGE, Rio de Janeiro

de Souza JS, Kliemann Neto FJ (2012) O impacto da incorporação da inflação na análise de projetos de investimentos. Produção 22(4):709–717

Dixit AK, Pindyck RS (1994) Investment under uncertainty. Princeton University Press, Princeton

Gitman LJ, Zutter CJ (2014) Principles of managerial finance, 14th edn. Hardcover

Gujarati DN (2002) Econometria básica, 3rd edn. Makron Books, São Paulo

Hertz DB (1964) Risk analysis in capital investment. Harvard Bus Rev 42(l):95–106

Lima JD de, Trentin MG, Oliveira GA, Batistus DR, Setti D (2014) Systematic analysis of economic viability: a proposal for investment projects. In: Joint conference CIO-ICIEOM-IIIE 2014—XX international conference on industrial engineering and operations management, 2014. Andalucia Tech. Universidad de Málaga, Málaga—ES, p 87

Lima JD de, Trentin MG, Oliveira GA, Batistus DR, Setti D (2015) A systematic approach for the analysis of the economic viability of investment projects. Int J Eng Manage Econ

Martins RA, Mello CHP, Turrioni JB (2014) Guia de Elaboração de Monografia e TCC em Engenharia de Produção, 1st edn. Atlas, São Paulo

Moore JH, Weatherford LR (2005) Tomada de decisão em administração com planilhas eletrônicas. Bookman, Porto Alegre

Rasoto A, Gnoatto AA, Oliveira AG de, Rosa CF da, Ishikawa G, Carvalho HA de, Lima IA de, Lima JD de, Trentin MG, Rasoto VI (2012) Gestão Financeira: enfoque em inovação, vol 10. Aymara, Curitiba, 148p (Série UTFinova)

Souza A, Clemente A (2008) Decisões Financeiras e Análise de Investimentos: Fundamentos, técnicas e aplicações, 6th edn. Atlas, São Paulo

Thermal Comfort Field Study Based on Adaptive Comfort Theory in Non-residential Buildings

Elena Barbadilla, José Guadix, Pablo Aparicio and Pablo Cortés

Abstract Indoor comfort has changed over the last years. Firstly, the key was the Indoor Air Quality (IAQ) inside a building but lately it shift away from IAQ towards greater expectations related to overall subjects' comfort. However, there should be a consistent and well-balanced relation between comfort and energy consumption due to the results on the environment, especially on climate change. This paper exposes a methodology to develop a field study in a non-residential building, based on adaptive comfort theory, which take into account the above desirable objectives.

Keywords Comfort · Fuzzy logic · Energy saving

1 Introduction

Three decades ago, indoor thermal comfort literature was mainly focused on Indoor Air Quality (IAQ) which depends on CO_2 emissions, workers' activities, buildings and the Heating, Ventilating and Air-Conditioning system (HVAC).

Nevertheless it is not enough and nowadays a comfortable indoor environment is expected. There has been an intensification of activity in thermal comfort researches over the last years driven by the connection of this topic to climate change, the urgency of decarbonizing the environment and reducing greenhouse gas emissions

E. Barbadilla (✉) · J. Guadix · P. Aparicio · P. Cortés
Dpto. de Organización Industrial y Gestión de Empresas II. Escuela Técnica Superior de Ingeniería, Universidad de Sevilla, Camino de los Descubrimientos S/N, 41092 Seville, Spain
e-mail: ebarbadilla@us.es

J. Guadix
e-mail: guadix@us.es

P. Aparicio
e-mail: pabloaparicio@us.es

P. Cortés
e-mail: pca@us.es

M. Amorim et al. (eds.), *Engineering Systems and Networks*, Lecture Notes in Management and Industrial Engineering, DOI 10.1007/978-3-319-45748-2_35

policies which are taken into account in international protocols (the Kyoto protocol to the united nations framework convention on climate change in 1980 and Montreal protocol in 1997) (de Dear et al. 2013).

Buildings account for 38.9 % of the total primary energy used in the United States and of this, 34.8 % is used by buildings for Heating, Ventilating and Air-Conditioning systems. In the European Union it increases up to 40 % and its reduction has become a mayor priority (de Dear et al. 2013).

International energy consumption and its distribution are represented in Fig. 1.

The European Council underlines the importance of energy efficiency and its increasing to get an outstanding energy saving in the European Union (2012/27/UE Directive). For that, buildings and their models are essential topics.

Users have more and more demanding requirements and thermal comfort has become a key related to health, productivity, and comfort in general. Moreover, it is probably that if users are not comfortable they do something to avoid it with effects on energy saving (UNE-EN 15251 2008). These days, a fixed temperature is not efficient and although HVAC systems usually have a control node, the information is not enough to change the operating conditions. They use a standard key which has not been proved and because of that it is usual a low satisfaction rate with thermal environment.

Selecting an indoor temperature based on users' preferences would lead to an improvement in comfort, avoiding fixed temperatures and a waste of energy.

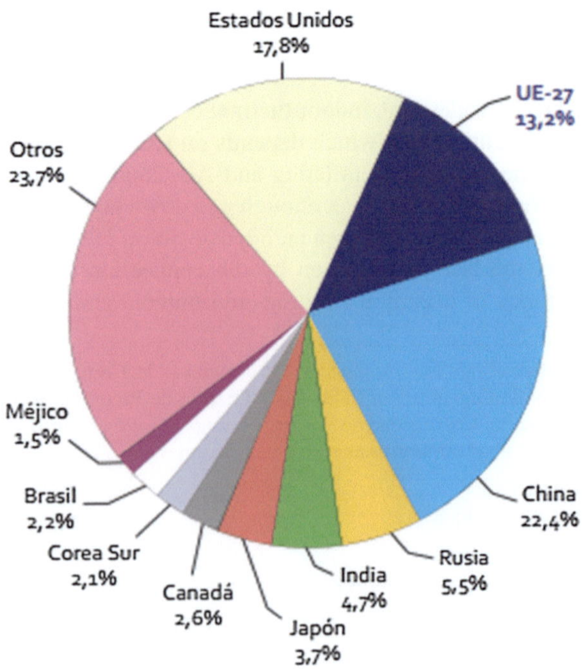

Fig. 1 International energy consumption and its distribution in 2014 (Industry, Energy and Tourism Spanish Ministry)

This paper shows, focusing on non-residential buildings, a methodology which has the aim to research adaptive methods to optimize energy efficiency for HVAC systems. It will also take into account users' comfort.

2 Comfort and Artificial Intelligence Methods

This section focuses on a general introduction to comfort, adaptive comfort theory and fuzzy logic. They define the principles of the methodology exposed in the paper.

2.1 Comfort and Buildings

American Society of Heating, Refrigerating and Air-Conditioning Engineers (ASHRAE) define thermal comfort as the condition of mind that expresses satisfaction with the thermal environment and is assessed by subjective evaluation. There are many others definitions in the literature related to this topic but they have in common the idea that comfort is not a worldwide standard. It is a concept with influences from a wide range of fields. Physical, physiological, psychological as well as cultural elements should be taken into account (Djongyang et al. 2010).

Over the last twenty years, one of the main paradigm changes was the conceptual reorientation in this field of study. There has been a shift away from the steady-state heat balance toward an adaptive comfort model (de Dear et al. 2013).

Rational Approach
The steady-state comfort theory uses values from researches that take place in climate chambers. The main advantage of this approach is that variables related to the thermal environment can be controlled.

One of the highlights was the development of the PMV model (Predicted Mean Vote) for evaluating indoor thermal comfort which was lately the basis of current thermal comfort standards.

Fanger defines PMV as the index which is able to predict the mean response related to users' subjective opinions about comfort of a large group of people according to a standard scale (the ASHRAE thermal sensation scale) and taking into account any combination of environmental values, activity levels and clothing insulation (Yau and Chew 2012). Moreover it will provide the deviation from presumed optimal thermal comfort or thermoneutrality (Van Hoof et al. 2010).

Since the design and development of the PMV model a huge number of studies have taken place in thermal comfort field using test chambers. Some of them, as Fanger in Denmark, Tanabe et al. in Japan, Chung Tong in Hong Kong, de Dear and Leow in Singapore support the strength of this model.

Whereas many others discuss its validity, reliability as well as the geographical applications range, its applications in another kind of buildings and the variables which define the input of the model as Oseland, Becker and Paciuck or de Dear and Brager who stated that the steady-state model overlooks thermal adaptive and its features (Djongyang et al. 2010).

Adaptive Comfort Theory
The adaptive approach is based on data from field studies with real people and buildings and allows analyzing another factors or elements that could not be simulated in a climate chamber since subjects are studied in their usual environment and wearing their regular clothing (Djongyang et al. 2010). Another advantage is that this model can deal with a great number of input data and has an experimental tested validity.

It is an adaptive model in which the subjects have the main responsibility to achieve comfort. Therefore they are supposed to have a range or degree of control on their environment. This control will provide a wider range of temperatures in which the users feel comfortable.

Nowadays both models coexist and each of them has their benefits and drawbacks. They are represented with their main features in the following Fig. 2.

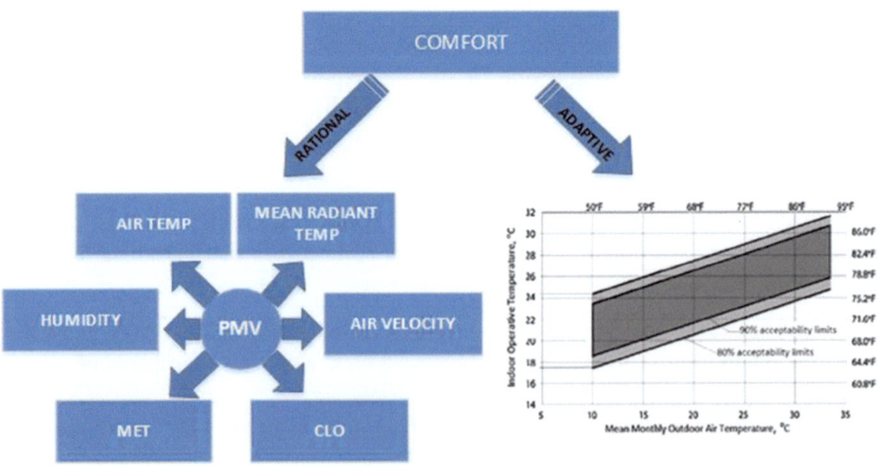

Fig. 2 Approaches in comfort field (Liang and Du 2008; ASHARE 55-2013)

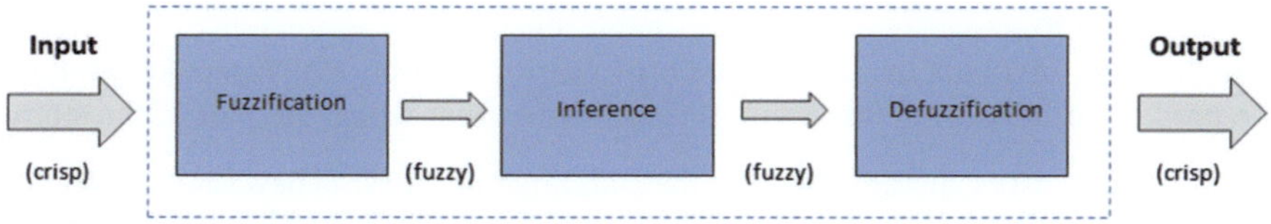

Fig. 3 Main stages in a fuzzy logic system

2.2 *Artificial Intelligence and Energy Efficiency*

Logic could be defined as the science that explains the laws, ways and forms of scientific knowledge.

Classical logic, also called logic of two values or bivalent, is based on that every proposition is true (taking the value one) or false (taking the zero value).

Fuzzy logic is one of the techniques classified within Artificial Intelligence and as all strategies belonging to this field emulates one of the smart features of the human being, in this case, reasoning with uncertainty.

It focuses on natural language joining the vagueness and the subjectivity of human activity in an automatic procedure (Pajares and Santos 2005).

It was exposed in the mid-sixties by the engineer Lofti A. Zadeh based mainly on the idea that people do not require accurate and numerical input information to control their environment and the Principle of Incompatibility. This principle takes into account the lack of a precise language in automation and control fields.

One of the aims will be to develop a system using fuzzy logic. Based on Mamdani approach, there will be possible to identify three main stages which are represented in Fig. 3.

- Fuzzification: associates every input with a degree of membership to a fuzzy set using the membership function to obtain a convenient fuzzy value.
- Inference: inference is the methodology which allows generating new knowledge from known information. This stage defines the control rules if-then that associate a fuzzy input with a fuzzy output.
- Defuzzification: obtains a crisp value from the results of the previous stage (inference stage).

3 Methodology

The methodology could split and resume into two tasks: development of instrumentation for the field study in a non-residential building in Seville and filling questionnaires related to perceived thermal comfort levels of building's users. This will extend over two years collecting data in winter season and in summer season.

Instrumentation
The first task involves the development of instrumentation which would automate the process of measuring the environmental parameters which are part of the study.

For that, sensors distributed over the buildings are used. They send, through a wireless technology, the measurements to a hub computer with connection to an external server for monitoring these input variables.

The office in the building in which the instrumentation will be developed and locations of the devices are represented in Fig. 4.

Environmental parameters relative to air temperature, relative humidity, air velocity and globe temperature have been selected because of they are the standard values which are required for most comfort indexes (such as the PMV) (McCartney and Nicol 2002).

Other parameters (CO_2 and illuminance) are measured to verify potential links between these magnitudes and perception of thermal comfort related to the users of the building.

Questionnaires
That perception of thermal comfort will be collected by questionnaires.

The questionnaires (checklists) designed for the study can be classified into three groups (Fig. 5): longitudinal, transverse and clothing. This idea is based on the methodology exposed in the SCATS project (McCartney and Nicol 2002).

- Longitudinal questionnaire: eight questions that should be answered after getting into the building (at the very least 15 min later). The users are asked about their comfort, activity level and clothing items daily.
- Transverse questionnaire: with questions related to illumination and air quality as an example.
- Clothing questionnaire: with questions related to clothing items.

Besides, users should fill in the longitudinal questionnaire every time they feel discomfort.

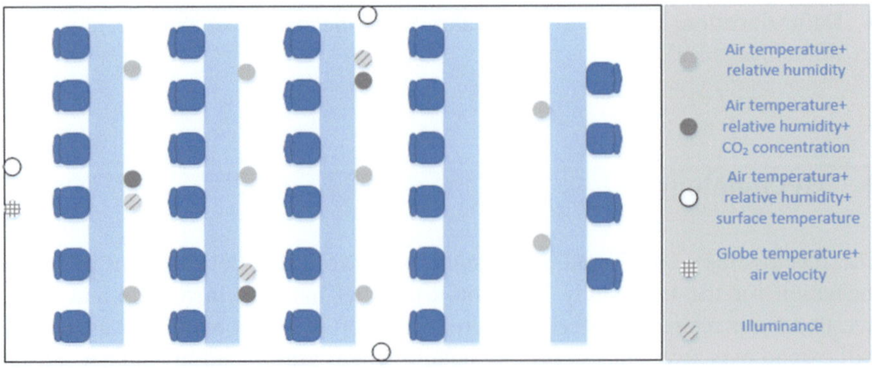

Fig. 4 Devices inside the building

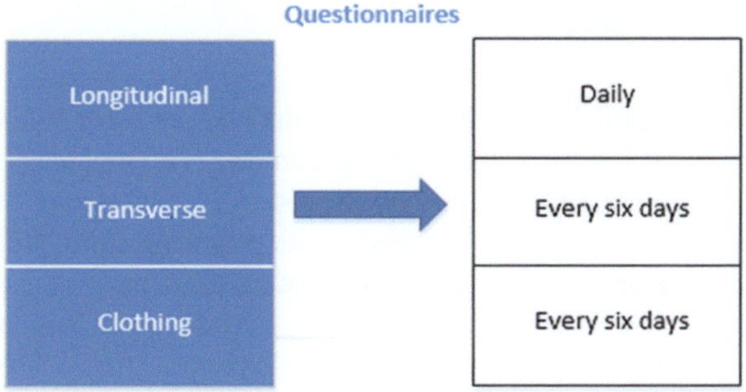

Fig. 5 Questionnaires' classification

The scheme in Fig. 6 will be the guide for the design and development of an adaptive method. Environmental parameters (from devices), subjects' opinions (from the questionnaires) and other public measurements concerning external environment (from a nearby weather station) will be the inputs of the system.

As a result of the fuzzy system an order will be obtained. This order will feed the HVAC system and therefore the temperature in the building will be modified, taking into account energy saving. It is supposed that this new temperature will lead to a change in the devices' measurements and a positive trend in users' comfort.

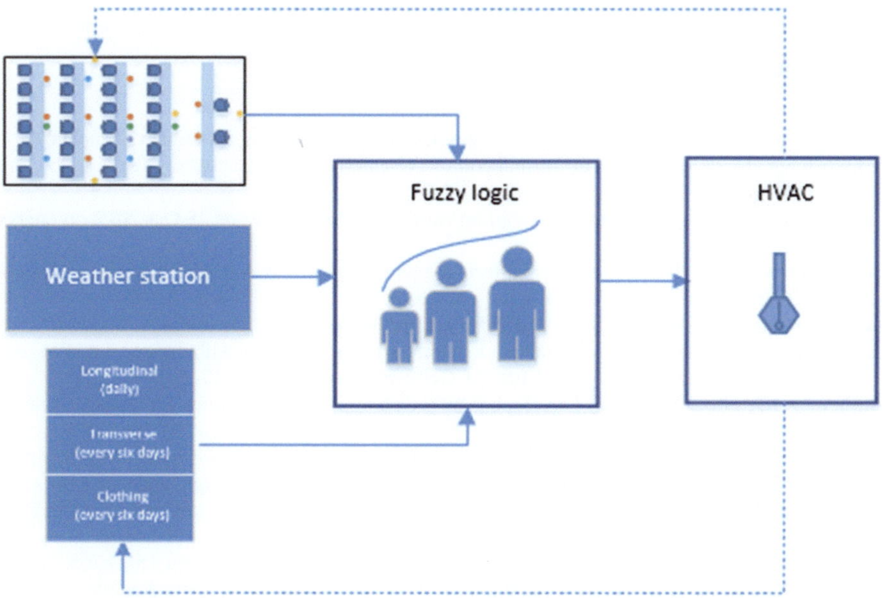

Fig. 6 Scheme for an adaptive method

4 Conclusions

PMV model is still recommended by comfort standards as the most important method for evaluating thermal comfort. The biggest advantages of that deterministic model are its fields of application. The adaptive comfort theory, based on real studies, has benefits related to its practical application, real human responses and the adaptive theory (Van Hoof et al. 2010).

Controlling indoor temperature by subjects has become an important guide in comfort area. In the future, it is expected that HVAC systems will be based on users' opinions thus the ranges of comfort would be known.

These systems, with uncertainty inputs, are difficult to model accurately so fuzzy logic could be an option instead of mathematical models. Fuzzy logic is not only based on a technical understanding of the system and pays attention to the experience of specialists and at the same time provides implementable results.

Taking into account the ideas above, the present methodology aims to develop an adaptive method that sets environmental conditions based on saving energy and subjects' comfort.

Final results must be tested to verify system's efficiency. Additionally, it would also be advisable extensively test the results in another kind of buildings to prevent an exclusive implementation in office environments.

References

ANSI/ASHRAE Standard 55-2013 Thermal environmental conditions for human occupancy

de Dear RJ et al (2013) Progress in thermal comfort research over the last twenty years. Indoor Air 23(6):442–461

Directive 2012/27/UE of the European Parliament and of the Council of 25 Oct 2012 on energy efficiency

Djongyang N et al (2010) Thermal comfort: a review paper. Renew Sustain Energy Rev 14 (9):2626–2640

Industry, Energy and Tourism Spanish Ministry. http://www.minetur.gob.es/

Liang J, Du R (2008) Design of intelligent comfort control system with human learning and minimum power control strategies. Energy Convers Manage 49(4):517–528

McCartney KJ, Nicol FJ (2002) Developing an adaptive control algorithm for Europe. Energy Build 34(6):623–635

Pajares G, Santos M (2005) Artificial intelligence and knowledge engineering. Ra-Ma, Spain

UNE-EN 15251:2008 Indoor environmental inputs parameters for design and assessment of energy performance of buildings addressing indoor air quality, thermal environment, lighting and acoustics

Van Hoof J et al (2010) Thermal comfort: research and practice. Front Biosci 15(2):765–788

Yau YH, Chew BT (2012) A review on predicted mean vote and adaptive thermal comfort models. Build Serv Eng Res Technol 35(1):23–35

Part IV
Education in Engineering Systems and Networks

Agents Playing the Beer Distribution Game: Solving the Dilemma Through the Drum-Buffer-Rope Methodology

José Costas, Borja Ponte, David de la Fuente, Jesús Lozano and José Parreño

Abstract The Beer Distribution Game (BDG) is a widely used experiential learning simulation game aimed at teaching the basic concepts around Supply Chain Management (SCM). The goal in this problem is to minimize inventory costs while avoiding stock-outs –hence the players face the dilemma between storage and shortage. Human players usually get confused giving rise to significant inefficiencies in the supply chain, such as the Bullwhip Effect. This research paper shows how artificial agents are capable of playing the BDG effectively. In order to solve the dilemma, we have integrated supply chain processes (*i.e.* a collaborative functioning) through the Drum-Buffer-Rope (DBR) methodology. This technique, from Goldratt's Theory of Constraints (TOC), is based on bottleneck management. In comparison to traditional alternatives, results bring evidence of the great advantages induced in the BDG by the systems thinking. Both the agent-based approach and the BDG exercise have proved to be very effective in illustrating managers the underlying structure of supply chain phenomenon.

Keywords Beer game · Supply chain management · Drum-buffer-rope methodology · Multi-agent system · Production

J. Costas
Polytechnic Institute of Viana do Castelo, School of Business Sciences of Valença, 49 30-678 Valença, Portugal
e-mail: josegual@esce.ipvc.pt

B. Ponte (✉) · D. de la Fuente · J. Lozano · J. Parreño
Department of Business Administration, University of Oviedo, Polytechnic School of Engineering, Gijón 33204, Spain
e-mail: ponteborja@uniovi.es

D. de la Fuente
e-mail: david@uniovi.es

J. Lozano
e-mail: lozano@uniovi.es

J. Parreño
e-mail: parreno@uniovi.es

M. Amorim et al. (eds.), *Engineering Systems and Networks*, Lecture Notes in Management and Industrial Engineering, DOI 10.1007/978-3-319-45748-2_36

1 Introduction

The Beer Distribution Game (BDG) is a role-playing simulation exercise of a simple supply chain that has been used in countless management courses since it was developed at the MIT Sloan School of Management (Jarmain 1963). It aims at teaching the main principles of Supply Chain Management (SCM), as provides practical evidence of the generation of inefficiencies along the production and distribution system (Sterman 1988). The experimental nature of this game has proved to be very effective in helping managers to understand the causal relationships between decision-making and supply chain behaviour (Goodwin and Franklin 1994). For this reason, it has been widely studied in the literature, especially in the last two decades given the strategic importance of SCM in the current competitive environment.

The BDG scenario is defined by a single-product linear supply chain, composed of four main levels: factory, distributor, wholesaler, and retailer (*e.g.* Kaminsky and Simchi-Levi 1998). At the beginning of each turn (a simulated week), the different levels receive the product from the previous level (material flow) and an order from the next level (information flow). Then, they try to ship the requested amount from its inventory. If it is not possible, a backorder is created and the remainder quantity will be fulfilled as soon as possible. Finally, they order an amount of beer from their supplier—in the case of the factory, these are manufacturing orders.

Under this context (*e.g.* Strozzi et al. 2007), there are only two sources inefficiencies: uncertainty in customer demand (simulated through random distributions) and lead time (fixed delay between orders and products are sent and received, *e.g.* three turns for both). It should be noted that production, storage, and transportation capacities are considered unconstrained.

The only goal for the participants in the BDG is to minimise costs. These are incurred in two ways (*e.g.* Chaharsooghi et al. 2008): carrying inventory and back-ordering, *e.g.* respectively $0.5 and $1 per unit of beer per period, creating an incentive to carry some inventory rather than back-order. As explained before, the only decision that each member must make is how much to order. Hence, the BDG dilemma emerges: if the purchase order is high (low), the stock-out risk reduces (surges) but the inventory costs tend to increase (decrease).

The results tend to be counterintuitive for the participants, as the small variations in consumer's demand are greatly amplified along the supply chain—this is the Bullwhip Effect (Lee et al. 1997). Sterman (1989) showed that the interaction of individual decisions produces aggregate dynamics in the supply chain which diverge from optimal behaviour. Therefore, the observed performance of human beings in the BDG is usually far from the optimal from a system-wide perspective (Lee and Whang 1999).

This paper shows how an electronic supply chain managed by artificial agents has been developed. The aim is to provide evidence that they can play the BDG effectively (*i.e.* outperforming classical alternatives) from a collaborative approach, which is the main contribution of this research. More specifically, the Drum-Buffer-Rope (DBR) technique, detailed in Sect. 2, has been used to implement the holistic functioning within the production and transportation system. The agent-based development of the system is described in Sect. 3. Section 4 shows the main results of this research study, while Sect. 5 presents a discussion of the results regarding the stated objectives.

2 Managing the Supply Chain Through the DBR Methodology

The Theory of Constraints (TOC) is a management philosophy coined by Goldratt (1990) that views any production system as being limited in achieving a higher level of performance by only one restriction. Thereby, it is aimed at achieving breakthrough improvements focusing on its constraint, which may vary over time. The logical thinking of this theory is based on a continuous improvement cycle (Goldratt and Cox 1992): (1) Identify the bottleneck; (2) Decide how to exploit the bottleneck; (3) Subordinate everything else in the system to the previous decision; (4) Implement measures to elevate the bottleneck; and (5) Evaluate if the bottleneck has been broken, and return to the beginning.

Within operations management, the TOC proposes the DBR methodology, named for its components (Goldratt and Cox 1992). The *drum* is the physical constraint of the system, *i.e.* the node that limits its performance. The other nodes follow the drum beat. Hence, the *buffer* protects the drum against variability, so that the full capacity in the bottleneck is exploited. The *rope* is required for subordinating the system to the drum—it is the release mechanism. That is, orders are released according to the buffer time before they are due. It should be underlined that the DBR configuration (planning state) must be complemented with the buffer management (monitoring stage). It implies administrating the buffer along the different nodes, in order to guide how the system is tuned for peak performance.

Although DBR technique was first oriented to the manufacturing process of the company, further development incorporates solutions for other areas. SCM is an emerging one of them. The early work deals with managing the whole system from a single company perspective (*e.g.* Cox and Spencer 1998). Later studies have used TOC to promote supply chain collaboration. Simatupang et al. (2004) provided a conceptual work for using this approach to assist the members to realise the potential benefits. Wu et al. (2010, 2014) developed two enhanced DBR-based replenishment models under capacity constraints, in order to facilitate plants and central warehouses to implement TOC. Costas et al. (2015) demonstrated that the DBR methodology induced signify operational improvements in the supply chain—as the Bullwhip Effect is greatly reduced, without any collateral damage.

When implementing the DBR method, the first question deals with identifying the bottleneck. In a real supply chain, it could be in any part of the system, depending on which one is limiting system performance. In the BDG, customer demand is the bottleneck—*i.e.* the other nodes cannot be the bottleneck due to the

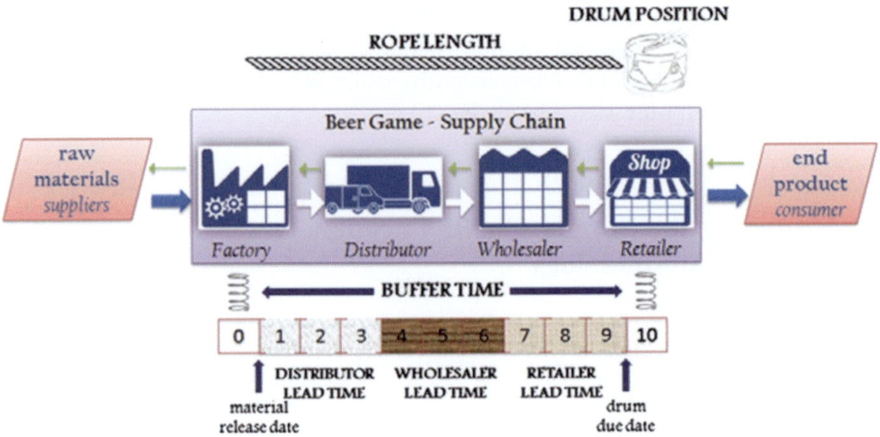

Fig. 1 Scheme of the supply chain when working according to the DBR methodology

assumption of infinite production, storage, and transportation capacities. Thus, the *drum* is placed at the retailer.

According to TOC logical thinking, the next step is to decide how to fully exploit the bottleneck. In the BDG, it is exploited by selling the product at the retailer. This node beats out the rate at which the whole supply chain can work at. To maximize the bottleneck flow means to decrease the missing sales.

This leads to the last key point[1]: factory, distributor, and wholesaler must work subordinated to the bottleneck. That is to say, the drum must be protected against process variation—*i.e.* from shortages. In other words, we need to buffer the constraint. Hence, the *buffer* is the material release duration (time interval by which the release of work is predated), while the *rope* is the release timing (it could be as a real-time feedback between the drum and the gating operation). Therefore, *e.g.* in the case of the factory the buffer is 9 turns (it is followed by three levels, and the lead time is 3 turns for every member). The rope length is the same as the buffer duration, and trying the rope ensures that excess flow cannot be admitted.

Thereby, at each turn, the factory decides the manufacturing orders that are issued, after analyzing the evolution of customer demand. Afterwards, each supply chain member (except the retailer) carries out buffer management. It means compensating the flow dissipated towards the following member. The DBR orders are dosage orders into the buffer, which are dissipative. Each agent decides what to dose subordinated to the bottleneck, so backorders creation does not make sense.

Figure 1 outlines the DBR implementation in the BDG explained above.[2]

[1]Note that, according to BDG assumptions, it is not possible to elevate the bottleneck. It is an external constraint, which is beyond the supply chain's sphere of influence. Hence the bottleneck will not be broken, and the continuous improvement cycle is reduced to three steps.

[2]Youngman (2009) elaborated an excellent guide that can be consulted to get more detail in the implementation process.

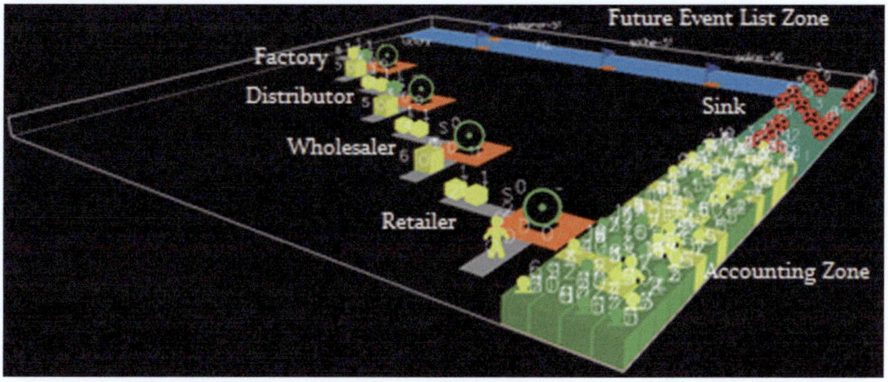

Fig. 2 Three-dimensional image of the graphical simulator

3 Description of the Multi-agent System

Agent-Based Modeling and Simulation (ABMS) is the computational method that enables to create, analyze, and experiment with models composed of agents that interact within an environment (Gilbert 2007). When the system is composed by intelligent agents that cooperate in order to achieve collective goals, it is called Multi-Agent System (MAS) (Wooldridge 2000). These emerging techniques are optimal for the study of SCM, because of the intrinsic characteristics of this problem (Ponte et al. 2014). They allow studying the effective of collaborative solutions in the supply chain, as many authors did—*e.g.* Moyaux et al. (2004) proposed the Quebec Wood Supply Game. Kimbrough et al. (2002) explored the concept of a supply chain managed by artificial agents, and demonstrated that they were capable of playing the BDG more effectively than humans.

We have used KAOS methodology to devise the conceptual model (Dardenne et al. 1993) robust software engineering techniques (Taguchi et al. 2000) to build the system, and NetLogo (Wilensky 1999) to implement it. An interface window provides the experimenter with the control to set-up parameters and to run each experimenter, as well as the graphics and monitoring to track the system. By way of example, Fig. 2 is a screenshot of the three-dimensional graphical simulator. It shows the product street (horizontal), through which the material flows from west to east—the patch closest to each node represent the on-hand inventory, and the orders street (vertical), through which the orders move from the south to the north. The other regions of the simulation are the future event list (space for agents who act as events), the sink for the system (for tangible facts), and the accounting zone (for statistical purposes.

Each supply chain node functions according to a finite state machine schema: (1) Idle state, where it waits until the drum triggers it; (2) Serve backorders state, where it completes past orders if remain at the node; (3) Shipping orders state, aimed at moving material downstream; (4) Sourcing state, when it takes care of the information flow; and (5) Reporting state, when it updates the results and exports information. In this way, the various supply chain members carry out the usual BDG operations according to the DBR methodology.

4 Numerical Results

As usual in the BDG (*e.g.* Ponte et al. 2013), we have simulated customer demand through a normal distribution with a mean of 12, while the standard deviation has been changed in order to evaluate the results in different scenarios: 1 (low variation), 3 (moderate variation), and 5 (high variation). We have carried out simulations of 330 periods. The first 30 are considered as warm-up period, hence they are not taken into account for the exposed results.

The first step is to check system stability according to common practices. We checked that the results behave stable using IR-charts, and replicated the same experiment to conduct ANOVA tests. We verified that *p*-value is greater than 5 % in the Levene test. Thereby, we failed to reject the null hypothesis (*H0: var (replica1) = var(replica2)*) and we consider the results to be statistically representative.

With the aim of simulating the human's behaviour in the BDG, we have implemented the order-up-to policy[3] in the system using three-period moving averages to forecast. The OUT was found to mimic real-life decisions made by players of the BDG (Sterman 1989).

Table 1 report the final results of the system in the three scenarios when using both methods. According to TOC's philosophy, we understand backlog costs as missing sales cost. That is to say, each level assumes a cost of $1 per unit of beer when there is shortage in the retailer. It must also be noted that inventory costs ($0.5 per unit of beer per period) considers the entire amount of product that is in the node's material flow (both on-hand inventory and product being transported).

Broadly speaking, Table 1 provides evidence of agents' high effectiveness when playing the BDG using the DBR methodology. They are able to dramatically reduce inventory costs at all levels without increasing missing sales.

[3]This replenishment method is based on a basic periodic review system for issuing orders depending on incoming demand and inventory position, aimed at bringing this position up to a defined level. See Chen et al. (2000) for more detail of this widely studied policy.

Table 1 Results of the tests

Supply chain performance	Demand—N(12, 1)		Demand—N(12, 3)		Demand—N(12, 5)	
	Agents	Humans	Agents	Humans	Agents	Humans
Inventory—factory	$3641	$21,662	$4258	$22,240	$4538	$39,636
Inventory—distributor	$1900	$8209	$2119	$14,405	$2225	$20,997
Inventory—wholesaler	$524	$2546	$696	$3939	$772	$5793
Inventory—retailer	$346	$452	$628	$719	$819	$1308
Missing sales—overall	$4	$652	$216	$496	$328	$344
Total supply chain costs	$6415	$33,521	$7917	$41,799	$8682	$68,078

Note that 'agents' refers to the DBR method, and 'humans' refers to the OUT policy

5 Discussion

The DBR method has demonstrated to be a highly effective technique to play the BDG. In other words, as expected, agents clearly outperform humans when facing this classic distribution game. It should be highlighted the huge difference: cost reduction is enormous. It is mainly due to the large reduction in the average inventory position of chain members—the Bullwhip Effect generated by human behaviour punishes the system, especially those nodes which are far from the client. Nevertheless—and paradoxically, even working with lower inventory levels, TOC philosophy displayed by agents leads to missing sales reduction.

How can this be explained? Playing the BDG with the OUT policy—even more with impulsive purchasing—causes large variations along the supply chain. Members tend to panic by shortage and thus contribute to information distortion. The peaks in orders received translate into larger peaks in orders issued (see left part of Fig. 3). These inefficiencies—provoked by the combined effect of lead times and demand uncertainty—lead to a poor performance. The more demand variability, the

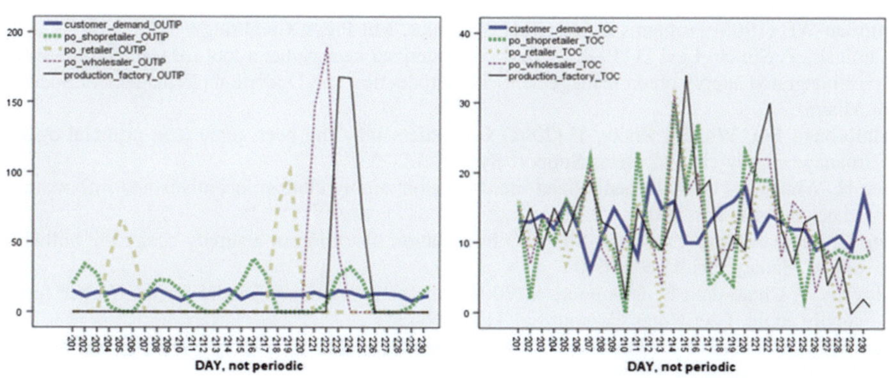

Fig. 3 Demand, orders, and production in the tests carried out with a N(12, 3) demand from turn 201 to 231 when the supply chain is managed by the OUT policy (*left*) and by the DBR (*right*)

more costs are assumed—as the members overprotect themselves. They reduce missing sales at the expense of unavoidable storage costs.

Conversely, this issue is under control when managing the supply chain from a systemic perspective. In this case, amplification also occurs along the system—and hence the factory supports the higher costs—but is much lower (see right part of Fig. 3 and note the difference between both scales). The supply chain is also damaged when demand variability increases, but the growth has been heavily damped (both in absolute and in relative terms). In summary, managing the system through the bottleneck results in a much better solution, in comparison with that one where the various levels selfishly seek the best solution for themselves.

Finally, we would like to insist on the importance of simulation as a way to educate students and managers in the comprehension process of the underlying effects of decision-making. The BDG draws an optimal context to study them.

References

Chaharsooghi SK, Heydari J, Zegordi SH (2008) A reinforcement learning model for supply chain ordering management: an application to the beer game. Decis Support Syst 45:948–959

Chen F, Drezner Z, Ryan JK, Simchi-Levi D (2000) Quantifying the bullwhip effect in a simple supply chain: the impact of forecasting, lead times and information. Manage Sci 46:436–443

Costas J, Ponte B, De la Fuente D, Pino R, Puche J (2015) Applying Goldratt's theory of constraints to reduce the bullwhip effect through agent-based modeling. Expert Syst App 42:2049–2060

Cox JF, Spencer MS (1998) The constraints management handbook. Lucie Press, Boca Raton

Dardenne A, Lamsweerde A, Fichas S (1993) Goal-directed requirements acquisition. Sci Comput Prog 20:3–50

Gilbert N (2007) Agent-based models. Sage Publications, London

Goldratt EM (1990) Theory of constraints. North River Press, Croton-on-Hudson, New York

Goldratt EM, Cox J (1992) The goal—a process of ongoing improvement. North River Press, Croton on-Hudson, New York

Goodwin JS, Franklin SG (1994) The beer distribution game: using simulation to teach systems thinking. J Manage 13:7–15

Jairman WE (1963) Problems in industrial dynamics. Mit Press, Cambridge

Kaminsky P, Simchi-Levi D (1998) A new computerized beer game: a tool for teaching the value of integrated supply chain management. The Production and Operations Management Society, Miami

Kimbrough SO, Wu DJ, Zhong F (2002) Computers play the beer game: can artificial agents manage supply chains? Decis Support Syst 33:323–333

Lee H, Whang S (1999) Decentralized multi-echelon supply chains: incentives and information. Manage Sci 6:475–490

Lee H, Padmanabhan V, Whang S (1997) Information distortion in a supply chain: the bullwhip effect. Manage Sci 45:633–640

Moyaux T, Chaib-draa B, D'Amours S (2004) An agent simulation model for the quebec forest supply chain. Lect Notes Comput Sci 3191:226–241

Ponte B, De la Fuente D, Pino P, Rosillo R, Fernández I (2013) Supply chain management by means of simulation. Polibits 48:55–60

Ponte B, Pino R, Fernández I, García N, Monterrey M (2014) Multiagent model for supply chain management. In: Managing complexity. Springer, London, pp 233–240

Simatupang TM, Wright AC, Sridharan R (2004) Applying the theory of constraints to supply chain collaboration. Supply Chain Manag 9:57–70
Sterman JD (1988) Deterministic chaos in models of human behaviour: Methodological issues and experimental results. Syst Dynam Rev 4:148–178
Sterman JD (1989) Modeling managerial behaviour: misperceptions of feedback in a dynamic decision making experiment. Manage Sci 35:321–339
Strozzi F, Bosch J, Zaldívar JM (2007) Beer game order policy optimization under changing customer demand. Decis Support Syst 42:2153–2163
Taguchi G, Chowdhury S, Taguchi S (2000) Robust engineering. Mc Graw-Hill, New York
Wilensky U (1999) NetLogo. In: The center for connected learning and computer-based modeling. Available via http://ccl.northwestern.edu/netlogo/. Cited 2 Jan 2015
Wooldridge M (2000) Reasoning about rational agents. MIT Press, Cambridge
Wu HH, Chen CP, Tsai CH, Tsai TP (2010) A study of an enhanced simulation model for TOC supply chain replenishment system under capacity constraint. Expert Syst App 37:6435–6440
Wu HH, Lee AHI, Tsai TP (2014) A two-level replenishment frequency model for TOC supply chain replenishment systems under capacity constraint. Comput Ind Eng 72:152–159
Youngman K (2009) A guide to implementing the theory of constraints (TOC). Available via http://www.dbrmfg.co.nz/. Cited 14 Feb 2015

Organizational Engineering: The Emerging Stage of Industrial Engineering

Javier Carrasco, Carlos Mataix and Ruth Carrasco-Gallego

Abstract Industrial Engineering (IE) has experienced a remarkable development since its inception in the early years of the past century. From the perspective of the present problematic situation of the world, including people and planet, a new space of opportunities opens to IE research and professional activities. In this respect, this paper reviews the two main previous stages of IE and proposes a characterization of its emergent current stage. The resulting IE concept is an extended one, which is proposed to be identified as Organizational Engineering.

Keywords Organizational engineering · Industrial engineering · Sustainability

1 Introduction

Industrial Engineering (IE) is a sound branch of the Engineering tree whose origins are commonly agreed to be found in the analytical studies on operational tasks by Taylor (1911) in the early years of the past century. Taylor's main focus on efficiency got a remarkable backing from the assembly line experience by Henri Ford who coupled a demand objective (people's accessibility to products) with an offer objective (cost efficiency). Since that time, Industrial Engineering has been evolving by the incorporation of several intertwining lines of advancement: new theories based on observed real phenomena; new mathematical techniques applied to challenging problems; practical experiences drawn by IE practitioners; increasing

J. Carrasco (✉) · C. Mataix · R. Carrasco-Gallego
Dpto. Ingeniería de Organización, Administración de Empresas y Estadística, Universidad Politécnica de Madrid, José Gutiérrez Abascal, 2, 28006 Madrid, Spain
e-mail: javier.carrasco@upm.es

C. Mataix
e-mail: carlos.mataix@upm.es

R. Carrasco-Gallego
e-mail: ruth.carrasco@upm.es

M. Amorim et al. (eds.), *Engineering Systems and Networks*, Lecture Notes in Management and Industrial Engineering, DOI 10.1007/978-3-319-45748-2_37

Table 1 Industrial Engineering Stages

	Traditional	Consolidated	Emergent
Environment			
Type of change	Laminar	Turbulent	Structural turbulence
Economic paradigm	Government drives	Market commands	Human development
Geo-politic paradigm	Bi-polarity	Mono-polarity	Multi-polarity
INDUSTRIAL ENG.			
Object of study	Manufacturing plant	Supply Networks	Sustainable Systems
Problem types	Operational	Configuration, design	Public-private design
Geographical scope	Local	International	Global
Organizational scope	Production and other	Firms network	Varied Organizations
Types of flows	Material	Information, finance	Knowledge
Actors	Firm	Network of firms	Private, public, CSOs
Decision criteria	Efficiency	Resilience	Sustainability

number of educational IE programs all over the world; diffusion of new knowledge by means of scientific conferences and reviews.

The objective of this paper is to elaborate insights on the new challenges faced by IE, reviewing its historical development.

We identify three stages in its evolution: the "traditional", the "consolidated" and the current "emergent" stage. The contextual changes that induced this evolution are outlined in a framework (Table 1). This is next used to understand the structural changes that IE is facing nowadays and to propose the lines of adaptation required. The resulting enlarged IE concept is such a distinctive approach that we propose to rename it as Organizational Engineering.

2 The Industrial Engineering Concept

Industrial Engineering is presently practiced all over the world, but named according to each particular country language and traditions, such as "Génie de Gestion" in France, "Ingegneria Gestionale" in Italy, "Engenharia de Produçâo" in Brazil and Portugal, "Ingeniería Industrial" in Latin America, "Ingeniería de Organización" in Spain. But in every case, its concept and approach are quite similar. In various countries there are professional or academic organizations devoted to IE development, such as the Institute of Industrial Engineers (IIE) in the USA, Associaçâo Brasileira de Engenharia de Produçâo (ABEPRO) in Brazil, Asociación para el Desarrollo de la Ingeniería de Organización (ADINGOR) in Spain.

The IIE (2015) official definition of IE is considered representative: "Industrial engineering is concerned with the design, improvement and installation of integrated systems of people, materials, information, equipment and energy. It draws upon specialized knowledge and skill in the mathematical, physical, and social sciences together with the principles and methods of engineering analysis and design, to specify, predict, and evaluate the results to be obtained from such systems."

Such systems are usually organizations and therefore "open systems", which implies that interactions with their environment are crucial to understand their behavior (Lewin 1936). Consequently, IE adopts an extended vision, including in its scope the particular organization and its relations with both external organizations and people (stakeholders, transactional environment) and the rest of the environment (contextual environment). The organization dynamics is dependent on the behavior of its people subsystem. Consequently IE knowledge is drawn not only from theoretical research, but also from experience, in particular, through proactive, collaborative processes first termed "action-research" by Lewin (1946).

3 Traditional Approach of IE

IE is considered to have reached consistency and momentum over the period from Taylor's time to the Second World War (WW2). Significant acquisitions in this period were: the "Gantt chart" by H.L. Gantt; the Economic Order Quantity (EOQ) by F.W. Harris; the works on motion study by F. and L. Gilbreth; the experiences on human motivation by E. Mayo. From WW2 to the 1970 decade, IE knew a steady development leading to what we consider the zenith of its traditional stage. Prominent impulse came from the inception in UK of Operational Research and its further development as Operations Research (OR) in the USA. This period raised relevant contributions such as linear programming, dynamic programming, graph theory or project programming; the incorporation of the incipient Informatics; new theories on human needs (Maslow 1943) and motivation (Herzberg 1968); the joint consideration of men and technology for the analysis of production systems by Ergonomics and the Socio-technical approach.

Considering jointly the development experienced throughout this first stage, the IE traditional approach can be characterized, in broad traits, as follows: the object of study is mainly the manufacturing plant; the geographical scope is local and the organizational scope is mainly the production function; problems studied are operational and deal with material transformation; actors concerned are usually firms looking for efficiency improvements (see Table 1, column 1).

4 From the Traditional to the Consolidated Stage

4.1 Contextual Changes Since the Seventies

The high growth rates of the economy in the 1960 decade (Emery and Trist 1965) led to remarkable "turbulences" in the first half of the seventies (the "oil crisis"). Huge raise in oil and raw material prices and financial costs depleted firm margins causing bankruptcy or severe financial difficulties.

This scenario required stretching responses from firms. On the one hand, from the supply side, firms had to find effective ways to reduce costs. Cost chapters such as inventory, storage space or transportation were then put into focus and the "logistic system" concept, brought from the military to the business field (Drucker 1962; Magee 1968), was applied to cut operational costs, to redesign the whole system and to redeploy vertically integrated firms as multi-firm systems, designated as supply networks in the 2000 decade. On the other hand, from the demand side, firms had to adapt to the new conditions of the competition arena, widening thus the product mix and innovating in products. Additionally, a pervading change in the business environment throughout this period was the development of the information and communication technologies, ICTs (see Table 1).

4.2 Traits of the IE Consolidated Stage

From the "oil crisis" of the 1970 decade up to the start of the great financial crisis of the last years (2008), turbulence deepened and IE was developing its "consolidated" stage, which got consistency in the years 2000–2007. Relevant facets of this development follow. In 1978 H.A. Simon was acknowledged with the Nobel Prize, representing the full recognition of the school of thought proposing a behavioral theory of the firm and decision processes based on bounded rationality. This implied the search for satisfactory alternatives instead of optimal solutions. In parallel, along the 1970 decade took place a significant debate on the appropriate focus of OR, questioning an excessive emphasis in optimization and pointing out a lack in dealing with problematic open system situations. In this respect, two papers from Ackoff (1979a, b) are fully representative. Even in the 1990 decade, the debate on OR appropriate focus stood (Corbett and van Wassenhove 1993), pointing out the gap between the managerial and the research foci. Already Simon had stated that model design had to follow two directions, either to optimize in a simplified context or to satisfy while retaining a richer set of real world properties. In this respect, the development of heuristic, simulation and multi-criteria decision techniques, powered by ICTs support, was the OR salient trait in the period, which allowed to address complex multi-actor strategic and system design problems.

Briefly, the main traits of the "IE consolidated stage" can be characterized as follows (see Table 1, column 2): the object of study is now the supply network; the geographical scope is international; the organizational scope is a network of firms; the flows studied are not only material (now including reverse logistics), but also informational and financial; the problems addressed are not only operational, but strategic and relative to network configuration and design, involving several actors cooperating in supply networks and wishing to reach resilient solutions. The economic activities are mainly ruled by the market offer side, which tends to form oligopolies and benefit from the governmental support, in particular, by private activity deregulation. The geopolitical scenario is shaped by the USA power and its liberal politics. This stage is not disjoint from the "traditional" one. On the contrary, it incorporates the knowledge acquired along the first one, which remains available to deal with "traditional" problems.

5 Macro-trends in the Present Scenario

The year 2008 marked the beginning of a huge financial crisis that, initiating in the USA, extended quickly all over the world and whose consequences are presumed to still last for several years. In such scenario, the present global trends become of highest significance and require tight attention. These trends can be grouped into two sets: global structural turbulence and changes in social values.

In respect to turbulence, the financial crisis of the last years shows that its depth is, not only global, but also structural. Relevant changes in the fundamentals of our current socio-economic world are: global warming induced by human activity and deterioration of ecosystem services (Rockström et al. 2009; Schultz et al. 2013); biodiversity loses (Lenzen et al. 2012; WWF 2015); emergence of a multi-polar world; raise of BRICS and other emergent countries; surge of pro-democracy movements (Tunisia, Egypt, Libya, Syria, …); extension of violent radical movements (Al Qaeda, Islamic State …); interconnected global risks (WEF 2015).

In respect to social values, significant changes are spreading among people all over the world. Relevant ones are: refusal of what is perceived as corruption practices from politicians and abuse from the financial industry; refusal of extreme poverty and economic, social and political inequity; emergence of aspirations from most disfavoured people, influenced by the "western" style of living; tensions between egoistic and solidarity values; consolidation of cooperation as a global value; social consensus fragility as a consequence of perceived insufficiency of current democracy practices; increasing ecological consciousness; revision of the economic concept of prosperity (DBR 2006).

To deal with this situation, limited initiatives have been undertaken. Significant examples follow: United Nations (UN) adopted a plan for the period 2000–2015, the Millennium Objectives, addressed to mitigate the situation of the poorest in the world; the World Bank and diverse National Governments finance development projects for the disadvantaged countries; most dynamic firms have adopted a *corporate social responsibility* paradigm (MIT-BCG 2012); civil society organizations (CSOs) work for the benefit of the most disfavoured people and the environment.

Remarkably, this year 2015, a global "post-2015 Agenda", focused on sustainability goals, is planned to be approved by United Nations (UN 2014).

6 Emergent Stage of IE: Organizational Engineering

Coming back to IE, some questions arise: in the present situation, what can be done by IE and how? Are there emergent opportunities for IE to contribute to the alleviation of the aforementioned problems? To answer these questions, it has to be noted that the new problems are interrelated network situations involving multiple actors, requiring appropriate "courses of action" undertaken by a coalition of different actors: people, private organizations (industrial or services firms, and others), CSOs, public bodies. In this respect, IE holds the great advantage of valuable accumulated knowledge and experience in dealing with supply networks (configuration, design, management of actor relations, knowledge transfer, etc.), which can be easily transferred to different kinds of problematic situations, constituting thus "emerging" opportunities for IE (see Table 1, column 3).

To succeed in such opportunities, five assumptions concerning the approach to be adopted by IE professionals and researchers are considered of great relevance:

- First, to play a role, not simply of problem solver in a confined situation, but one of "change agent" or "promoter", collaborating with a variety of stakeholders on "designing and improving" roads to deal with open unsatisfactory situations, promoting thus cross-sector partnerships.
- Second, to be prepared to work in relation, not only to a variety of organization types, but also to people from diverse cultures and socio-economic contexts, being aware of the crucial importance of people perceptions and attitudes.
- Third, to extend the usual offer-side viewpoint of previous IE approaches (traditional and consolidated) to a wider focus including sustainable human needs and well-being for present and future generations. This refers not only to world's poorest people, but also to health, education and social services for people in developing and developed countries; correspondingly, this requires reviewing and re-orientating the demand side towards sustainable consumption patterns.
- Fourth, to conceive solutions and courses of action within the constraints originated by the judicious preservation of natural resources and the correct functioning of the planet ecological systems (Rockström et al. 2009), in line with paths opened by concepts as "closed-loop supply chains", "circular economy", "industrial ecology" or "ecological footprint".
- Fifth, to adopt a wider appropriate perspective (Ghoshal 2005) and become familiar with concepts, theories and methods such as: "wicked problems" (Churchman 1967; Conklin 2005); "persistent problems" (Loorbach 2009); "network society" (Castells 2000); "complex situations" (Kovacic and Sousa-Poza 2013); "transitions for sustainable development" (Loorbach 2009);

"governance for sustainable development" (Zeijl-Rozema et al. 2008); "metagovernance" (Meuleman 2008); "complex systems theory" (Midgley 2003); "embedded case studies" (Scholz and Tietje 2002); "cognitive mapping" (Kitchin 1994); "sense making" (Weick 1995); "consensus building processes" (Innes and Booher 1999); "problem structuring methods" (Kato 2011).

We propose to refer to this IE "emerging" stage as Organizational Engineering. This stage, including the above-mentioned five assumptions, is not considered to be disjoint from the previous ones, but a deeper and extended stage containing the precedent ones. The resulting Organizational Engineering (OE) concept constitutes an emerging paradigm for IE that deserves to be clearly identified. The term "organizational" in the OE wording highlights thus the focus of this approach on designing and improving interlinked organizations devoted to cope with complex, real-world problematic situations.

7 Conclusions

Industrial Engineering (IE), since its inception in the early years of the past century, has developed a sound trajectory, having reached two significant stages that we characterize as the "traditional" and the "consolidated" ones. Since the 2008 financial crises, the world context is changing rapidly, which implies that a new perspective deserves to be adopted by IE in order to preserve its usefulness for coping with the new problematic situations. In this respect, this paper tries to contribute to the corresponding reflection proposing an "emergent" approach for IE.

The 2050 planet scenario seems incompatible with a global extension of the current model of living standardized in the developed countries. In order to preserve planet sustainability and a peaceful mankind, a transition towards a new societal model is necessary. This requires world governance and organizations working for the benefit, not only of customers able to pay, but of the entire world population. As a consequence, more complex organizational configurations and goals are needed, which represent emerging opportunities for IE, but require an appropriate vision.

The resulting IE concept is a new, deeper and extended one, which is proposed to be identified as Organizational Engineering.

References

Ackoff RL (1979a) The future of operations research is past. J Opl Res Soc 30(2):93–104
Ackoff RL (1979b) Resurrecting the future of operational research. J Opl Res Soc 30(3):189–199
Castells M (2000) The rise of the network society. Wiley, New York
Conklin J (2005) Dialogue mapping: shared understanding of wicked problems. Wiley, New York
Corbett CJ, van Wassenhove LN (1993) The natural drift: what happened to operations research? Oper Res 41(4):625–640

Churchman C (1967) Wicked problems. Manage Sci 14(B):141–142
DBR (2006) Measures of well-being, current issues, 08.09.2006, Deutsche Bank Research. https://www.dbresearch.com
Drucker P (1962) Economy's dark continent. Fortune 65(103):265–270
Emery FE, Trist EL (1965) The causal texture of organizational environments. Hum Relat 18: 21–32
Ghoshal S (2005) Bad management theories are destroying good management practices. Acad Manage Learn Educ 4(1):75–91
Herzberg F (1968) One more time: how do you motivate your employees? Harvard Bus Rev, pp 53–62
IIE (2015) About IIE. Institute of Industrial Engineers website. Available online at: https://www.iienet2.org/Details.aspx?id=282. (consulted 21 Apr 2015)
Innes JE, Booher DE (1999) Consensus building and complex adaptive systems. J Am Plan Assoc 65(4):412–423
Kato H (2011) Problem-structuring methods based on a cognitive mapping approach. In: Komiyama et al (eds) Sustainability science. United Nations Press, Tokio
Kitchin RM (1994) Cognitive maps: what are they and why study them? J Environ Psychol 14(1): 1–19
Kovacic SF, Sousa-Poza A (eds) (2013) Managing and engineering in complex situations. Springer, Berlin
Lenzen M et al (2012) International trade drives biodiversity threats in developing nations. Nature 486:109–112
Lewin K (1936) Principles of topological psychology. McGraw-Hill, New York
Lewin K (1946) Action research and minority problems. J Soc Issues 2(4):34–46
Loorbach D (2009) Transition management for sustainable development: a prescriptive, complexity-based governance framework. Int J Policy, Adm Inst 23(1):161–183
Magee JF (1968) Industrial logistics. McGraw-Hill, New York
Maslow AH (1943) A theory of human motivation. Psychol Rev 50(4):370–396
Meuleman L (2008) Public management and the metagovernance of hierarchies, networks and markets. Physica-Verlag, Heidelberg
Midgley G (ed) (2003) Systems thinking. SAGE Publications, USA
MIT-BCG (2012) Sustainability nears a tipping point. MIT Sloan Man Rev, winter, Heidelberg
Rockström J et al (2009) A safe operating space for humanity. Nature 461:472–475
Simon HA (1978). http://www.nobelprize.org/nobel_prizes/economic-sciences/laureates/1978/simon-lecture.pdf
Scholz RW, Tietje O (2002) Embedded case studies: integrating quantitative and qualitative knowledge. Sage Publications, USA
Schultz M et al (2013) Human prosperity requires global sustainability. Stockholm Resilience Centre, Stockholm University, Stockholm
Taylor FW (1911) Principles of scientific management. Harper and brother, NY and London
United Nations (2014) The road to dignity. https://sustainabledevelopment.un.org/content/documents/5527SR_advance%20unedited_final.pdf
WEF (2015) Global risks report 2015 world economic forum. http://www.weforum.org/reports/global-risks-report-20152013
Weick K (1995) Sense making in organizations. Thousand Oaks, CA
WWF (2015) Living planet report, world wildlife Fund 12, N.W. Washington, DC 20037
Zeijl-Rozema A et al (2008) Governance for sustainable development: a framework. Sustain Dev 16:410–421

Perception of the Evolution of Industrial Engineering Areas Based on the Brazilian ENADE-INEP Assessment System

Eduardo Guilherme Satolo, Renato Luis Garrido Monaro, Milton Vieira Junior and Daniel Luis Garrido Monaro

Abstract Due to increased competitiveness, Brazilian industry, in general, deal with constant factory changes and with the insertion of new models to plan, schedule and control production. All these aspects together make them to face a new challenge: the lack of skilled manpower at all hierarchical levels. For Engineering, especially at Industrial Engineering, the professional profile changes continuously and then, the engineering education in universities has the necessity to reinvent themselves the whole time to meet market demands. Thus, evaluation methods shall be deployed to quantify and provide information that help in analysis of vocational training of the 21st century Engineer. Therefore, the purpose of this paper is to analyze the evolution of Industrial Engineering areas based on the Brazilian ENADE-INEP assessment system applied on 2005, 2008, 2011 and 2014, presenting a diagnosis of the knowledge areas along these years. From the results it is possible the perception of a trend to equality the areas on examination, besides the possibility of an interdisciplinary approach of the questions.

Keywords Industrial Engineering courses · Evolution · Thematic areas · Students performance evaluation · ENADE system

E.G. Satolo (✉)
Post-Graduate Program in Agribusiness and Development (PGAD),
UNESP – Univ Estadual Paulista, Faculdade de Ciência e Engenharia
(FCE/UNESP-Tupã), Tupã, SP, Brazil
e-mail: eduardo@tupa.unesp.br

R.L.G. Monaro
FAJ – Faculdade de Jaguariúna, Centro Universitário Católico de Vitória, Vitória, ES, Brazil
e-mail: prof.renatomonaro@gmail.com

M. Vieira Junior
UNINOVE – Universidade Nove de Julho, São Paulo, SP, Brazil
e-mail: mvieirajr@uninove.br

D.L.G. Monaro
UNIMEP - Universidade Metodista de Piracicaba; FAJ - Faculdade de Jaguariúna,
Jaguariúna, SP, Brazil
e-mail: danielmonaro@yahoo.com.br

M. Amorim et al. (eds.), *Engineering Systems and Networks*, Lecture Notes
in Management and Industrial Engineering, DOI 10.1007/978-3-319-45748-2_38

1 Introduction

In industrial scenario, which is part of a globalized economic environment, companies seek to increase the competitiveness to face the market and new competitors that appear every day. This causes a constantly changing in the direction of productivity of production systems through the rearrangement of factories layouts and the addition of new models to plan, schedule and control production. These changes aim to improve the information and production flow, reduce the lead time, reduce costs and enhance the competitiveness advantage through increased efficiency of processes (Elmaraghy et al. 2012; Smith et al. 1996).

However, one of the most critical problems that Brazilian industry deals with is the lack of skilled manpower at all levels. This has been aggravated especially on engineer's formation, once that recent graduates are employed and need to be trained intensively to present performance results in the shortest time possible.

Since Industrial Revolution the changes on industry requirements occurs dynamically. These changes constantly transform the engineer's profile and become the main challenge for engineering education in universities into the 21st century (Gwynne 2012; Nose and Rebelatto 2001; Tryggvason and Apelian 2006).

To reach these changes, as well new proposals for Engineering Education based on the levelled distribution between theory and practice, a lot of challenges arise for the Universities, including the development of interdisciplinary skills.

Thus, methods to measure and evaluate educational systems are being proposed aiming the quantification of data obtained from the Industrial Engineering educational process. In Brazilian case, the National Examination of Students Performance (ENADE-INEP) is the main assessment instrument known and recognized by the educational system, and it allows guiding the knowledge of Industrial Engineering areas.

In this context, this paper aims to analyze the evolution of Industrial Engineering areas based on questions from ENADE-INEP examination applied in 2005, 2008, 2011 and 2014, bringing a diagnosis of the influence of each area on questions.

2 Industrial Engineering Education and Assessment System Evolution

Within a historical context related to changes in engineer's profile, it is highlighted that only after the Second World War education was considered as a focus for academic research and this insertion was motivated by Grinter report (1956) and the launch of Sputnik (1957) (Lamancusa 2006). However, only in the late '80s, new proposals for management of production systems, brought closer the academic engineer's profile to the needs of industry, dealing with abilities such as: entrepreneurship, creativity, management and their integration into social, political and economic environment (Gwynne 2012; Lamancusa 2006; Rojter 2010), beyond the honesty, integrity and communication skills (Lamancusa and Simpson 2004).

Recently, in 2005/2006, the Bologna Agreement restructured higher education courses on 2 stages of education. The first consists of 3 years (Bachelor Degree), and the second consists of 2 years of specialization in which the student obtains the diploma of B.Sc.

Parallel to the transformation of the engineer's profile in developed countries, in 1958 emerged in Brazil the Undergraduation in Industrial Engineering as a choice of Mechanical Engineering course at the Polytechnic School of the University of São Paulo (USP), thus meeting the changes of the engineer's profile on the market, which sought the engineer to manager profile and with a holistic view of the production system (Oliveira et al. 2010).

But it was only in the mid-1970s that the Ministry of Education of Brazil (MEC) turned the course into a secondary formation linked to already existing six basic areas of Engineering—Civil, Electrical, Mechanical, Materials, Metallurgy and Mining (Borchardt et al. 2009). Finally, in 2002, with the growing of industry needs, aligned with the growth of Industrial Engineering courses in Brazil (see Fig. 1), MEC sets the Industrial Engineering as a basic Engineering area. This definition creates a new professional profile, based on a generalist, humanist, critical and reflective formation, with capacity to absorb new technologies and develop solutions to problems relevant to the society (Oliveira et al. 2005).

It is still observed in Fig. 1 that the definition of Industrial Engineering as a basic area by MEC was a correct decision, once there was a wide acceptance of this professional and the number of courses had grown exponentially, currently having

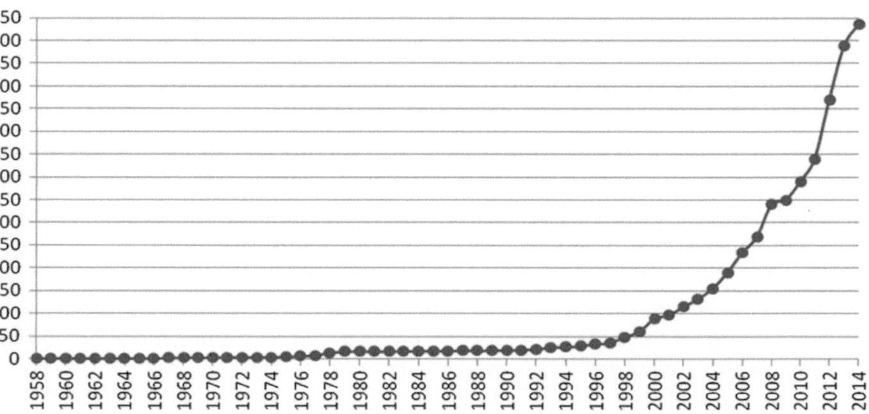

Fig. 1 Cumulative number of industrial engineering courses in Brazil. *Source* Prepared by the authors from data of Oliveira et al. (2010), ABEPRO (2014) and INEP (2014)

736 registered courses, according to the National Institute for Studies and Researches in Education—Anísio Teixeira (INEP 2014).

With the growth of courses offered in Brazil, assessment systems arise during the '90s. The main initiatives that appear were the students examination and on-site assessment of courses, but they were not integrated and had different goals. Since 2002 the assessment procedures for accreditation and renewal of recognition started to cover also the evaluation of supply conditions, and the process became known as the Assessment Condition of Education (ACE) held by INEP/MEC.

Thus, the National System of Higher Education Assessment (SINAES) was established with the publication of the Federal Law number 10,861 of April 14th, 2004, known as the SINAES Law. With this Law the purpose had become the integration between ways and formats of existing evaluation: students' examination, on-site assessments and other types. The aim was to obtain data for the generation of educational policies respecting the diversity of existing higher education institutions in different regions of the country. The SINAES has, among others, the purpose of improving the quality of higher education based on some fundamental principles, evaluation processes and regulatory acts that support the regulatory process performed by MEC (Aguiar Filho 2013).

Institutional reviews would have as inspiration the experience previously performed in public higher education institutions (Institutional Evaluation Program of the Brazilian Universities—PAIUB) and, in its internal aspect should provide the participation of the entire community (teachers, coaches, administrators and students) of the IES and the creation of Own Assessment Committees (known as CPA). In the foreign aspect the effectuation of an evaluation mode was seeking that, in fact, had not previously gotten to operate, with specific rules, appropriate tools and training. The INEP identified itself existing difficulty in operating with a large number of assessment tools. Then, the starting point was to introduce a new assessment tool, unique for all courses, which brought a concern: preventing the reappearance of subjective interpretation of evaluators.

Finally, ENADE assessment was regulated by MEC Ordinance number 2051 of July 9th, 2004, and it would bring out the new character of evaluation through the students examination, which should be held from sample size (depending on the operating costs identified with ENC), and were proposed to evaluate the universe of areas, not the universe of students in each area. In this assessment was sought to identify the value gain ("change analysis") obtained by the teaching provided by IES, based on the assessment of students who would be in different moments: at the input and output of the education process, in other words, freshmen and graduating. The examination would be performed with default time interval (3 years) and should eliminate the previously existing "logic of ranking" between IES.

For Engineering Courses, the ENADE began in 2005, and nowadays it consists of an assessment by means of a test of (i) general knowledge, common to all groups; (ii) basic engineering background, common to all groups; and (iii) specific education in each area of engineering course. It aims also to assess the evolution of the student development during the years of study by means of comparison between freshmen and seniors performance in the test.

3 Research Method

The qualitative analysis was employed as a research method to achieve the proposal of this paper. This method, according to Miguel (2011), confirm that the subjective reality of the individuals involved in this research characterized by the performance of students in ENADE assessment is relevant and contributes to the development of research. Another point that supports the use of qualitative research is the fact that the survey results will be used to attempt to describe and decode understanding, not the frequency of occurrence of the variables under study.

There are in literature a set of techniques linked to qualitative research, among which may include: documentary research, case study and ethnography (Neves 1996).

Thus, for this research we used the document research technique, which is characterized by the search for information in documents which received no scientific treatment, such as reports, newspaper articles, magazines, cards, films, recordings, photographs, and other materials of disclosure (Sá-Silva et al. 2009). For this research the main source of data is linked to the area of documents issued by the MEC, through summary reports and course performance reports. The conduction of documentary research, according to Silva et al. (2005), occurs by the use of three steps which are described in Fig. 2.

Stage 1 - Pre-analysis
Identify pertinent materials to study in governmental sites (MEC and INEP), scientific journals (national and international databases), articles in newspapers and magazines, National Guidelines for Industrial Engineering courses.

⇩

Stage 2 - Analytical description
Study and analyze summary reports and ENADE reviews. Identify key points in student performance in assessing ENADE. Create and analyze graphs and comparative tables.

⇩

Stage 3 - Referential interpretation
Reflection on the points raised in the analytical description and action plan of establishment, seeking to answer the research questions and identifying opportunities.

Fig. 2 Stages of documentary research. *Source* Prepared by the authors

4 Results

The ENADE test applied to the Industrial Engineering courses has been of interest for researchers like Souza et al. (2013) and Monaro et al. (2014), with studies aiming different aspects of the tests applied in 2005, 2008 and 2011. For the present study the analysis of 2014 test has been included in order to extend other studies conducted until this moment.

Therefore, the initial analysis was conducted from the ENADE tests applied in 2005, 2008, 2011 and 2014, by understanding the focus and sorting the 81 questions that composed all of these tests. Figure 3 shows the percentage distribution of Industrial Engineering areas identified on questions.

A general analysis of the assessment cycles denotes that 5 of the 10 areas defined by ABEPRO (Brazilian Association of Industrial Engineering) were considered on 69.1 % of the questions: Operations and Processes Engineering (18.8 %); Quality Engineering (13.6 %); Economic Engineering (13.6 %); Product Development Engineering (12.1 %); and Organizational Engineering (10.4 %).

However, this analysis can be deeper and with a better understanding if consider the distribution of the 4 tests applied, according to Table 1.

The analysis of Table 1 allows denoting that, along the years of the examination, a trend of levelling of importance of the Industrial Engineering areas contemplate on the questions, even though the areas of "Operations and Processes Engineering" and "Logistics" had greater percentage of questions on the last ENADE test.

Such fact can be a result of a change on the Committee responsible for the structure and the guidelines of the test occurred in 2014. This new assessment cycle can be a trend or can denote a transition process towards to a distribution of questions according to the importance of the areas on the Industrial Engineering educational process.

Other important inferences can arise for each area:

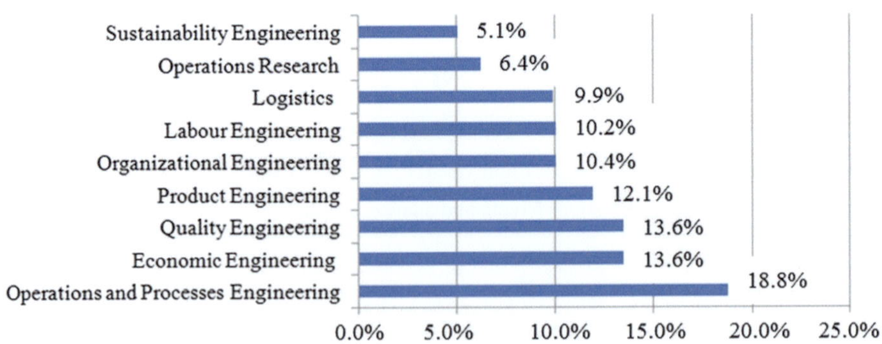

Fig. 3 Distribution of the industrial engineering areas identified on the questions of ENADE tests (2005, 2008, 2011 and 2014). *Source* Prepared by the authors

Table 1 Percentage distribution of questions over the triennial assessment

Industrial engineering areas	2005	2008	2011	2014	Average	Total (%)
Operations and processes engineering	25.0 %	17.6 %	14.8 %	17.6 %	19.2 %	18.5
Economic engineering	10.0 %	17.6 %	14.8 %	11.8 %	14.2 %	13.6
Quality engineering	10.0 %	17.6 %	14.8 %	11.8 %	14.2 %	13.6
Product engineering	10.0 %	11.8 %	14.8 %	11.8 %	12.2 %	12.3
Organizational engineering	15.0 %	5.9 %	14.8 %	5.9 %	11.9 %	11.1
Labor engineering	10.0 %	11.8 %	7.4 %	11.8 %	9.7 %	9.9
Logistics	5.0 %	5.9 %	11.1 %	17.6 %	7.3 %	9.9
Operations research	10.0 %	5.9 %	3.7 %	5.9 %	6.5 %	6.2
Sustainability engineering	5.0 %	5.9 %	3.7 %	5.9 %	4.9 %	4.9
Total questions/year	20	20	27	17		100.0

Source Prepared by the authors

1. The area of "Operations and Processes Engineering" had the reduction in its importance in the distribution of questions from 25 % in 2008 to 19.2 % in 2014.
2. The areas of "Product Engineering", "Economic Engineering" and "Quality Engineering" had strong oscillations, with representation from 10 % in 2005, rising from 17.6 % in 2008, decreasing to 14.8 % of the questions in 2011, and finally to 11.8 % in 2014.
3. The "Organizational Engineering" area presented the greater oscillating level among the areas, with a percentage of 15 % on the test of 2005, a small percentage on the test of 2008 (5.9 %), another growth in 2011 (14.8 %) and another fall in 2014 (5.9 %).
4. The "Logistics" area has been presented an increasing on the representative of the evaluation questions, jumping from 5 % in 2005 to 17.6 % in 2014.
5. The "Operational Research" area is presented as the area with greatest loss of representativeness and decreased from 10 % in 2005 to 3.7 % in 2011, with a little growth in 2014 (5.9 %).
6. The "Sustainability Engineering" area performed practically stabilized with an average percentage of 5.9 % of the questions.
7. Finally, the "Labour Engineering" area, after successive falls on the tests of 2008 and 2011, presented a growth on the test of 2014 (11.8 %).

5 Conclusions

Industrial Engineering is the course that has the highest number of requests for new courses at Brazilian government agencies. This occurs due to the lack in demand for professionals who possess the skills and competencies inherent in the formation of the Industrial Engineer market. This fact brings the necessity of conducting assessments, enabling the courses evaluation and thus the quality of education and the professionals who are formed in Brazil.

However, the formation of skilled professionals is not only the implementation of an assessment that quantifies the knowledge that was acquired by the student throughout his graduation, but a complex set of issues, including: the level of knowledge and ongoing training of course in each university; the involvement of IES and responsible directors by constant updating of pedagogical proposal of disciplines and content; class associations conducting disclosure of the course, assisting in the preparation of these reviews to envisage the profile of the desired market egress; and government agencies that oversee and carry out charging an ethical and professional manner with quality education IES.

It is important to denote the efforts of the responsible for the test elaboration (the Area Committee) to make the assessment of Industrial Engineering courses with an interdisciplinary approach, leveling the distribution of the areas, adding value to the educational process that brings a multidisciplinary formation, with knowledge on the diverse areas of Engineering.

It is verified therefore that there is a strong tendency in the construction of the ENADE evaluation for Industrial Engineering courses in trying to equalize the distribution of questions for the areas of knowledge of the course. Since the specific knowledge assessment consists of 17 questions addressing the 10 areas of Industrial Engineering, the trend is to have 2 questions by assessment area.

Souza et al. (2013) highlights an important aspect also occurred regarding the evolution of the questions in the construction process, at the beginning the questions were more objective and direct whereas currently only one question can address various contents thereby emphasizing the multidisciplinary characteristic.

References

ABEPRO (2014) Relatório da Consulta Avançada—Sistema e-MEC. Available via http://www.abepro.org.br/arquivos/websites/1/relatorio_consulta_publica_avancada_curso_06_06_2014_15_46_03.xls.pdf

Aguiar Filho AS, Batista ME, Hickson RS (2013) Implementação de ações voltadas ao ENADE a partir dos resultados da autoavaliação institucional: um estudo de caso. In: Coloquio Internacional De Gestión Universitaria En América Del Sur, Buenos Aires, Argentina, 13. Available via https://repositorio.ufsc.br/bitstream/handle/123456789/114856/2013314%20-%20Implementa%c3%a7%c3%a3o%20de%20a%c3%a7%c3%b5es.pdf?sequence=1&isAllowed=y

Borchardt M, Vaccaro GL, Azevedo D, Ponte J Jr (2009) O perfil do engenheiro de produção: visão de empresas da região metropolitana de Porto Alegre. Produção19(2):230–248. doi:10.1590/S0103-65132009000200002

Elmaraghy W, Elmaraghy H, Yomiyama T, Monostori L (2012) Complexity in engineering design and manufacturing. CIRP Ann Manuf Technol 61(2):793–814. doi:org/10.1016/j.cirp.2012.05.001

Grinter LE (1956) Report on the evaluation of engineering education. Eng Educ 46(1):25–63. Available via https://www.asee.org/papers-and-publications/publications/The-Grinter-Report-PDF.pdf

Gwynne P (2012) Engineering a revolution in eng. education. Res Technol Manage 55(4):8–9. Available via http://search.proquest.com/docview/1321119682/fulltextPDF/14DC9E5949C64 8E6PQ/1?accountid=8112

INEP (2014) Instituições de educação superior e cursos cadastrados: consulta avançada. Available via: http://emec.mec.gov.br/

Lamancusa JS (2006) The reincarnation of the engineering "shop". In: ASME 2006 International Design Engineering Technical Conferences and Computers and Information in Engineering Conference. Philadelphia, Pennsylvania, USA, p 849–857. doi:10.1115/DETC2006-99723

Lamancusa JS, Simpson TW (2004) The learning factory: 10 years of impact at Penn State. In International Conference On Engineering Education. Gainesville, Florida, USA, p 1–8. Available via http://www.ineer.org/Events/ICEE2004/Proceedings/Papers/329_PSU_Learning_Factory_(1).pdf

Miguel PAC (2011) Metodologia de pesquisa em engenharia de produção e gestão de operações. Elsevier, Rio de Janeiro

Monaro RLG, Satolo EG, Vieira M Jr, Monaro DLG (2014) Análise dos cursos de engenharia de produção no Brasil a partir dos métodos avaliativos do sistema SINAES com foco no ENADE. In: Simpósio de Engenharia de Produção, Bauru, São Paulo, Brazil. Available via http://www.simpep.feb.unesp.br/abrir_arquivo_pdf.php?tipo=artigo&evento=9&art=997&cad=384&opcao=com_id

Neves JL (1996) Pesquisa qualitativa: características, usos e possibilidade. Caderno de Pesquisa em Administração 1(3):2–5. Available via http://www.dcoms.unisc.br/portal/upload/com_arquivo/pesquisa_qualitativa_caracteristicas_usos_e_possibilidades.pdf

Nose MM, Rebelatto DA (2001) O perfil do engenheiro segundo as empresas. In: Congresso Brasileiro de ensino de engenharia. Porto Alegre, Rio Grande do Sul, Brazil, p 1–6. Available via http://www.abenge.org.br/CobengeAnteriores/2001/trabalhos/DTC007.pdf

Oliveira V, Barbosa C, Chrispim E (2005) Cursos de Engenharia de Produção: crescimento e projeções. In Encontro Nacional de Engenharia de Produção. Porto Alegre, Rio Grande do Sul, Brazil, p 1–8. Available via http://www.abepro.org.br/biblioteca/ENEGEP2005_Enegep1101_0328.pdf

Oliveira VF, Vieira Junior M, Cunha GDD (2010) Trajetória e estado da arte da formação em engenharia, arquitetura e agronomia: volume VII: engenharia de produção. CONFEA, Brasília. Available via http://www.ufjf.br/observatorioengenharia/files/2012/01/vol07.pdf

Rojter J (2010) The allocation to the study of humanities and social sciences at Australian engineering education. In International IGIP-SEFI Annual Conference. Trnava, Slovakia, p 1–9. Available via http://www.sefi.be/wp-content/papers2010/papers/1246.pdf

Sá-Silva JR, Almeida CD, Guindani JF (2009) Pesquisa documental: pistas teóricas e metodológicas. Revista Brasileira de História & Ciências Sociais 1(1):1–15

Silva CR, Gobbi BC, Simão AA (2005) O uso da análise de conteúdo como uma ferramenta para a pesquisa qualitativa: descrição e aplicação do método. Organizações Rurais & Agroindustriais 7(1):70–81. Available via http://revista.dae.ufla.br/index.php/ora/article/view/210/207

Smith RP, Barton RR, Nowack CA, Castro JZ (1996) Concurrent engineering: a partnership approach. In: ASEE annual meeting, 1.120.1 – 1.120.5. Washington, D.C., USA. Available via http://citeseerx.ist.psu.edu/viewdoc/download?doi=10.1.1.8.9447&rep=rep1&type=pdf

Sousa FR, Silva DM, Mello JAVB, Mello AJR (2013) Caminhos formacionais na Engenharia de Produção: uma análise do conteúdo das provas específicas do ENADE nos anos 2005, 2008 e 2011. Revista Eletrônica Produção & Engenharia 4(1):365–373. Available via http://www.fmepro.org/ojs/index.php/rpe/article/download/23/24

Tryggvason G, Apelian D (2006) Re-engineering engineering education for the challenges of the 21st century. JOM J Minerals Metals Mat Soc 58(10):14–17. doi:10.1007/s11837-006-0194-6

The Skateboard Manufacturing Company: A Teaching Case on Production Planning and Control

Marco Aurélio de Mesquita, Leonel José Girotti and João Vitor Tomotani

Abstract In manufacturing companies, it is possible to identify different production environments, each suited to a given production strategy. The capability of future managers to deal with these differences is critical and education should provide them a broad and critical view of possible planning models and methods. The aim of this paper is to present a teaching case that enables students to differentiate between two main approaches in the context of production scheduling, namely: MRP and APS. The teaching case was applied in a course of Production Planning and Control in an undergraduate degree in Production Engineering in Brazil. In its current version, the case comprises of five stages, covering the topics of demand forecasting, master scheduling, material requirements planning, finite capacity scheduling, and systems integration. A survey was conducted to evaluate the effectiveness of the teaching plan. The results indicate that students gain a better understanding of the differences between the approaches and now the skateboard factory case is an integral part of the course. Future applications will help to improve and expand the case, including new features related to the discipline contents.

Keywords Teaching case · Case-based learning · Production planning and control

---

The original version of this chapter was revised: The first author's name has been modified. The erratum to this chapter is available at 10.1007/978-3-319-45748-2_40

M.A. de Mesquita (✉)
Engenharia de Produção. Escola Politécnica, Universidade de São Paulo. Av. Prof. Almeida Prado, 531, Cid. Universitária, 05508-900 São Paulo, SP, Brazil
e-mail: marco.mesquita@poli.usp.br

L.J. Girotti · J.V. Tomotani
Engenharia de Produção. Escola Politécnica, Universidade de São Paulo, São Paulo, Brazil
e-mail: leo_girotti@hotmail.com

J.V. Tomotani
e-mail: t.jvitor@gmail.com

M. Amorim et al. (eds.), *Engineering Systems and Networks*, Lecture Notes in Management and Industrial Engineering, DOI 10.1007/978-3-319-45748-2_39

1 Introduction

Engineering Education is often criticized for its excessive and rigorous teaching of engineering sciences and mathematics, which are unlikely to be later used and thus disconnect students from practice (Smith 2005). Empirical studies support this claims, indicating a gap between the skills developed in engineering schools and those demanded by employers (Basnet 2000; Abraham and Carns 2009).

Studies have been conducted in many countries intending to determine the technical and personal skills required of engineers by today's industries, and how these requirements are addressed by engineering schools. Some critical points of these studies are summarized by Mills and Treagust (2003), such as engineering courses being excessively focused on engineering science and techniques without providing sufficient integration of these topics. Still according to the authors, a mixed approach combining project-based learning and traditional lectures has been successfully adopted by several institutions as a way to satisfy industry needs, without sacrificing the knowledge of engineering basics.

An important point in learning is that students must be able to recognize meaningful patterns of information, understand such information, apply it in a given situation, evaluate the results of their actions and develop new, coherent learning based on their experience (Krathwohl 2002). Case-based instruction addresses these issues, helping students to understand 'when, where, and why' to use the knowledge they are learning, and requiring them to integrate multiple sources of information to deal with real problems.

Vivas and Allada (2006) suggest that the gap between theory and practice in engineering education can be reduced to some extent through the use of case studies. Cases can be problem-based, historical in nature, present exemplary scenario, dilemma-based, and/or illustrate critical issues in the field (Yadav et al. 2010; Yadav and Barry 2009).

A survey conducted by Girotti and Mesquita (2015) evaluated teaching methodologies used by PPC professors of Production Engineering courses in Brazil. The results reveal that 85 % of the professors considered that case studies supported by computer applications could be a relevant contribution to the learning in PPC, but, as some of them highlighted, it is difficult to find teaching cases and software for use on the topic.

This paper presents and evaluates a teaching case designed to explain the main concepts and methods for Production Planning and Control (PPC), from the perspective of two production environments, namely, Make-to-Stock (MTS) and Make-to-Order (MTO). The case-based approach tries to bring the reality of the PPC into the classroom. Through a more iterative approach, it is believed the students would be able handle different tools, developing professional production engineering skills, as well as being more motivated due to the immersion in a realistic case. In addition to evaluating the effectiveness of the plan, the aim is to improve the case from each new application for use in teaching PPC.

The methodology adopted to develop and evaluate the teaching plan was the action research. Case and Light (2011) claim this methodology to be particularly effective for engineering courses that seek not only to study their educational practices, but also an effective way to improve them.

The structure of this paper is as following. The first section is the introduction and formulation of the problem and objectives. Section 2 presents the teaching case and its implementation, and Sect. 3 presents the teaching plan applied to the discipline. The results of the applications are presented and discussed in Sect. 4. Section 5 ends the paper with conclusions and future developments.

2 The Teaching Case

2.1 The Skateboard Factory

The case considers two fictitious companies, each with a different production strategy. The first company is the “Sk8 MRP”, a small to medium-sized skate-board manufacturer operating under Make-to-Stock (MTS) strategy. It is the main company of the case and is where the main character works. He is a young Engineer who joins Sk8 MRP to work on the PPC. At different stages, he should solve some problems related to the production planning and scheduling. The second company is called Metal Casting Inc. (MCI) which provides casted and/or machined parts upon demand (Make-to-Order, MTO). Specifically, in the teaching case, it provides the trucks of the skateboards to Sk8 MRP.

The Sk8 MRP manufactures decks and wheels, buying the other components to skateboard assembly (Fig. 1). Other products they sell separately are boxes of decks, sets of wheels and trucks. In order to configure a realistic teaching case, demand, cost and technical data were generated. We estimated a demand level for finished products based on Brazilian skateboards market and did a research in machine manufacturing websites, which provided the technical data required.

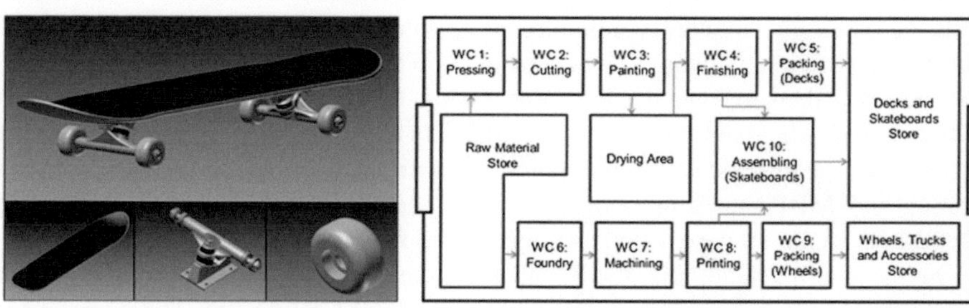

Fig. 1 The product and plant layout

2.2 The Five Stages of the Case

To follow the chronology of themes of the subject, the teaching case was divided into stages, aiming at encompassing the main topics related to the discipline. In order to complete the case, students were divided in groups of five.

Below is described in greater detail the five stages of the case, including the software used for their resolution. Although the teaching case makes use of certain software, the professor can replace them by others he considers more appropriate. The teaching case details and documents are provided in Girotti (2013), and in the website developed by the authors (https://sk8factory.wordpress.com/).

The aim of the first stage is that the students develop a demand forecasting process and, later, critically evaluate the results of such process. For this, the monthly sales of each product were given to students in standard text file format. The time series provided covers a period of almost five years. Each sales history has some peculiarity, such as trends, seasonality or the presence of outliers. Afterwards, the students are requested to provide forecasts for the next three months, using the most appropriate forecasting method among the ones presented in class and justifying their choice. The aim is to place the student in a situation similar to business reality, where both the uncertainty about the best model to use as the presence of outliers and inconsistencies in the collected data are common.

On the second stage, the objective is to review and improve the production planning method adopted in the Sk8 MRP. Currently, the company decides how much of each item to produce based on the finished product inventory level and standard lot sizing constraints. The item with the lowest relative inventory level (measured in days of supply) has priority in production. The question is how to improve the production planning process. The student´s tasks are: (1) to elaborate in a spreadsheet a master production schedule for the next three months, using the demand forecasts from the previous stage and the capacity constrains; and (2) to undertake a critical analysis of the solution and assess the feasibility of deploying the solution procedure in the routine of the production planning and control.

The MRP approach to production scheduling is considered in the third stage. Using the results from the previous stage and additional information about the items (bills of materials, inventory policies etc.), students perform the calculation of material requirements, using a VBA application based on MRP logic. In addition, students are asked to evaluate both the quality of the solution generated and the software used. The main outputs of the MRP run are: the material requirements and the capacity requirements in the planning horizon (Arnold et al. 2012). After running the MRP, students should take the output in table format and, using spreadsheet tools, prepare a better report of the material and capacity plans.

On the fourth stage, the focus shifts to the second approach to production scheduling, namely the APS, employed by the Metal Casting Inc. The company is facing problems in meeting the promised due dates and, because of that, decides to hire a consultant to evaluate and improve the performance of the Detailed Scheduling module of its Advanced Production Scheduling System (APS/DS).

So, the student′s task is to generate a schedule for the open orders and critically evaluate the available software. For this stage, different kinds of software can be employed: there are both commercial software and educational applications developed in spreadsheets that could be applied. In present case study, it was chosen the Lekin software (academic version), developed by the Stern School of Business—New York University (Pinedo et al. 2002). However, in 2015, a VBA application was developed to replace Lekin (Fig. 2).

Finally, in the fifth stage, the demand for the students is to elaborate a report explaining the differences between the MRP and APS approaches, looking for additional information on books, papers and the internet. The groups should explain whether it is possible to use the two approaches together or if one replaces the other. The aim of this stage summarizes the goal of the educational plan and the teaching case presented, i.e., to help understanding the difference between the two approaches of production scheduling.

3 Teaching Plan

The teaching case presented in the previous section is the main component of the teaching plan prepared for the discipline in question. According to Seno and Belhot (2009), the teaching plan should cover all activities that will be developed by both the teacher and students. As such, before starting the course, the teacher must analyse the reality where it will be deployed, what are the goals to be achieved,

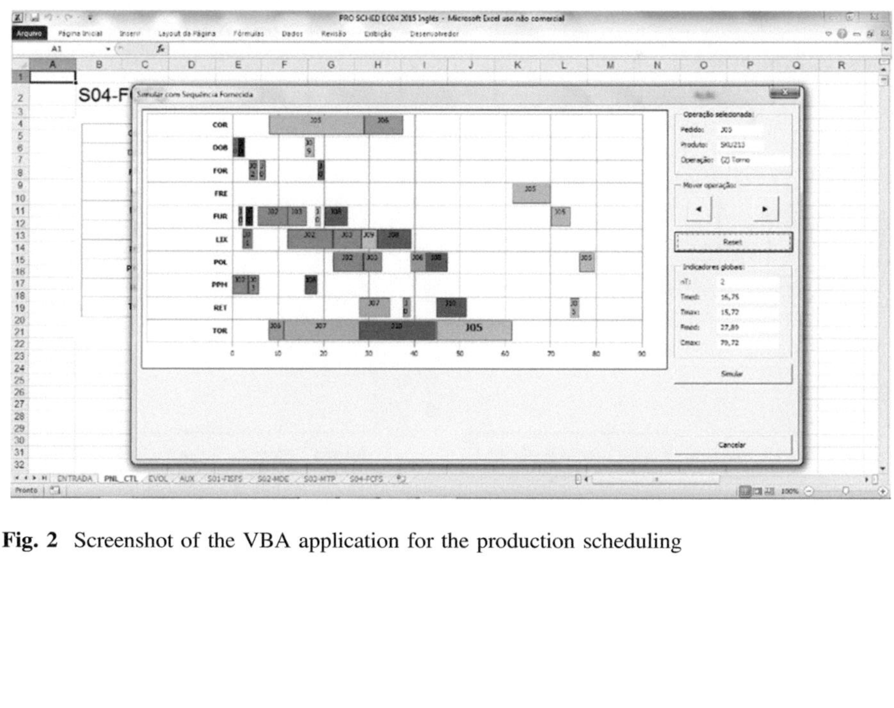

Fig. 2 Screenshot of the VBA application for the production scheduling

which resources he needs and which he already has. The ultimate goal is to make the process of teaching/learning as efficient and effective as possible.

The new syllabus of the PPC discipline was applied in an undergraduate course of Production Engineering at the University of São Paulo, Brazil. The course is offered once a year and the plan was applied in the last two classes.

In addition to the teaching case, at the end of each topic, lists of exercises and test are assigned to the students. Lectures notes and additional readings are provided by the Virtual Learning Environment (VLE). The lists have questions and numerical exercises, like regular textbooks, and are corrected by a teaching assistant and handed back to the students. In the first application, the case had little weight in the final grade. In the second, the case gained more weight and lists became facultative. Also, with the intention of evaluating student learning and assisting them, fast and facultative assessments (10 min duration) were made at the end of each subject in the first edition of the course.

4 Case and Survey Results

4.1 Students' Grades in the Teaching Case

This paper presents a teaching case supported by software and its first two applications in undergraduate classes (two different years) of a Production Engineering course were evaluated. Figure 3 presents the results of student's performance in solving the teaching case.

When comparing the grades of the two applications, there are some noticeable differences. However, such a comparison is not fully appropriate given that the lower weight of the teaching case in the first application made only a few groups to conclude all stages (and thus being included in the analysis).

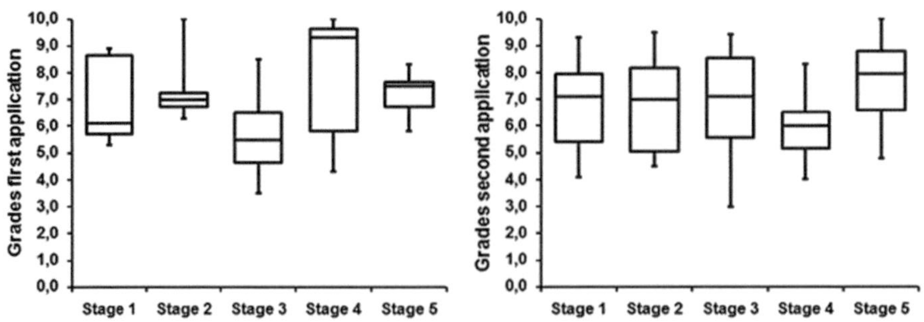

Fig. 3 Boxplot of the grades by stage of the first and second application

4.2 Students Evaluation of the Teaching Case

A survey of students was conducted to assess the effectiveness of the teaching case. There were 28 responses from a total of 79 students on the first application of the survey and 29 out of 74 students on the second application; i.e., a response rate of 35.4 % on the first application and 39.2 % on the second. Some of the most relevant results are presented below. Figure 4 concerns the participation of the students in the resolution of the teaching case and if the differences between the MRP and APS approaches were clear.

Another question (Fig. 5) evaluated the contribution of the teaching case in the learning process. Further results of the survey are detailed in Girotti (2013).

4.3 Discussion

From the results presented in Sects. 4.1 and 4.2, it is possible to evaluate the effects of the teaching case.

Adding up the responses of the two surveys, only 5.3 % of the students checked the lowest values (1 or 2) when asked whether the differences between the MRP and APS approaches were clear by the end of the case. As such, the majority of respondents believed to have assimilated the concepts covered.

On the question about the importance of the teaching case, the result is more positive in the second application, with a much larger number of students giving the highest score and a much smaller number of students assigning the lower scores. The results indicate that the students agree that the teaching case is important for understanding the key question it seeks to answer.

Promoting the participation in the teaching case by increasing its weight on the final grade resulted in greater participation and a more positive assessment on its usefulness for the learning process.

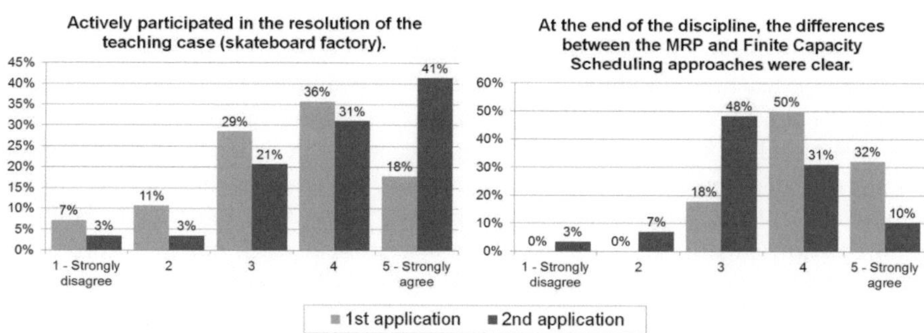

Fig. 4 Self-assessment: participation in solving the teaching case

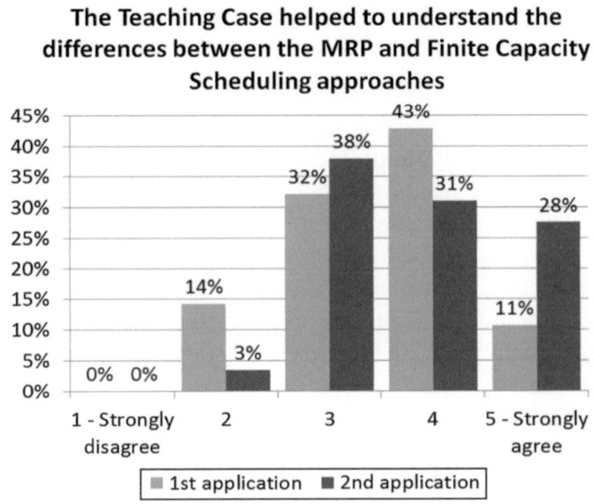

Fig. 5 Perception of the students on the usefulness of the teaching case

The existence of a correlation between active participation in the resolution of the teaching case and the perception of learning through the case was verified (r = 0.629). Such results are consistent with the answers given by the students for the open question of the questionnaire, where they could make suggestions or critics. Of all the 57 respondents in both surveys, 22 answered this question, with half of them commenting on the teaching case. On these comments, 100 % praised the proposal, recommended that the case had a greater weight on the final grade, and/or suggested changes in the case. These responses show that the teaching case approach contributes to the course through the motivational factor and team working. These results are consistent with the ones suggested by Yadav et al. (2010) and Garcia et al. (2012).

5 Conclusion

In the present paper, we propose and validate the effectiveness of a case-based approach in teaching Production Planning and Control (PPC). Specifically, a teaching plan which mainly includes the use of a teaching case was evaluated. The plan was tested on two groups of different years of PPC classes of a Production Engineering undergraduate program in Brazil. A highlight of the teaching case is the exposure of the students to open situations, where they must apply both the acquired knowledge in the subject and the discernment when making decisions.

The first two experiences have left some lessons. The professor who wishes to deploy a teaching case should deal with planning issues such as: additional activities assigned to students; analyze the alternative software; among others. Due to the encouraging preliminary results, the test case will be applied to new classes, including small improvements with each new application. In future editions, the case may include more stages, advancing on topics such as Cost Management, Inventory Control and Pull Production, including synergies with other subjects.

Acknowledgments The authors express their sincere thanks to Umberto Crepaldi, for helping with the development of the VBA applications used throughout the teaching case.

References

Abraham SE, Carns LA (2009) Do business schools value the competencies that business value? J Educ Bus 84(6):350–356
Arnold JR, Chapman SN, Clive LM (2012) Introduction to materials management, 7th edn. Prentice-Hall, Upper Saddle River
Basnet C (2000) Production management in New Zealand: is education relevant to practice? Int J Oper Prod Manage 20(6):730–744
Case JM, Light G (2011) Emerging methodologies in engineering education research. J Eng Educ 100(1):186–210
Garcia J, Sinfield J, Yadav A, Robin A (2012) Learning through entrepreneurially oriented case-based instruction. Int J Eng Educ 28(2):448–457
Girotti LJ (2013) A fábrica de skates: um caso didático em planejamento e controle da produção (Master's thesis). Universidade de São Paulo, São Paulo
Girotti LJ, Mesquita MA (2015) Production planning and control: a survey of teachers in production engineering. Production, Ahead of print
Krathwohl DR (2002) A revision of bloom's taxonomy. Theor Pract 41(4):212–218
Mills JE, Treagust DF (2003) Engineering education—is problem-based or project-based learning the answer? Australas J Eng Educ 3(2):2–16
Pinedo ML, Chao X, Leung J (2002) LEKIN—flexible job-shop scheduling system [software], Retrieved 10 Dec 2014 from http://community.stern.nyu.edu/om/software/lekin/
Seno WP, Belhot RV (2009) Delimitando a fronteira de competências para a capacitação de professores de engenharia para o ensino a distância. Gestão & Produção 16(3):502–514
Smith G (2005) Problem-based learning: can it improve managerial thinking? J Manage Educ 29 (2):357–378
Vivas JF, Allada V (2006) Enhancing engineering education using thematic case-based learning. Int J Eng Educ 22(2):236–246
Yadav A, Berry BE (2009) Using case-based instruction to increase ethical understanding in engineering: what do we know? What do we need? Int J Eng Educ 25(1):138–143
Yadav A, Shaver GM, Meckl P (2010) Lessons learned: implementing the case teaching method in a mechanical engineering course. J Eng Educ 99(1):55–69

Erratum to: The Skateboard Manufacturing Company: A Teaching Case on Production Planning and Control

Marco Aurélio de Mesquita, Leonel José Girotti and João Vitor Tomotani

Erratum to:
Chapter "The Skateboard Manufacturing Company: A Teaching Case on Production Planning and Control" in: M. Amorim et al. (eds.), *Engineering Systems and Networks*, Lecture Notes in Management and Industrial Engineering, DOI 10.1007/978-3-319-45748-2_39

In the original version of Chapter 39, the first author name "Macro Aurélio de Mesquita" was inadvertently published with typo error and it should read as "Marco Aurélio de Mesquita". The erratum chapter and the book have been updated with the changes.

The updated original online version for this chapter can be found at 10.1007/978-3-319-45748-2_39

M.A. de Mesquita (✉)
Engenharia de Produção. Escola Politécnica, Universidade de São Paulo. Av. Prof. Almeida Prado, 531, Cid. Universitária, 05508-900 São Paulo, SP, Brazil
e-mail: marco.mesquita@poli.usp.br

L.J. Girotti · J.V. Tomotani
Engenharia de Produção. Escola Politécnica, Universidade de São Paulo, São Paulo, Brazil
e-mail: leo_girotti@hotmail.com

J.V. Tomotani
e-mail: t.jvitor@gmail.com

M. Amorim et al. (eds.), *Engineering Systems and Networks*, Lecture Notes in Management and Industrial Engineering, DOI 10.1007/978-3-319-45748-2_40

MIX
Papier aus verantwortungsvollen Quellen
Paper from responsible sources
FSC® C105338

If you have any concerns about our products,
you can contact us on
ProductSafety@springernature.com

In case Publisher is established outside the EU,
the EU authorized representative is:
Springer Nature Customer Service Center GmbH
Europaplatz 3, 69115 Heidelberg, Germany

Printed by Libri Plureos GmbH
in Hamburg, Germany